Springer Undergraduate Texts
in Mathematics and Technology

Mario Lefebvre

Basic Probability Theory with Applications

 Springer

Mario Lefebvre
Département de mathématiques et de génie industriel
École Polytechnique de Montréal, Québec
C.P. 6079, succ. Centre-ville
Montréal H3C 3A7
Canada
mlefebvre@polymtl.ca

Series Editors
Jonathan M. Borwein
Faculty of Computer Science
Dalhousie University
Halifax, Nova Scotia B3H 1W5
Canada
jborwein@cs.dal.ca

Helge Holden
Department of Mathematical Sciences
Norwegian University of Science and
  Technology
Alfred Getz vei 1
NO-7491 Trondheim
Norway
holden@math.ntnu.no

ISBN 978-1-4614-2923-4   e-ISBN 978-0-387-74995-2
DOI 10.1007/978-0-387-74995-2
Springer Dordrecht Heidelberg London New York

Mathematics Subject Classification (2000): 60-01

Springer is part of Springer Science+Business Media (www.springer.com)

To the memory of my father

I will never believe that God plays dice with the universe.

Albert Einstein

*Then they gave lots to them, and the lot fell upon Matthias,
and he was counted with the eleven apostles.*

Acts 1: 26

Preface

The main intended audience for this book is undergraduate students in pure and applied sciences, especially those in engineering. Chapters 2 to 4 cover the probability theory they generally need in their training. Although the treatment of the subject is surely sufficient for non-mathematicians, I intentionally avoided getting too much into detail. For instance, topics such as mixed type random variables and the Dirac delta function are only briefly mentioned.

Courses on probability theory are often considered difficult. However, after having taught this subject for many years, I have come to the conclusion that one of the biggest problems that the students face when they try to learn probability theory, particularly nowadays, is their deficiencies in basic differential and integral calculus. Integration by parts, for example, is often already forgotten by the students when they take a course on probability. For this reason, I have decided to write a chapter reviewing the basic elements of differential calculus. Even though this chapter might not be covered in class, the students can refer to it when needed. In this chapter, an effort was made to give the readers a good idea of the use in probability theory of the concepts they should already know.

Chapter 2 presents the main results of what is known as *elementary probability*, including Bayes' rule and elements of combinatorial analysis. Although these notions are not mathematically complicated, it is often a chapter that the students find hard to master. There is no trick other than doing a lot of exercises to become comfortable with this material.

Chapter 3 is devoted to the more technical subject of random variables. All the important models for the applications, such as the binomial and normal distributions, are introduced. In general, the students do better when examined on this subject and feel that their work is more rewarded than in the case of combinatorial analysis, in particular.

Random vectors, including the all-important central limit theorem, constitute the subject of Chapter 4. I have endeavored to present the material as simply as possible. Nevertheless, it is obvious that double integrals cannot be simpler than single integrals.

Applications of Chapters 2 to 4 are presented in Chapters 5 to 7. First, Chapter 5 is devoted to the important subject of reliability theory, which is used in most engineering disciplines, in particular in mechanical engineering. Next, the basic queueing models are studied in Chapter 6. Queueing theory is needed for many computer science engineering students, as well as for those in industrial engineering. Finally, the last application considered, in Chapter 7, is the concept of time series. Civil engineers, notably those specialized in hydrology, make use of stochastic processes of this type when they want to model various phenomena and forecast the future values of a given variable, such as the flow of a river. Time series are also widely used in economy and finance to represent the variations of certain indices.

No matter the level and the background of the students taking a course on probability theory, one thing is always true: as mentioned above, they must try to solve many exercises before they can feel that they have mastered the theory. To this end, the book contains more than 400 exercises, many of which are multiple part questions. At the end of each chapter, the reader will find some solved exercises, whose solutions can be found in Appendix C, followed by a large number of unsolved exercises. Answers to the even-numbered questions are provided in Appendix D at the end of the book. There are also many multiple choice questions, whose answers are given in Appendix E.

It is my pleasure to thank all the people I worked with over the years at the École Polytechnique de Montréal and who provided me with interesting exercises that were included in this work.

Finally, I wish to express my gratitude to Vaishali Damle, and the entire publishing team at Springer, for their excellent support throughout this book project.

Mario Lefebvre
Montréal, July 2008

Contents

List of Tables

List of Figures

1

Review of differential calculus

This chapter presents the main elements of differential calculus needed in probability theory. Often, students taking a course on probability theory have problems with concepts such as integrals and infinite series. In particular, the integration by parts technique is recalled.

1.1 Limits and continuity

The first concept that we recall is that of the *limit* of a function, which is defined formally as follows.

Definition 1.1.1. *Let f be a real-valued function. We say that $f(x)$ tends to f_0 ($\in \mathbb{R}$) as x tends to x_0 if for any positive number ϵ there exists a positive number δ such that*

$$0 < |x - x_0| < \delta \quad \Longrightarrow \quad |f(x) - f_0| < \epsilon.$$

*We write that $\lim_{x \to x_0} f(x) = f_0$. That is, f_0 is the **limit** of the function $f(x)$ as x tends to x_0.*

Remarks. (i) The function $f(x)$ need not be defined at the point x_0 for the limit to exist.

(ii) It is possible that $f(x_0)$ exists, but $f(x_0) \neq f_0$.

(iii) We write that $\lim_{x \to x_0} f(x) = \infty$ if, for any $M > 0$ (as large as we want), there exists a $\delta > 0$ such that

$$0 < |x - x_0| < \delta \quad \Longrightarrow \quad f(x) > M.$$

Similarly for $\lim_{x \to x_0} f(x) = -\infty$.

(iv) In the definition, x_0 is assumed to be a real number. However, the definition can actually be extended to the case when $x_0 = \pm\infty$.

M. Lefebvre, *Basic Probability Theory with Applications*, Springer Undergraduate Texts in Mathematics and Technology, DOI: 10.1007/978-0-387-74995-2_1,
© Springer Science + Business Media, LLC 2009

Sometimes, we are interested in the limit of the function $f(x)$ as x *decreases* or *increases* to a given real number x_0. The *right-hand limit* (resp., *left-hand limit*) of the function $f(x)$ as x decreases (resp., increases) to x_0 is denoted by $\lim_{x \downarrow x_0} f(x)$ [resp., $\lim_{x \uparrow x_0} f(x)$]. Some authors write $\lim_{x \to x_0^+} f(x)$ [resp., $\lim_{x \to x_0^-} f(x)$]. If the limit of $f(x)$ as x tends to x_0 exists, then

$$\lim_{x \downarrow x_0} f(x) = \lim_{x \uparrow x_0} f(x) = \lim_{x \to x_0} f(x).$$

Definition 1.1.2. *The real-valued function $f(x)$ is said to be* **continuous** *at the point $x_0 \in \mathbb{R}$ if (i) it is defined at this point, (ii) the limit as x tends to x_0 exists, and (iii) $\lim_{x \to x_0} f(x) = f(x_0)$. If f is continuous at every point $x_0 \in [a, b]$ [or (a, b), etc.], then f is said to be* continuous *on this interval.*

Remarks. (i) In this textbook, a *closed* interval is denoted by $[a, b]$, whereas (a, b) is an *open* interval. We also have, of course, the intervals $[a, b)$ and $(a, b]$.

(ii) If we rather write, in the definition, that the limit $\lim_{x \downarrow x_0} f(x)$ [resp., $\lim_{x \uparrow x_0} f(x)$] exists and is equal to $f(x_0)$, then the function is said to be *right-continuous* (resp., *left-continuous*) at x_0. A function that is continuous at a given point x_0 such that $a < x_0 < b$ is both right-continuous and left-continuous at that point.

(iii) A function f is said to be *piecewise continuous* on an interval $[a, b]$ if this interval can be divided into a *finite* number of subintervals on which f is continuous and has right- and left-hand limits.

(iv) Let $f(x)$ and $g(x)$ be two real-valued functions. The *composition* of the two functions is denoted by $g \circ f$ and is defined by

$$(g \circ f)(x) = g[f(x)].$$

A result used in Chapter 3 states that the composition of two continuous functions is itself a continuous function.

Example 1.1.1. Consider the function

$$u(x) = \begin{cases} 0 \text{ if } x < 0, \\ 1 \text{ if } x \geq 0, \end{cases} \tag{1.1}$$

which is known as the Heaviside or unit step function. In probability, this function corresponds to the *distribution function* of the constant 1. It is also used to indicate that the possible values of a certain *random variable* are the set of nonnegative real numbers. For example, writing that

$$f_X(x) = e^{-x} u(x) \quad \text{for all } x \in \mathbb{R}$$

is tantamount to writing that

$$f_X(x) = \begin{cases} 0 & \text{if } x < 0, \\ e^{-x} & \text{if } x \geq 0, \end{cases}$$

where $f_X(x)$ is called the *density function* of the random variable X.

The function $u(x)$ is defined for all $x \in \mathbb{R}$. In other contexts, $u(0)$ is chosen differently as above. For instance, in some applications $u(0) = 1/2$. At any rate, the unit step function is continuous everywhere, except at the origin, because (for us)

$$\lim_{x \downarrow 0} u(x) = 1 \quad \text{and} \quad \lim_{x \uparrow 0} u(x) = 0.$$

However, with the choice $u(0) = 1$, we may assert that $u(x)$ is right-hand continuous at $x = 0$.

The previous definitions can be extended to the case of real-valued functions of two (or more) variables. In particular, the function $f(x, y)$ is *continuous* at the point (x_0, y_0) if

$$\lim_{\substack{x \to x_0 \\ y \to y_0}} f(x, y) = f(\lim_{x \to x_0} x, \lim_{y \to y_0} y).$$

This formula implies that the function $f(x, y)$ is defined at (x_0, y_0) and that the limit of $f(x, y)$ as (x, y) tends to (x_0, y_0) exists and is equal to $f(x_0, y_0)$.

1.2 Derivatives

Definition 1.2.1. *Suppose that the function $f(x)$ is defined at $x_0 \in (a, b)$. If*

$$f'(x_0) := \lim_{x \to x_0} \frac{f(x) - f(x_0)}{x - x_0} \equiv \lim_{\Delta x \to 0} \frac{f(x_0 + \Delta x) - f(x_0)}{\Delta x}$$

*exists, we say that the function $f(x)$ is **differentiable** at the point x_0 and that $f'(x_0)$ is the **derivative** of $f(x)$ (with respect to x) at x_0.*

Remarks. (i) For the function $f(x)$ to be differentiable at x_0, it must at least be continuous at that point. However, this condition is not sufficient, as can be seen in the example below.

(ii) If the limit is taken as $x \downarrow x_0$ (resp., $x \uparrow x_0$) in the previous definition, then the result (if the limit exists) is called the right-hand (resp., left-hand) derivative of $f(x)$ at x_0 and is sometimes denoted by $f'(x_0^+)$ [resp., $f'(x_0^-)$]. If $f'(x_0)$ exists, then $f'(x_0^+) = f'(x_0^-)$.

(iii) The derivative of f at an arbitrary point x is also denoted by $\frac{d}{dx}f(x)$, or by $Df(x)$. If we set $y = f(x)$, then

$$f'(x_0) \equiv \frac{dy}{dx}\bigg|_{x = x_0}.$$

(iv) If we differentiate $f'(x)$, we obtain the *second-order derivative* of the function f, denoted by $f''(x)$ or $\frac{d^2}{dx^2}f(x)$. Similarly, $f'''(x)$ [or $f^{(3)}(x)$, or $\frac{d^3}{dx^3}f(x)$] is the *third-order derivative* of f, and so on.

(v) One way to find the values of x that maximize or minimize the function $f(x)$ is to calculate the first-order derivative $f'(x)$ and to solve the equation $f'(x) = 0$. If $f'(x_0) = 0$ and $f''(x_0) < 0$ [resp., $f''(x_0) > 0$], then f has a *relative* maximum (resp., minimum) at $x = x_0$. If $f'(x) \neq 0$ for all $x \in \mathbb{R}$, we can check whether the function $f(x)$ is always increasing or decreasing in the interval of interest.

Example 1.2.1. The function $f(x) = |x|$ is continuous everywhere, but is not differentiable at the origin, because we find that

$$f'(x_0^+) = 1 \quad \text{and} \quad f'(x_0^-) = -1.$$

Example 1.2.2. The function $u(x)$ defined in Example 1.1.1 is obviously not differentiable at $x = 0$, because it is not continuous at that point.

Example 1.2.3. The function

$$F_X(x) = \begin{cases} 0 \text{ if } x < 0, \\ x \text{ if } 0 \leq x \leq 1, \\ 1 \text{ if } x > 1 \end{cases}$$

is defined and continuous everywhere. It is also differentiable everywhere, except at $x = 0$ and $x = 1$. We obtain that the derivative $F_X'(x)$ of $F_X(x)$, which is denoted by $f_X(x)$ in probability theory, is given by

$$f_X(x) = \begin{cases} 1 \text{ if } 0 \leq x \leq 1, \\ 0 \text{ elsewhere.} \end{cases}$$

Note that $f_X(x)$ is discontinuous at $x = 0$ and $x = 1$. The function $F_X(x)$ is an example of what is known in probability as a *distribution function*.

Remark. Using the *theory of distributions*, we may write that the derivative of the Heaviside function $u(x)$ is the *Dirac delta function* $\delta(x)$ defined by

$$\delta(x) = \begin{cases} 0 \text{ if } x \neq 0, \\ \infty \text{ if } x = 0. \end{cases} \tag{1.2}$$

The Dirac delta function is actually a *generalized function*. It is, by definition, such that

$$\int_{-\infty}^{\infty} \delta(x)\,dx = 1.$$

We also have, if $f(x)$ is continuous at $x = x_0$, that

$$\int_{-\infty}^{\infty} f(x)\delta(x - x_0)\, dx = f(x_0).$$

We do not recall the basic differentiation rules, except that for the *derivative of a quotient*:

$$\frac{d}{dx}\frac{f(x)}{g(x)} = \frac{g(x)f'(x) - f(x)g'(x)}{g^2(x)} \quad \text{if } g(x) \neq 0.$$

Remark. Note that this formula can also be obtained by differentiating the product $f(x)h(x)$, where $h(x) := 1/g(x)$.

Likewise, the formulas for calculating the derivatives of elementary functions are assumed to be known. However, a formula that is worth recalling is the so-called *chain rule* for derivatives.

Proposition 1.2.1. (Chain rule) *Let the real-valued function $h(x)$ be the composite function $(g \circ f)(x)$. If f is differentiable at x and g is differentiable at the point $f(x)$, then h is also differentiable at x and*

$$h'(x) = g'[f(x)]f'(x).$$

Remark. If we set $y = f(x)$, then the chain rule may be written as

$$\frac{d}{dx}g(y) = \frac{d}{dy}g(y) \cdot \frac{dy}{dx} = g'(y)f'(x).$$

Example 1.2.4. Consider the function $h(x) = \sqrt{x^2 + 1}$. We may write that $h(x) = (g \circ f)(x)$, with $f(x) = x^2 + 1$ and $g(x) = \sqrt{x}$. We have:

$$g'(x) = \frac{1}{2\sqrt{x}} \quad \Longrightarrow \quad g'[f(x)] = \frac{1}{2\sqrt{f(x)}} = \frac{1}{2\sqrt{x^2 + 1}}.$$

Then,

$$f'(x) = 2x \quad \Longrightarrow \quad h'(x) = \frac{1}{2\sqrt{x^2 + 1}} \cdot (2x) = \frac{x}{\sqrt{x^2 + 1}}.$$

Finally, another useful result is known as *l'Hospital's rule*.

Proposition 1.2.2. (L'Hospital's rule) *Suppose that*

$$\lim_{x \to x_0} f(x) = \lim_{x \to x_0} g(x) = 0$$

or that

$$\lim_{x \to x_0} f(x) = \lim_{x \to x_0} g(x) = \pm\infty.$$

If (i) $f(x)$ and $g(x)$ are differentiable in an interval (a, b) containing the point x_0, except perhaps at x_0, and (ii) the function $g(x)$ is such that $g'(x) \neq 0$ for all $x \neq x_0$ in the interval (a, b), then

$$\lim_{x \to x_0} \frac{f(x)}{g(x)} = \lim_{x \to x_0} \frac{f'(x)}{g'(x)}$$

holds. If the functions $f'(x)$ and $g'(x)$ satisfy the same conditions as $f(x)$ and $g(x)$, we can repeat the process. Moreover, the constant x_0 may be equal to $\pm\infty$.

Remark. If $x_0 = a$ or b, we can replace the limit as $x \to x_0$ by $\lim x \downarrow a$ or $\lim x \uparrow b$, respectively.

Example 1.2.5. In probability theory, one way of defining the *density function* $f_X(x)$ of a *continuous random variable* X is by calculating the limit of the ratio of the *probability* that X takes on a value in a small interval about x to the length of this interval:

$$f_X(x) := \lim_{\epsilon \downarrow 0} \frac{\text{Probability that } X \in [x - (\epsilon/2), x + (\epsilon/2)]}{\epsilon}.$$

The probability in question is actually equal to zero (in the limit as ϵ decreases to 0). For example, we might have that $\{\text{Probability that } X \in [x - (\epsilon/2), x + (\epsilon/2)]\} = \exp\{-x + (\epsilon/2)\} - \exp\{-x - (\epsilon/2)\}$, for $x > 0$ and ϵ small enough.

By making use of l'Hospital's rule (with ϵ as variable), we find that

$$f_X(x) := \lim_{\epsilon \downarrow 0} \frac{\exp\{-x + (\epsilon/2)\} - \exp\{-x - (\epsilon/2)\}}{\epsilon}$$

$$= \lim_{\epsilon \downarrow 0} \frac{\exp\{-x + (\epsilon/2)\}(1/2) - \exp\{-x - (\epsilon/2)\}(-1/2)}{1}$$

$$= \exp\{-x\} \quad \text{for } x > 0.$$

In two dimensions, we define the *partial derivative* of the function $f(x, y)$ with respect to x by

$$\frac{\partial}{\partial x} f(x, y) \equiv f_x(x, y) = \lim_{\Delta x \to 0} \frac{f(x + \Delta x, y) - f(x, y)}{\Delta x},$$

when the limit exists. The partial derivative of $f(x, y)$ with respect to y is defined similarly. Note that even if the partial derivatives $f_x(x_0, y_0)$ and $f_y(x_0, y_0)$ exist, the function $f(x, y)$ is not necessarily continuous at the point (x_0, y_0). Indeed, the limit of the function $f(x, y)$ must not depend on the way (x, y) tends to (x_0, y_0). For instance, let

$$f(x, y) = \begin{cases} \dfrac{xy}{x^2 + y^2} & \text{if } (x, y) \neq (0, 0), \\ \\ 0 & \text{if } (x, y) = (0, 0). \end{cases} \tag{1.3}$$

Suppose that (x, y) tends to $(0, 0)$ along the line $y = kx$, where $k \neq 0$. We then have:

$$\lim_{\substack{x \to 0 \\ y \to 0}} f(x, y) = \lim_{x \to 0} \frac{x(kx)}{x^2 + (kx)^2} = \frac{k}{1 + k^2} \quad (\neq 0).$$

Therefore, we must conclude that the function $f(x, y)$ is discontinuous at $(0, 0)$. However, we can show that $f_x(0, 0)$ and $f_y(0, 0)$ both exist (and are equal to 0).

1.3 Integrals

Definition 1.3.1. *Let $f(x)$ be a continuous (or piecewise continuous) function on the interval $[a, b]$ and let*

$$I = \sum_{k=1}^{n} f(\xi_k)(x_k - x_{k-1}),$$

where $\xi_k \in [x_{k-1}, x_k]$ for all k and $a = x_0 < x_1 < \cdots < x_{n-1} < x_n = b$ is a partition of the interval $[a, b]$. The limit of I as n tends to ∞, and $x_k - x_{k-1}$ decreases to 0 for all k, exists and is called the (**definite**) **integral** *of $f(x)$ over the interval $[a, b]$. We write:*

$$\lim_{\substack{n \to \infty \\ (x_k - x_{k-1}) \downarrow 0 \, \forall k}} I = \int_a^b f(x) \, dx.$$

Remarks. (i) The limit must not depend on the choice of the partition of the interval $[a, b]$.

(ii) The function $f(x)$ is called the *integrand*.

(iii) If $f(x) \geq 0$ in the interval $[a, b]$, then the integral of $f(x)$ over $[a, b]$ gives the area between the curve $y = f(x)$ and the x-axis from a to b.

(iv) If the interval $[a, b]$ is replaced by an infinite interval, or if the function $f(x)$ is not defined or not bounded at at least one point in $[a, b]$, then the integral is called *improper*. For example, we define the integral of $f(x)$ over the interval $[a, \infty)$ as follows:

$$\int_a^\infty f(x) \, dx = \lim_{b \to \infty} \int_a^b f(x) \, dx.$$

If the limit exists, the improper integral is said to be *convergent*; otherwise, it is *divergent*. When $[a, b]$ is the entire real line, we should write that

$$I_1 := \int_{-\infty}^\infty f(x) \, dx = \lim_{a \to -\infty} \int_a^0 f(x) \, dx + \lim_{b \to \infty} \int_0^b f(x) \, dx$$

and *not*

$$I_1 = \lim_{b \to \infty} \int_{-b}^{b} f(x)\, dx.$$

This last integral is actually the *Cauchy principal value* of the integral I_1. The Cauchy principal value may exist even if I_1 does not.

Definition 1.3.2. *Let $f(x)$ be a real-valued function. Any function $F(x)$ such that $F'(x) = f(x)$ is called a **primitive** (or **indefinite integral** or **antiderivative**) of $f(x)$.*

Theorem 1.3.1. (Fundamental Theorem of Calculus) *Let $f(x)$ be a continuous function on the interval $[a, b]$ and let $F(x)$ be a primitive of $f(x)$. Then,*

$$\int_{a}^{b} f(x)\, dx = F(b) - F(a).$$

Example 1.3.1. The function

$$f_X(x) = \frac{1}{\pi(1 + x^2)} \quad \text{for } x \in \mathbb{R}$$

is the *density function* of a particular *Cauchy distribution*. To obtain the *average value* of the *random variable X*, we can calculate the improper integral

$$\int_{-\infty}^{\infty} x f_X(x)\, dx = \int_{-\infty}^{\infty} \frac{x}{\pi(1 + x^2)}\, dx.$$

A primitive of the integrand $g(x) := x f_X(x)$ is

$$G(x) = \frac{1}{2\pi} \ln(1 + x^2).$$

We find that the improper integral diverges, because

$$\lim_{a \to -\infty} \int_{a}^{0} g(x)\, dx = \lim_{a \to -\infty} [G(0) - G(a)] = -\infty$$

and

$$\lim_{b \to \infty} \int_{0}^{b} g(x)\, dx = \lim_{b \to \infty} [G(b) - G(0)] = \infty.$$

Because $\infty - \infty$ is indeterminate, the integral indeed diverges. However, the Cauchy principal value of the integral is

$$\lim_{b \to \infty} \int_{-b}^{b} g(x)\, dx = \lim_{b \to \infty} [G(b) - G(-b)] = \lim_{b \to \infty} 0 = 0.$$

There are many results on integrals that could be recalled. We limit ourselves to mentioning a couple of techniques that are helpful to find indefinite integrals or evaluate definite integrals.

1.3.1 Particular integration techniques

(1) First, we remind the reader of the technique known as *integration by substitution*. Let $x = g(y)$. We can write that

$$\int f(x)\, dx = \int f[g(y)]g'(y)\, dy.$$

This result actually follows from the chain rule. In the case of a definite integral, assuming that

(i) $f(x)$ is continuous on the interval $[a, b]$,

(ii) the inverse function $g^{-1}(x)$ exists and

(iii) $g'(y)$ is continuous on $[g^{-1}(a), g^{-1}(b)]$ (resp., $[g^{-1}(b), g^{-1}(a)]$) if $g(y)$ is an increasing (resp., decreasing) function,

we have:

$$\int_a^b f(x)\, dx = \int_c^d f[g(y)]g'(y)\, dy,$$

where $a = g(c) \Leftrightarrow c = g^{-1}(a)$ and $b = g(d) \Leftrightarrow d = g^{-1}(b)$.

Example 1.3.2. Suppose that we want to evaluate the definite integral

$$I_2 := \int_0^4 x^{-1/2} e^{x^{1/2}}\, dx.$$

Making the substitution $x = g(y) = y^2$, so that $y = g^{-1}(x) = x^{1/2}$ (for $x \in [0, 4]$), we can write that

$$\int_0^4 x^{-1/2} e^{x^{1/2}}\, dx \overset{y = x^{1/2}}{=} \int_0^2 y^{-1} e^y\, 2y\, dy = 2e^y \Big|_0^2 = 2(e^2 - 1).$$

Remark. If we are only looking for a primitive of the function $f(x)$, then after having found a primitive of $f[g(y)]g'(y)$ we must replace y by $g^{-1}(x)$ (assuming it exists) in the function obtained. Thus, in the above example, we have:

$$\int x^{-1/2} e^{x^{1/2}}\, dx = \int y^{-1} e^y\, 2y\, dy = 2e^y = 2e^{x^{1/2}}.$$

(2) Next, a very useful integration technique is based on the formula for the derivative of a product:

$$\frac{d}{dx} f(x)g(x) = f'(x)g(x) + f(x)g'(x).$$

Integrating both sides of the previous equation, we obtain:

$$f(x)g(x) = \int f'(x)g(x)\,dx + \int f(x)g'(x)\,dx$$

$$\Longleftrightarrow \quad \int f(x)g'(x)\,dx = f(x)g(x) - \int f'(x)g(x)\,dx.$$

Setting $u = f(x)$ and $v = g(x)$, we can write that

$$\int u\,dv = uv - \int v\,du.$$

This technique is known as *integration by parts*. It is often used in probability to calculate the *moments* of a *random variable*.

Example 1.3.3. To obtain the *expected value* of the square of a *standard normal random variable*, we can try to evaluate the integral

$$I_3 := \int_{-\infty}^{\infty} cx^2 e^{-x^2/2}\,dx,$$

where c is a positive constant. When applying the integration by parts technique, one must decide which part of the integrand to assign to u. Of course, u should be chosen so that the new (indefinite) integral is easier to find. In the case of I_3, we set

$$u = cx \quad \text{and} \quad dv = xe^{-x^2/2}\,dx.$$

Because

$$v = \int dv = \int xe^{-x^2/2}\,dx = -e^{-x^2/2},$$

it follows that

$$I_3 = cx(-e^{-x^2/2})\Big|_{-\infty}^{\infty} + \int_{-\infty}^{\infty} ce^{-x^2/2}\,dx.$$

The constant c is such that the above improper integral is equal to 1 (see Chapter 3). Furthermore, making use of l'Hospital's rule, we find that

$$\lim_{x \to \infty} xe^{-x^2/2} = \lim_{x \to \infty} \frac{x}{e^{x^2/2}} = \lim_{x \to \infty} \frac{1}{e^{x^2/2} \cdot x} = 0.$$

Similarly,

$$\lim_{x \to -\infty} xe^{-x^2/2} = 0.$$

Hence, we may write that $I_3 = 0 + 1 = 1$.

Remarks. (i) Note that there is no elementary function that is a primitive of the function $e^{-x^2/2}$. Therefore, we could not have obtained a primitive (in terms of elementary functions) of $x^2 e^{-x^2/2}$ by proceeding as above.

(ii) If we set $u = ce^{-x^2/2}$ (and $dv = x^2 dx$) instead, then the resulting integral will be more complicated. Indeed, we would have:

$$I_3 = ce^{-x^2/2} \frac{x^3}{3} \Big|_{-\infty}^{\infty} + \int_{-\infty}^{\infty} c \frac{x^4}{3} e^{-x^2/2} dx = \int_{-\infty}^{\infty} c \frac{x^4}{3} e^{-x^2/2} dx.$$

A particular improper integral that is important in probability theory is defined by

$$F(\omega) = \int_{-\infty}^{\infty} e^{j\omega x} f(x) dx,$$

where $j := \sqrt{-1}$ and ω is a real parameter, and where we assume that the real-valued function $f(x)$ is *absolutely integrable*; that is,

$$\int_{-\infty}^{\infty} |f(x)| dx < \infty.$$

The function $F(\omega)$ is called the *Fourier transform* (up to a constant factor) of the function $f(x)$. It can be shown that

$$f(x) = \frac{1}{2\pi} \int_{-\infty}^{\infty} e^{-j\omega x} F(\omega) d\omega.$$

We say that $f(x)$ is the *inverse Fourier transform* of $F(\omega)$.

In probability theory, a *density function* $f_X(x)$ is, by definition, nonnegative and such that

$$\int_{-\infty}^{\infty} f_X(x) dx = 1.$$

Hence, the function $F(\omega)$ is well defined. In this context, it is known as the *characteristic function* of the *random variable* X and is often denoted by $C_X(\omega)$. By differentiating it, we can obtain the *moments* of X (generally more easily than by performing the appropriate integrals).

Example 1.3.4. The characteristic function of a *standard normal random variable* is

$$C_X(\omega) = e^{-\omega^2/2}.$$

We can show that the *expected value* of the square of X is given by

$$-\frac{d^2}{d\omega^2} e^{-\omega^2/2} \Big|_{\omega=0} = -e^{-\omega^2/2}(\omega^2 - 1) \Big|_{\omega=0} = 1,$$

which agrees with the result found in the previous example.

Finally, to obtain the *moments* of a *random variable* X, we can also use its *moment-generating function*, which, in the *continuous case*, is defined (if the integral exists) by

$$M_X(t) = \int_{-\infty}^{\infty} e^{tx} f_X(x)\,dx,$$

where we may assume that t is a real parameter. When X is nonnegative, the moment-generating function is actually the *Laplace transform* of the function $f_X(x)$.

1.3.2 Double integrals

In Chapter 4, on *random vectors*, we have to deal with double integrals to obtain various quantities of interest. A *joint density function* is a nonnegative function $f_{X,Y}(x,y)$ such that

$$\int_{-\infty}^{\infty} \int_{-\infty}^{\infty} f_{X,Y}(x,y)\,dx\,dy = 1.$$

Double integrals are needed, in general, to calculate the probability $P[A]$ that the *continuous random vector* (X,Y) takes on a value in a given subset A of $\mathbb{R}^2$. We have:

$$P[A] = \int \int_A f_{X,Y}(x,y)\,dx\,dy.$$

One has to describe the region A with the help of functions of x or of y (whichever is easier or more appropriate in the problem considered). That is,

$$P[A] = \int_a^b \int_{g_1(x)}^{g_2(x)} f_{X,Y}(x,y)\,dy\,dx = \int_a^b \left\{ \int_{g_1(x)}^{g_2(x)} f_{X,Y}(x,y)\,dy \right\} dx \qquad (1.4)$$

or

$$P[A] = \int_c^d \int_{h_1(y)}^{h_2(y)} f_{X,Y}(x,y)\,dx\,dy = \int_c^d \left\{ \int_{h_1(y)}^{h_2(y)} f_{X,Y}(x,y)\,dx \right\} dy. \qquad (1.5)$$

Remarks. (i) If the functions $g_1(x)$ and $g_2(x)$ [or $h_1(y)$ and $h_2(y)$] are constants, and if the function $f_{X,Y}(x,y)$ can be written as

$$f_{X,Y}(x,y) = f_X(x) f_Y(y),$$

then the double integral giving $P[A]$ can be expressed as a product of single integrals:

$$P[A] = \int_a^b f_X(x)\,dx \int_{\alpha}^{\beta} f_Y(y)\,dy,$$

where $\alpha = g_1(x) \; \forall x$ and $\beta = g_2(x) \; \forall x$.

(ii) Conversely, we can write the product of two single integrals as a double integral:

$$\int_a^b f(x)\,dx \int_c^d g(y)\,dy = \int_a^b \int_c^d f(x)g(y)\,dy\,dx.$$

Example 1.3.5. Consider the function

$$f_{X,Y}(x,y) = \begin{cases} e^{-x-y} & \text{if } x \geq 0 \text{ and } y \geq 0, \\ 0 & \text{elsewhere} \end{cases}$$

(see Figure 1.1). To obtain the probability that (X,Y) takes on a value in the region

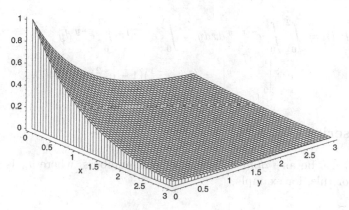

Fig. 1.1. Joint density function in Example 1.3.5.

$$A := \{(x,y) \in \mathbb{R}^2 : x \geq 0, y \geq 0, 0 \leq 2x + y \leq 1\}$$

(see Figure 1.2), we calculate the double integral

$$P[A] = \int_0^{1/2} \int_0^{1-2x} e^{-x-y}\,dy\,dx = \int_0^{1/2} \left\{ -e^{-x-y} \Big|_{y=0}^{y=1-2x} \right\} dx$$

$$= \int_0^{1/2} \left\{ e^{-x} - e^{x-1} \right\} dx = -e^{-x} - e^{x-1} \Big|_0^{1/2}$$

$$= -e^{-1/2} - e^{-1/2} + 1 + e^{-1} \simeq 0.1548.$$

We may also write that

$$P[A] = \int_0^1 \int_0^{(1-y)/2} e^{-x-y}\,dx\,dy = \int_0^1 \left\{ -e^{-x-y} \Big|_{x=0}^{x=(1-y)/2} \right\} dy$$

$$= \int_0^1 \left\{ e^{-y} - e^{-(1+y)/2} \right\} dy = -e^{-y} + 2e^{-(1+y)/2} \Big|_0^1$$

$$= -e^{-1} + 2e^{-1} + 1 - 2e^{-1/2} \simeq 0.1548.$$

Remark. In this example, the functions $g_i(x)$ and $h_i(y)$, $i = 1, 2$, that appear in (1.4) and (1.5) are given by

$$g_1(x) \equiv 0, \quad g_2(x) = 1 - 2x, \quad h_1(y) \equiv 0 \quad \text{and} \quad h_2(y) = (1 - y)/2.$$

If B is the rectangle defined by

$$B = \{(x, y) \in \mathbb{R}^2 : 0 \le x \le 1, 0 \le y \le 2\},$$

then we have:

$$P[B] = \int_0^2 \int_0^1 e^{-x-y} \, dx \, dy = \int_0^1 e^{-x} \, dx \int_0^2 e^{-y} \, dy$$

$$= -e^{-x} \Big|_0^1 - e^{-y} \Big|_0^2 = (1 - e^{-1})(1 - e^{-2}) \simeq 0.5466.$$

1.4 Infinite series

Let $a_1, a_2, \ldots, a_n, \ldots$ be an *infinite sequence* of real numbers, where a_n is given by a certain formula or rule, for example,

$$a_n = \frac{1}{n+1} \quad \text{for } n = 1, 2, \ldots .$$

We denote the infinite sequence by $\{a_n\}_{n=1}^{\infty}$ or simply by $\{a_n\}$. An infinite sequence is said to be *convergent* if $\lim_{n \to \infty} a_n$ exists; otherwise, it is *divergent*.

Next, from the sequence $\{a_n\}_{n=1}^{\infty}$ we define a new infinite sequence by

$$S_1 = a_1, \quad S_2 = a_1 + a_2, \quad \ldots, \quad S_n = a_1 + a_2 + \cdots + a_n, \quad \ldots .$$

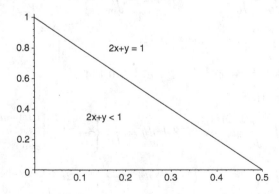

Fig. 1.2. Integration region in Example 1.3.5.

Definition 1.4.1. *The infinite sequence* $S_1, S_2, \ldots, S_n, \ldots$ *is represented by* $\sum_{n=1}^{\infty} a_n$ *and is called an* **infinite series**. *Moreover,* $S_n := \sum_{k=1}^{n} a_k$ *is called the nth* **partial sum** *of the series. Finally, if the limit* $\lim_{n\to\infty} S_n$ *exists (resp., does not exist), we say that the series is* **convergent** *(resp.,* **divergent***).*

In probability, the set of possible values of a *discrete random variable* X may be finite or *countably infinite* (see p. 27). In the latter case, the *probability function* p_X of X is such that

$$\sum_{k=1}^{\infty} p_X(x_k) = 1,$$

where $x_1, x_2, \ldots$ are the possible values of X. In the most important cases, the possible values of X are actually the integers $0, 1, \ldots$.

1.4.1 Geometric series

A particular type of infinite series encountered in Chapter 3 is known as a *geometric series*. These series are of the form

$$\sum_{n=1}^{\infty} ar^{n-1} \quad \text{or} \quad \sum_{n=0}^{\infty} ar^n, \tag{1.6}$$

where a and r are real constants.

Proposition 1.4.1. *If* $|r| < 1$, *the geometric series* $S(a, r) := \sum_{n=0}^{\infty} ar^n$ *converges to* $a/(1-r)$. *If* $|r| \geq 1$ *(and* $a \neq 0$*), then the series is* divergent.

To prove the above results, we simply have to consider the nth partial sum of the series:

$$S_n = \sum_{k=1}^{n} ar^{k-1} = a + ar + ar^2 + \cdots + ar^{n-1}.$$

We have:

$$rS_n = ar + ar^2 + ar^3 + \cdots + ar^{n-1} + ar^n,$$

so that

$$S_n - rS_n = a - ar^n \stackrel{r\neq 1}{\Longrightarrow} S_n = \frac{a(1-r^n)}{1-r}.$$

Hence, we deduce that

$$S := \lim_{n\to\infty} S_n = \begin{cases} \dfrac{a}{1-r} & \text{if } |r| < 1, \\ \text{does not exist if } |r| > 1. \end{cases}$$

If $r = 1$, we have that $S_n = na$, so that the series diverges (if $a \neq 0$). Finally, we can show that the series is also divergent if $r = -1$.

Other useful formulas connected with geometric series are the following: if $|r| < 1$, then

$$\sum_{k=1}^{\infty} ar^k = \frac{ar}{1-r}$$

and

$$\sum_{k=0}^{\infty} ak\,r^k = \frac{ar}{(1-r)^2}. \tag{1.7}$$

In probability, we also use *power series*, that is, series of the form

$$S(x) := a_0 + a_1 x + a_2 x^2 + \cdots + a_n x^n + \cdots,$$

where a_k is a constant, for $k = 0, 1 \ldots$. In particular, we use the fact that it is possible to express functions, for instance, the exponential function e^{cx}, as a power series:

$$e^{cx} = 1 + cx + \frac{c^2}{2!}x^2 + \cdots + \frac{c^n}{n!}x^n + \cdots \quad \text{for all } x \in \mathbb{R}. \tag{1.8}$$

This power series is called the *series expansion* of e^{cx}.

Remark. Note that a geometric series $S(a, r)$ is a power series $S(r)$ for which all the constants a_k are equal to a.

In general, a power series converges for all values of x in an interval around 0. If a given series expansion is valid for $|x| < R$ (> 0), we say that R is the *radius of convergence* of the series. For $|x| < R$, the series can be differentiated and integrated term by term:

$$S'(x) = a_1 + 2a_2 x + \cdots + na_n x^{n-1} + \cdots \tag{1.9}$$

and

$$\int_0^x S(t)\,dt = a_0 x + a_1 \frac{x^2}{2} + \cdots + a_n \frac{x^{n+1}}{n+1} + \cdots .$$

The *interval of convergence* of a power series having a radius of convergence $R > 0$ is at least $(-R, R)$. The series may or may not converge for $x = -R$ and $x = R$. Because

$$S(0) = a_0 \in \mathbb{R},$$

any power series converges for $x = 0$. If the series does not converge for any $x \neq 0$, we write that $R = 0$. Conversely, if the series converges for all $x \in \mathbb{R}$, then $R = \infty$.

To calculate the radius of convergence of a power series, we can make use of *d'Alembert's ratio test*: suppose that the limit

$$L := \lim_{n \to \infty} \left| \frac{u_{n+1}}{u_n} \right| \tag{1.10}$$

exists. Then, the series $\sum_{n=0}^{\infty} u_n$

(a) converges *absolutely* if $L < 1$;

(b) diverges if $L > 1$.

Remarks. (i) If the limit L in (1.10) does not exist, or if $L = 1$, then the test is inconclusive. There exist other criteria that can be used, for instance, Raabe's test.

(ii) An infinite series $\sum_{n=0}^{\infty} u_n$ converges *absolutely* if $\sum_{n=0}^{\infty} |u_n|$ converges. A series that converges absolutely is also convergent.

(iii) In the case of a power series, we calculate

$$\lim_{n \to \infty} \left| \frac{a_{n+1} x^{n+1}}{a_n x^n} \right| = \lim_{n \to \infty} \left| \frac{a_{n+1}}{a_n} \right| |x|.$$

For example, the series expansion of the exponential function e^{cx} given in (1.8) is valid for all $x \in \mathbb{R}$. Indeed, $a_n = c^n/n!$, so that

$$\lim_{n \to \infty} \left| \frac{a_{n+1} x^{n+1}}{a_n x^n} \right| = \lim_{n \to \infty} \left| \frac{c}{n+1} \right| |x| = 0 < 1 \quad \forall x \in \mathbb{R}.$$

Example 1.4.1. To obtain the *mean* of a *geometric random variable*, we can compute the infinite sum

$$\sum_{k=1}^{\infty} k(1-p)^{k-1} p = \frac{p}{1-p} \sum_{k=0}^{\infty} k(1-p)^k \stackrel{(1.7)}{=} \frac{p}{1-p} \frac{1-p}{p^2} = \frac{1}{p},$$

where $0 < p < 1$. To prove Formula (1.7), we can use (1.9).

Example 1.4.2. A *Poisson random variable* is such that its *probability function* is given by

$$p_X(x) = e^{-\lambda} \frac{\lambda^x}{x!} \quad \text{for } x = 0, 1, \ldots,$$

where λ is a positive constant. We have:

$$\sum_{x=0}^{\infty} p_X(x) = \sum_{x=0}^{\infty} e^{-\lambda} \frac{\lambda^x}{x!} = e^{-\lambda} \sum_{x=0}^{\infty} \frac{\lambda^x}{x!} \stackrel{(1.8)}{=} e^{-\lambda} e^{\lambda} = 1,$$

as required.

Example 1.4.3. The power series

$$S_k(x) := 1 + kx + \frac{k(k-1)}{2!} x^2 + \cdots + \frac{k(k-1)(k-2) \cdots (k-n+1)}{n!} x^n + \cdots$$

is the series expansion of the function $(1+x)^k$, and is called the *binomial series*. In probability, k will be a natural integer. It follows that the series has actually a finite number of terms and thus converges for all $x \in \mathbb{R}$. Moreover, we can then write that

$$\frac{k(k-1)(k-2)\cdots(k-n+1)}{n!} = \frac{k!}{(k-n)!n!}$$

for $n = 1, \ldots, k$.

To conclude this review of calculus, we give the main logarithmic formulas:

$$\ln ab = \ln a + \ln b; \quad \ln a/b = \ln a - \ln b; \quad \ln a^b = b \ln a.$$

We also have:

$$\ln e^{cx} = e^{\ln(cx)} = cx$$

and

$$e^{f(x)\ln x} = x^{f(x)}.$$

1.5 Exercises for Chapter 1

Solved exercises[1]

Question no. 1
   Calculate $\lim_{x \downarrow 0} x \sin(1/x)$.

Question no. 2
   For what values of x is the function

$$f(x) = \begin{cases} \dfrac{\sin x}{x} & \text{if } x \neq 0, \\ 1 & \text{if } x = 0 \end{cases}$$

continuous?

Question no. 3
   Differentiate the function $f(x) = \sqrt{3x+1}(2x^2+1)^2$.

Question no. 4
   Find the limit $\lim_{x \downarrow 0} x \ln x$.

Question no. 5
   Evaluate the definite integral

$$I_5 := \int_1^e \frac{\ln x}{x}\, dx.$$

---

[1] The solutions can be found in Appendix C.

Question no. 6

Find the value of the definite integral

$$I_6 := \int_{-\infty}^{\infty} x^3 e^{-x^2/2} dx.$$

Question no. 7

Find the Fourier transform of the function

$$f(x) = ce^{-cx} \quad \text{for } x \geq 0,$$

where c is a positive constant.

Question no. 8

Let

$$f(x, y) = \begin{cases} x + y & \text{if } 0 \leq x \leq 1, 0 \leq y \leq 1, \\ 0 & \text{elsewhere.} \end{cases}$$

Calculate

$$I_8 := \int_A \int f(x, y)\, dx dy,$$

where $A := \{(x, y) \in \mathbb{R}^2 : 0 \leq x \leq 1, 0 \leq y \leq 1, x^2 < y\}$ (see Figure 1.3).

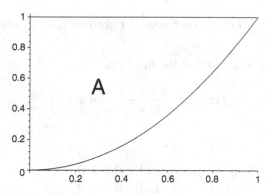

Fig. 1.3. Region A in solved exercise no. 8.

Question no. 9

Find the value of the infinite series

$$S_9 := \frac{1}{8} + \frac{1}{16} + \frac{1}{32} + \cdots .$$

Question no. 10
    Calculate

$$S_{10} := \sum_{k=1}^{\infty} k^2(1-p)^{k-1}p,$$

where $0 < p < 1$.

Exercises

Question no. 1
    Let

$$F(x) = \begin{cases} 0 & \text{if } x < 0, \\ 1/2 & \text{if } x = 0, \\ 1 - (1/2)e^{-x} & \text{if } x > 0. \end{cases}$$

Calculate (a) $\lim_{x \uparrow 0} F(x)$, (b) $\lim_{x \downarrow 0} F(x)$, and (c) $\lim_{x \to 0} F(x)$.

Question no. 2
    Consider the function

$$F(x) = \begin{cases} 0 & \text{if } x < 0, \\ 1/3 & \text{if } 0 \leq x < 1, \\ 2/3 & \text{if } 1 \leq x < 2, \\ 1 & \text{if } x \geq 2. \end{cases}$$

For what values of x is $F(x)$ left-continuous? right-continuous? continuous?

Question no. 3
    Find the limit as x tends to 0 of the function

$$f(x) = \frac{x^2 \sin(1/x)}{\sin x} \quad \text{for } x \in \mathbb{R}.$$

Question no. 4
    Is the function

$$f(x) = \begin{cases} e^{1/x} & \text{if } x \neq 0, \\ 0 & \text{if } x = 0 \end{cases}$$

continuous or discontinuous at $x = 0$? Justify.

Question no. 5
    Find the fourth-order derivative of the function $F(\omega) = e^{-\omega^2/2}$, for any $\omega \in \mathbb{R}$, and evaluate the derivative at $\omega = 0$.

Question no. 6

Find the following limit:

$$\lim_{\epsilon \downarrow 0} \frac{\left(1 + x - \frac{\epsilon}{2}\right) e^{-x + (\epsilon/2)} - \left(1 + x + \frac{\epsilon}{2}\right) e^{-x - (\epsilon/2)}}{\epsilon}.$$

Question no. 7

Determine the second-order derivative of $f(x) = \sqrt[3]{2x^2}$.

Question no. 8

Calculate the derivative of

$$f(x) = 1 + \sqrt[3]{x^2} \quad \text{for } x \in \mathbb{R}$$

and find the value of x that minimizes this function.

Question no. 9

Use the fact that

$$\int_0^\infty x^{n-1} e^{-x} dx = (n-1)! \quad \text{for } n = 1, 2, \ldots$$

to evaluate the integral

$$\int_0^\infty x^{n-1} e^{-cx} dx,$$

where c is a positive constant.

Question no. 10

Use the following formula:

$$\int_{-\infty}^\infty e^{-(x-m)^2/2} dx = \sqrt{2\pi},$$

where m is a real constant, to calculate the definite integral

$$\int_{-\infty}^\infty x^2 e^{-x^2/2} dx.$$

Question no. 11

Evaluate the improper integral

$$\int_0^\infty \frac{e^{-x}}{x} dx.$$

Question no. 12

Find a primitive of the function

$$f(x) = e^{-x}\sin x \quad \text{for } x \in \mathbb{R}.$$

Question no. 13

Find the Fourier transform of the function

$$f(x) = \frac{c}{2}e^{-c|x|} \quad \text{for } x \in \mathbb{R},$$

where c is a positive constant.

Question no. 14

Let

$$f(x,y) = \begin{cases} \frac{3}{2}x & \text{if } 1 \le x \le 2, 1 \le y \le 2, x \le y, \\ 0 & \text{elsewhere.} \end{cases}$$

Calculate the double definite integral

$$\int\int_A f(x,y)\,dxdy,$$

where

$$A := \{(x,y) \in \mathbb{R}^2 : 1 \le x \le 2, 1 \le y \le 2, x^2 < y\}.$$

Question no. 15

The *convolution* of two functions, f and g, is denoted by $f * g$ and is defined by

$$(f * g)(x) = \int_{-\infty}^{\infty} f(y)g(x-y)\,dy.$$

Let

$$f(x) = \begin{cases} ce^{-cx} & \text{if } x \ge 0, \\ 0 & \text{if } x < 0, \end{cases}$$

where c is a positive constant, and assume that $g(x) \equiv f(x)$ (i.e., g is *identical* to f). Find $(f * g)(x)$.

Remark. In probability theory, the convolution of f_X and f_Y is the *probability density function* of the sum $Z := X + Y$ of the *independent random variables* X and Y. The result of the above exercise implies that the sum of two independent *exponentially distributed* random variables with the same *parameter c* has a *gamma distribution* with parameters $\alpha = 2$ and $\lambda = c$.

Question no. 16
  Prove that

$$I := \int_{-\infty}^{\infty} \frac{1}{\sqrt{2\pi}} e^{-z^2/2} dz = 1$$

by first writing that

$$I^2 = \frac{1}{2\pi} \int_{-\infty}^{\infty} e^{-z^2/2} dz \int_{-\infty}^{\infty} e^{-w^2/2} dw = \frac{1}{2\pi} \int_{-\infty}^{\infty} \int_{-\infty}^{\infty} e^{-(z^2+w^2)/2} dz\, dw$$

and then using *polar coordinates*. That is, set $z = r\cos\theta$ and $w = r\sin\theta$ (with $r \geq 0$), so that

$$r = \sqrt{z^2 + w^2} \quad \text{and} \quad \theta = \tan^{-1}(w/z).$$

Remark. We find that $I^2 = 1$. Justify why this implies that $I = 1$ (and not $I = -1$).

Question no. 17
  Determine the value of the infinite series

$$\frac{1}{2!}x^2 - \frac{1}{3!}x^3 + \cdots + \frac{(-1)^n}{n!}x^n + \cdots .$$

Question no. 18
  Let

$$S(q) = \sum_{n=1}^{\infty} q^{n-1},$$

where $0 < q < 1$. Calculate

$$\int_0^{1/2} S(q)\, dq.$$

Question no. 19
(a) Calculate the infinite series

$$M(t) := \sum_{k=0}^{\infty} e^{tk} e^{-\alpha} \frac{\alpha^k}{k!},$$

where $\alpha > 0$.

(b) Evaluate the second-order derivative $M''(t)$ at $t = 0$.

Remark. The function $M(t)$ is the *moment-generating function* of a random variable X having a *Poisson distribution* with *parameter* α. Moreover, $M''(0)$ gives us the *expected value* of X^2.

Question no. 20

(a) Determine the value of the power series

$$G(z) := \sum_{k=0}^{\infty} z^k (1-p)^k p,$$

where $z \in \mathbb{R}$ and $p \in (0,1)$ are such that $|z| < (1-p)^{-1}$.

(b) Calculate

$$\frac{d^k}{dz^k} G(z) \quad \text{for } k = 0, 1, \ldots$$

at $z = 0$.

Remark. The function $G_X(z) := \sum_{k=0}^{\infty} z^k p_X(k)$ is called the *generating function* of the *discrete random variable* X taking its values in the set $\{0, 1, \ldots\}$ and having *probability mass function* $p_X(k)$. Furthermore,

$$\frac{1}{k!} \frac{d^k}{dz^k} G_X(z) \Big|_{z=0}$$

yields $p_X(k)$.

Multiple choice questions

Question no. 1

   Calculate the limit

$$\lim_{x \to 1} \frac{\sqrt{x} - 1}{1 - x}.$$

(a) -1 (b) $-1/2$ (c) $1/2$ (d) 1 (e) does not exist

Question no. 2

   Let $u(x)$ be the Heaviside function (see p. 2) and define $h(x) = e^{u(x)/2}$, for $x \in \mathbb{R}$. Calculate the limit of the function $h(x)$ as x tends to 0.

(a) $1/2$ (b) 1 (c) $(1 + e^{1/2})/2$ (d) $e^{1/2}$ (e) does not exist

Question no. 3

   Evaluate the second-order derivative of the function

$$F(\omega) = \frac{2(2 + j\omega)}{(4 + \omega^2)} \quad \text{for } \omega \in \mathbb{R}$$

at $\omega = 0$.

a) $-1/4$ (b) $-1/2$ (c) 0 (d) $1/4$ (e) $1/2$

Question no. 4

Find the following limit:

$$\lim_{x \to \infty} \left(1 + \frac{1}{x}\right)^x.$$

Indication. Take the natural logarithm of the expression first and then use the fact that $\ln x$ is a continuous function.

(a) 0   (b) 1   (c) e (d) ∞ (e) does not exist

Question no. 5

Use the formula

$$\int_{-\infty}^{\infty} \frac{1}{\sqrt{2\pi}b} \exp\left\{-\frac{(x-a)^2}{2b^2}\right\} dx = 1,$$

where a and b (> 0) are constants, to evaluate the definite integral

$$\int_0^{\infty} \frac{1}{\sqrt{2\pi}} e^{-2x^2} dx.$$

(a) 1/4   (b) 1/2   (c) 1   (d) 2   (e) 4

Question no. 6

Calculate the definite integral

$$\int_0^1 x \ln x \, dx.$$

(a) $-\infty$ (b) -1 (c) $-1/2$ (d) $-1/4$ (e) 0

Question no. 7

Suppose that

$$f(x) = \begin{cases} 1 \text{ if } 0 \le x \le 1, \\ 0 \text{ elsewhere} \end{cases}$$

and that $g(x) = f(x)$ for all $x \in \mathbb{R}$. Find $(f * g)(3/2)$, where $*$ denotes the *convolution* of f and g (see no. 15, p. 22).

(a) 0   (b) 1/2   (c) 1   (d) 3/2   (e) 2

Question no. 8

Let

$$f(x, y) = \begin{cases} 2 - x - y \text{ if } 0 < x < 1, 0 < y < 1, \\ 0 \quad \text{elsewhere.} \end{cases}$$

Calculate the following double integral:

$$\int \int_A f(x, y) \, dx dy,$$

where
$$A := \{(x, y) \in \mathbb{R}^2 : 0 < x < 1, 0 < y < 1, x + y > 1\}.$$

(a) 1/6   (b) 1/4   (c) 1/3   (d) 1/2   (e) 5/6

Question no. 9
  Find the value of the natural logarithm of the *infinite product*

$$\prod_{n=1}^{\infty} e^{1/2^n} = e^{1/2} \times e^{1/4} \times e^{1/8} \times \cdots .$$

(a) −1   (b) −1/2   (c) 0   (d) 1/2   (e) 1

Question no. 10
  We define $a_0 = 0$ and

$$a_k = e^{-c} \frac{c^{|k|}}{2|k|!} \quad \text{for } k = \ldots, -2, -1, 1, 2, \ldots,$$

where $c > 0$ is a constant. Find the value of the infinite series $\sum_{k=-\infty}^{\infty} a_k$.

(a) $1 - e^{-c}$ (b) 0   (c) 1   (d) $(1/2)e^{-c}$ (e) $1 + e^{-c}$

2

Elementary probability

This first chapter devoted to probability theory contains the basic definitions and concepts in this field, without the formalism of *measure theory*. However, the range of problems that can be solved by using the formulas of *elementary probability* is very broad, particularly in *combinatorial analysis*. Therefore, it is necessary to do numerous exercises in order to master these basic concepts.

2.1 Random experiments

A *random experiment* is an experiment that, at least theoretically, may be repeated as often as we want and whose outcome cannot be predicted, for example, the roll of a die. Each time the experiment is repeated, an *elementary outcome* is obtained. The set of all elementary outcomes of a random experiment is called the *sample space*, which is denoted by Ω.

Sample spaces may be discrete or continuous.

(a) Discrete sample spaces. (i) Firstly, if the number of possible outcomes is finite. For example, if a die is rolled and the number that shows up is noted, then $\Omega = \{1, 2, \ldots, 6\}$.

ii) Secondly, if the number of possible outcomes is *countably infinite*, which means that there is an infinite number of possible outcomes, but these outcomes can be put in a one-to-one correspondence with the positive integers. For example, if a die is rolled until a "6" is obtained, and the number of rolls made before getting this first "6" is counted, then we have that $\Omega = \{0, 1, 2, \ldots\}$. This set is *equivalent* to the set of all natural integers $\{1, 2, \ldots\}$, because we can associate the natural number $k+1$ with each element $k = 0, 1, \ldots$ of Ω.

(b) Continuous sample spaces. If the sample space contains one or many intervals, the sample space is then *uncountably infinite*. For example, a die is rolled until a "6" is obtained and the *time* needed to get this first "6" is recorded. In this case, we have that $\Omega = \{t \in \mathbb{R} : t > 0\}$ [or $\Omega = (0, \infty)$].

M. Lefebvre, *Basic Probability Theory with Applications*, Springer Undergraduate Texts in Mathematics and Technology, DOI: 10.1007/978-0-387-74995-2_2,
© Springer Science + Business Media, LLC 2009

2.2 Events

Definition 2.2.1. *An* **event** *is a set of elementary outcomes. That is, it is a subset of the sample space Ω. In particular, every elementary outcome is an event, and so is the sample space itself.*

Remarks. (i) An elementary outcome is sometimes called a *simple* event, whereas a *compound* event is made up of at least two elementary outcomes.

(ii) To be precise, we should distinguish between the elementary outcome ω, which is an element of Ω, and the *elementary event* $\{\omega\} \subset \Omega$.

(iii) The events are denoted by A, B, C, and so on.

Definition 2.2.2. *Two events, A and B, are said to be* **incompatible** *(or* **mutually exclusive***) if their intersection is empty. We then write that $A \cap B = \emptyset$.*

Example 2.2.1. Consider the experiment that consists in rolling a die and recording the number that shows up. We have that $\Omega = \{1, 2, 3, 4, 5, 6\}$. We define the events

$$A = \{1, 2, 4\}, \quad B = \{2, 4, 6\} \quad \text{and} \quad C = \{3, 5\}.$$

We have:

$$A \cup B = \{1, 2, 4, 6\}, \quad A \cap B = \{2, 4\} \quad \text{and} \quad A \cap C = \emptyset.$$

Therefore, A and C are incompatible events. Moreover, we may write that $A' = \{3, 5, 6\}$, where the symbol $'$ denotes the *complement* of the event.

To represent a sample space and some events, we often use a *Venn diagram* as in Figure 2.1. In general, for three events we have the diagram in Figure 2.2.

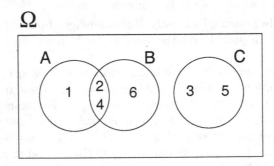

Fig. 2.1. Venn diagram for Example 2.2.1.

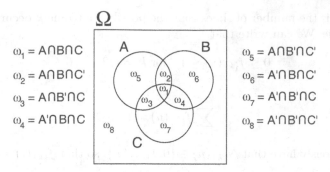

$\omega_1 = A \cap B \cap C$

$\omega_2 = A \cap B \cap C'$

$\omega_3 = A \cap B' \cap C$

$\omega_4 = A' \cap B \cap C$

$\omega_5 = A \cap B' \cap C'$

$\omega_6 = A' \cap B \cap C'$

$\omega_7 = A' \cap B' \cap C$

$\omega_8 = A' \cap B' \cap C'$

Fig. 2.2. Venn diagram for three arbitrary events.

2.3 Probability

Definition 2.3.1. *The* **probability** *of an event $A \subset \Omega$, denoted by $P[A]$, is a real number obtained by applying to A the function P that possesses the following properties:*

(i) $0 \le P[A] \le 1$;

(ii) if $A = \Omega$, then $P[A] = 1$;

(iii) if $A = A_1 \cup A_2 \cup \cdots \cup A_n$, where $A_1, \ldots, A_n$ are incompatible events, then we may write that

$$P[A] = \sum_{i=1}^{n} P[A_i] \quad for \ n = 2, 3, \ldots, \infty.$$

Remarks. (i) Actually, we only have to write that $P[A] \ge 0$ in the definition, because we can show that

$$P[A] + P[A'] = 1,$$

which implies that $P[A] = 1 - P[A'] \le 1$.
(ii) We also have the following results:

$$P[\emptyset] = 0 \quad \text{and} \quad P[A] \le P[B] \quad \text{if } A \subset B.$$

(iii) The definition of the probability of an event is motivated by the notion of *relative frequency*. For example, suppose that the random experiment that consists in rolling a die is repeated a very large number of times, and that we wish to obtain the probability of any of the possible outcomes of this experiment, namely, the integers $1, 2, \ldots, 6$. The relative frequency of the elementary event $\{k\}$ is the quantity $f_{\{k\}}(n)$ defined by

$$f_{\{k\}}(n) = \frac{N_{\{k\}}(n)}{n},$$

where $N_{\{k\}}(n)$ is the number of times that the possible outcome k occurred among the n rolls of the die. We can write that

$$0 \leq f_{\{k\}}(n) \leq 1 \quad \text{for } k = 1, 2, \ldots, 6$$

and that

$$\sum_{k=1}^{6} f_{\{k\}}(n) = 1.$$

Indeed, we obviously have that $N_{\{k\}}(n) \in \{0, 1, \ldots, n\}$, so that $f_{\{k\}}(n)$ belongs to $[0, 1]$, and

$$\sum_{k=1}^{6} f_{\{k\}}(n) = \frac{N_{\{1\}}(n) + \cdots + N_{\{6\}}(n)}{n} = \frac{n}{n} = 1.$$

Furthermore, if A is an event containing two possible outcomes, for instance "1" and "2," then

$$f_A(n) = f_{\{1\}}(n) + f_{\{2\}}(n),$$

because the outcomes 1 and 2 cannot occur on the *same* roll of the die.

Finally, the probability of the elementary event $\{k\}$ can theoretically be obtained by taking the limit of $f_{\{k\}}(n)$ as the number n of rolls tends to infinity:

$$P[\{k\}] = \lim_{n \to \infty} f_{\{k\}}(n).$$

The probability of an arbitrary event can be expressed in terms of the relative frequency of this event, thus it is logical that the properties of probabilities more or less mimic those of relative frequencies.

Sometimes, the probability of an elementary outcome is simply equal to 1 divided by the total number of elementary outcomes. In this case, the elementary outcomes are said to be *equiprobable* (or *equally likely*). For example, if a *fair* (or *unbiased*) die is rolled, then we have that $P[\{1\}] = P[\{2\}] = \cdots = P[\{6\}] = 1/6$.

If the elementary outcomes r_i are *not* equiprobable, we can (try to) make use of the following formula:

$$P[A] = \sum_{r_i \in A} P[\{r_i\}].$$

However, this formula is only useful if we know the probability of all the elementary outcomes r_i that constitute the event A.

Now, if A and B are incompatible events, then we deduce from the third property of $P[\cdot]$ that $P[A \cup B] = P[A] + P[B]$. If A and B are not incompatible, we can show (see Figure 2.3) that

$$P[A \cup B] = P[A] + P[B] - P[A \cap B].$$

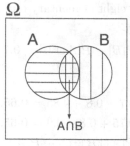

$$P[A \cup B] = P[A] + P[B] - P[A \cap B]$$

Fig. 2.3. Probability of the union of two arbitrary events.

Similarly, in the case of three arbitrary events, we have:

$$P[A \cup B \cup C] = P[A] + P[B] + P[C] - P[A \cap B] - P[A \cap C] - P[B \cap C] + P[A \cap B \cap C].$$

Example 2.3.1. The three most popular options for a certain model of new car are A: automatic transmission, B: V6 engine, and C: air conditioning. Based on the previous sales data, we may suppose that $P[A] = 0.70$, $P[B] = 0.75$, $P[C] = 0.80$, $P[A \cup B] = 0.80$, $P[A \cup C] = 0.85$, $P[B \cup C] = 0.90$, and $P[A \cup B \cup C] = 0.95$, where $P[A]$ denotes the probability that an arbitrary buyer chooses option A, and so on. Calculate the probability of each of the following events:
(a) the buyer chooses at least one of the three options;
(b) the buyer does not choose any of the three options;
(c) the buyer chooses only air conditioning;
(d) the buyer chooses exactly one of the three options.

Solution. (a) We seek $P[A \cup B \cup C] = 0.95$ (by assumption).
(b) We now seek $P[A' \cap B' \cap C'] = 1 - P[A \cup B \cup C] = 1 - 0.95 = 0.05$.
(c) The event whose probability is requested is $A' \cap B' \cap C$. We can write that

$$P[A' \cap B' \cap C] = P[A \cup B \cup C] - P[A \cup B] = 0.95 - 0.8 = 0.15.$$

(d) Finally, we want to calculate

$$P[(A \cap B' \cap C') \cup (A' \cap B \cap C') \cup (A' \cap B' \cap C)]$$
$$\stackrel{\text{inc.}}{=} P[A \cap B' \cap C'] + P[A' \cap B \cap C'] + P[A' \cap B' \cap C]$$
$$= 3P[A \cup B \cup C] - P[A \cup B] - P[A \cup C] - P[B \cup C]$$
$$= 3(0.95) - 0.8 - 0.85 - 0.9 = 0.3.$$

Remarks. (i) The indication "inc." above the "=" sign means that the equality is true because of the *incompatibility* of the events. We use this type of notation often in this book to justify the passage from an expression to another.

(ii) The probability of each of the eight elementary outcomes is indicated in the diagram of Figure 2.4. First, we calculate

$$P[A \cap B] = P[A] + P[B] - P[A \cup B] = 0.7 + 0.75 - 0.8 = 0.65.$$

Likewise, we have:

$$P[A \cap C] = 0.7 + 0.8 - 0.85 = 0.65,$$
$$P[B \cap C] = 0.75 + 0.8 - 0.9 = 0.65,$$
$$P[A \cap B \cap C] = P[A \cup B \cup C] - P[A] - P[B] - P[C]$$
$$+ P[A \cap B] + P[A \cap C] + P[B \cap C]$$
$$= 0.95 - 0.7 - 0.75 - 0.8 + 3(0.65) = 0.65.$$

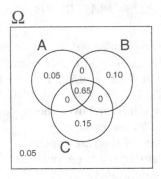

Fig. 2.4. Venn diagram for Example 2.3.1.

2.4 Conditional probability

Definition 2.4.1. *The* **conditional probability** *of event A, given that event B occurred, is defined (and denoted) by (see Figure 2.5)*

$$P[A \mid B] = \frac{P[A \cap B]}{P[B]} \quad \text{if } P[B] > 0. \tag{2.1}$$

From the above definition, we obtain the *multiplication rule:*

$$P[A \cap B] = P[A \mid B]P[B] \quad \text{if } P[B] > 0 \tag{2.2}$$

and

$$P[A \cap B] = P[B \mid A]P[A] \quad \text{if } P[A] > 0.$$

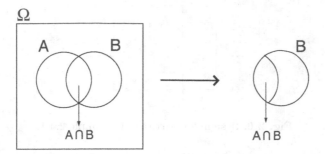

Fig. 2.5. Notion of conditional probability.

Definition 2.4.2. *Let A and B be two events such that $P[A]P[B] > 0$. We say that A and B are **independent** events if*

$$P[A \mid B] = P[A] \quad or \quad P[B \mid A] = P[B]. \tag{2.3}$$

We deduce from the multiplication rule that A and B are independent if and only if (iff)

$$P[A \cap B] = P[A]P[B]. \tag{2.4}$$

Actually, this equation is the definition of independence of events A and B in the general case when we can have that $P[A]P[B] = 0$. However, Definition 2.4.2 is more intuitive, whereas the general definition of independence given by Formula (2.4) is purely mathematical.

In general, the events $A_1, A_2, \ldots, A_n$ are independent iff

$$P[A_{i_1} \cap \cdots \cap A_{i_k}] = \prod_{j=1}^{k} P[A_{i_j}]$$

for $k = 2, 3, \ldots, n$, where $A_{i_l} \neq A_{i_m}$ if $l \neq m$.

Remark. If A and B are two incompatible events, then they *cannot* be independent, unless $P[A]P[B] = 0$. Indeed, in the case when $P[A]P[B] > 0$, we have:

$$P[A \mid B] = \frac{P[A \cap B]}{P[B]} = \frac{P[\emptyset]}{P[B]} = \frac{0}{P[B]} = 0 \neq P[A].$$

Example 2.4.1. A device is constituted of two components, A and B, subject to failures. The components are connected in parallel (see Figure 2.6) and are not independent. We estimate the probability of a failure of component A to be 0.2 and that of a failure of component B to be 0.8 if component A is down, and to be 0.4 if component A is not down.

(a) Calculate the probability of a failure (i) of component B and (ii) of the device.

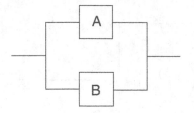

Fig. 2.6. System for part (a) of Example 2.4.1.

Solution. Let A (resp., B) be the event "component A (resp., B) is down." By assumption, we have that $P[A] = 0.2$, $P[B \mid A] = 0.8$, and $P[B \mid A'] = 0.4$.

(i) We may write (see Figure 2.7) that

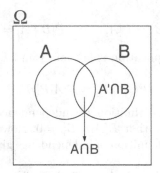

Fig. 2.7. Venn diagram for part (a) of Example 2.4.1.

$$P[B] = P[A \cap B] + P[A' \cap B] = P[B \mid A]P[A] + P[B \mid A']P[A']$$
$$= (0.8)(0.2) + (0.4)(0.8) = 0.48.$$

(ii) We seek $P[\text{Device failure}] = P[A \cap B] = P[B \mid A]P[A] = 0.16$.

(b) In order to increase the reliability of the device, a third component, C, is added to the system in such a way that components A, B, and C are connected in parallel (see Figure 2.8). The probability that component C fails is equal to 0.2, independently from the state (up or down) of components A and B. Calculate the probability that the device made up of components A, B, and C breaks down.

Solution. By assumption, $P[C] = 0.2$ and C is independent of A and B. Let F be the event "the subsystem made up of components A and B fails." We can write that

$$P[F \cap C] \stackrel{\text{ind.}}{=} P[A \cap B]P[C] \stackrel{\text{(a)\_(ii)}}{=} (0.16)(0.2) = 0.032.$$

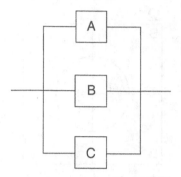

Fig. 2.8. System for part (b) of Example 2.4.1.

2.5 Total probability

Let $B_1, B_2, \ldots, B_n$ be *incompatible* and *exhaustive* events; that is, we have:

$$B_i \cap B_j = \emptyset \quad \text{if } i \neq j \quad \text{and} \quad \bigcup_{i=1}^{n} B_i = \Omega.$$

We say that the events B_i constitute a *partition* of the sample space Ω. It follows that

$$P\left[\bigcup_{i=1}^{n} B_i\right] = \sum_{i=1}^{n} P[B_i] = P[\Omega] = 1.$$

Now, let A be an arbitrary event. We can write that (see Figure 2.9)

$$P[A] = \sum_{i=1}^{n} P[A \cap B_i] = \sum_{i=1}^{n} P[A \mid B_i] P[B_i] \tag{2.5}$$

(the second equality above being valid when $P[B_i] > 0$, for $i = 1, 2, \ldots, n$).

Remark. This formula is sometimes called the *law of total probability*.

Finally, suppose that we wish to calculate $P[B_i \mid A]$, for $i = 1, \ldots, n$. We have:

$$P[B_i \mid A] = \frac{P[B_i \cap A]}{P[A]} = \frac{P[A \mid B_i]P[B_i]}{\sum_{j=1}^{n} P[A \cap B_j]} = \frac{P[A \mid B_i]P[B_i]}{\sum_{j=1}^{n} P[A \mid B_j]P[B_j]}. \tag{2.6}$$

This formula is called *Bayes' formula*.

Remark. We also have (*Bayes' rule*):

$$P[B \mid A] = \frac{P[A \mid B]P[B]}{P[A]} \quad \text{if } P[A]P[B] > 0. \tag{2.7}$$

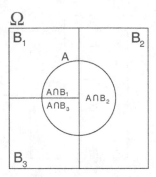

$$P[A]=P[A \cap B_1]+P[A \cap B_2]+P[A \cap B_3]$$

Fig. 2.9. Example of the law of total probability with $n = 3$.

Example 2.5.1. Suppose that machines M_1, M_2, and M_3 produce, respectively, 500, 1000, and 1500 parts per day, of which 5%, 6%, and 7% are defective. A part produced by one of these machines is taken at random, at the end of a given workday, and it is found to be defective. What is the probability that it was produced by machine M_3?

Solution. Let A_i be the event "the part taken at random was produced by machine M_i," for $i = 1, 2, 3$, and let D be "the part taken at random is defective." We seek

$$P[A_3 \mid D] = \frac{P[D \mid A_3]P[A_3]}{\sum_{i=1}^{3} P[D \mid A_i]P[A_i]} = \frac{(0.07)\left(\frac{1500}{3000}\right)}{(0.05)\left(\frac{1}{6}\right) + (0.06)\left(\frac{1}{3}\right) + (0.07)\left(\frac{1}{2}\right)}$$
$$= \frac{105}{190} \simeq 0.5526.$$

2.6 Combinatorial analysis

Suppose that we perform a random experiment that can be divided into two steps. On the first step, outcome A_1 *or* outcome A_2 may occur. On the second step, either of outcomes B_1, B_2, or B_3 may occur. We can use a *tree diagram* to describe the sample space of this random experiment, as in Figure 2.10.

Example 2.6.1. Tests conducted with a new breath alcohol analyzer enabled us to establish that (i) 5 times out of 100 the test proved positive even though the person subjected to the test was not intoxicated; (ii) 90 times out of 100 the test proved positive and the person tested was really intoxicated. Moreover, we estimate that 1% of the persons subjected to the test are really intoxicated. Calculate the probability that
(a) the test will be positive for the next person subjected to it;
(b) a given person is intoxicated, given that the test is positive.

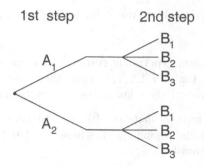

Fig. 2.10. Example of a tree diagram.

Solution. Let A be the event "the test is positive" and let E be "the person subjected to the test is intoxicated." From the above assumptions, we can construct the tree diagram in Figure 2.11, where the *marginal* probabilities of events E and E' are written above the branches, as well as the conditional probabilities of events A and A', given that E or E' occurred. Furthermore, we know by the multiplication rule that the product of these probabilities is equal to the probability of the intersections $E \cap A$, $E \cap A'$, and so on.

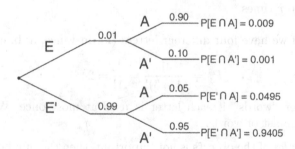

Fig. 2.11. Tree diagram in Example 2.6.1.

(a) We have:
$$P[A] = P[E \cap A] + P[E' \cap A] = 0.0585.$$

(b) We calculate
$$P[E \mid A] = \frac{P[E \cap A]}{P[A]} \stackrel{\text{(a)}}{=} \frac{0.009}{0.0585} \simeq 0.1538.$$

Note that this probability is very low. If we assume that 60% of the individuals subjected to the test are intoxicated (rather than 1%), then we find that $P[A]$ becomes 0.56 and $P[E \mid A] \simeq 0.9643$, which is much more reasonable. Therefore, this breath alcohol analyzer is only efficient if we use it for individuals who are suspected of being intoxicated.

Remark. In general, if a random experiment comprises k steps and if there are n_j possible outcomes on the jth step, for $j = 1, \ldots, k$, then there are $n_1 \times \cdots \times n_k$ elementary outcomes in the sample space. This is known as the *multiplication principle.*

Suppose now that we have n *distinct* objects and that we take, at random and *without* replacement, r objects among them, where $r \in \{(0,)1, \ldots, n\}$. The number of possible *arrangements* is given by

$$n \times (n-1) \times \cdots \times [n - (r-1)] = \frac{n!}{(n-r)!} := P_r^n. \tag{2.8}$$

The symbol P_r^n designates the number of *permutations* of n distinct objects taken r at a time. The *order* of the objects is important.

Remarks. (i) Reminder. We have that $n! = 1 \times 2 \times \cdots \times n$, for $n = 1, 2, 3, \ldots,$ and $0! = 1$, by definition.

(ii) Taking r objects *without* replacement means that the objects are taken one at a time and that a given object cannot be chosen more than once. This is equivalent to taking the r objects all at once. In the case of sampling *with* replacement, any object can be chosen up to r times.

Example 2.6.2. If we have four different letters (for instance, a, b, c, and d), then we can form

$$P_3^4 = \frac{4!}{(4-3)!} = \frac{4!}{1!} = 24$$

different three-letter "words" if each letter is used at most once. We can use a tree diagram to draw the list of words.

Finally, if the *order* of the objects is not important, then the number of ways to take, at random and *without* replacement, r objects among n distinct objects is given by

$$\frac{n \times (n-1) \times \cdots \times [n - (r-1)]}{r!} = \frac{n!}{r!(n-r)!} := C_r^n \equiv \binom{n}{r} \tag{2.9}$$

for $r \in \{(0,)1, \ldots, n\}$. The symbol C_r^n, or $\binom{n}{r}$, designates the number of *combinations* of n distinct objects taken r at a time.

Remarks. (i) Each combination of r objects enables us to form $r!$ different permutations, because

$$P_r^r = \frac{r!}{(r-r)!} = \frac{r!}{0!} = r!.$$

(ii) Moreover, it is easy to check that $C_r^n = C_{n-r}^n$.

Example 2.6.3. Three parts are taken, at random and *without* replacement, among 10 parts, of which 2 are defective. What is the probability that at least 1 defective part is obtained?

Solution. Let F be the event "at least one part is defective among the three parts taken at random." We can write that

$$P[F] = 1 - P[F'] = 1 - \frac{C_0^2 \cdot C_3^8}{C_3^{10}}$$

$$= 1 - \frac{1 \cdot \frac{8!}{3!5!}}{\frac{10!}{3!7!}} = 1 - \frac{6 \times 7}{9 \times 10} = \frac{8}{15} = 0.5\bar{3}.$$

2.7 Exercises for Chapter 2

Solved exercises

Question no. 1

We consider the following random experiment: a fair die is rolled; if (and only if) a "6" is obtained, the die is rolled a second time. How many elementary outcomes are there in the sample space Ω?

Question no. 2

Let $\Omega = \{e_1, e_2, e_3\}$, where $P[\{e_i\}] > 0$, for $i = 1, 2, 3$. How many different partitions of Ω, excluding the partition $\emptyset$, Ω can be formed?

Question no. 3

A fair die is rolled twice, independently. Knowing that an even number was obtained on the first roll, what is the probability that the sum of the two numbers obtained is equal to 4?

Question no. 4

Suppose that $P[A] = P[B] = 1/4$ and that $P[A \mid B] = P[B]$. Calculate $P[A \cap B']$.

Question no. 5

A system is made up of three independent components. It operates if at least two of the three components operate. If the reliability of each component is equal to 0.95, what is the reliability of the system?

Question no. 6

Suppose that $P[A \cap B] = 1/4$, $P[A \mid B'] = 1/8$, and $P[B] = 1/2$. Calculate $P[A]$.

Question no. 7

Knowing that we obtained at least once the outcome "heads" in three independent tosses of a fair coin, what is the probability that we obtained "heads" three times?

Question no. 8

Suppose that $P[B \mid A_1] = 1/2$ and that $P[B \mid A_2] = 1/4$, where A_1 and A_2 are two equiprobable events forming a partition of Ω. Calculate $P[A_1 \mid B]$.

Question no. 9

Three horses, a, b, and c, enter in a race. If the outcome bac means that b finished first, a second, and c third, then the set of all possible outcomes is

$$\Omega = \{abc, acb, bac, bca, cab, cba\}.$$

We suppose that $P[\{abc\}] = P[\{acb\}] = 1/18$ and that each of the other four elementary outcomes has a 2/9 probability of occurring. Moreover, we define the events

$$A = \text{``}a \text{ finishes before } b\text{''} \quad \text{and} \quad B = \text{``}a \text{ finishes before } c.\text{''}$$

(a) Do the events A and B form a partition of Ω?

(b) Are A and B independent events?

Question no. 10

Let ε be a random experiment for which there are three elementary outcomes: A, B, and C. Suppose that we repeat ε indefinitely and independently. Calculate, in terms of $P[A]$ and $P[B]$, the probability that A occurs before B.

Hints. (i) You can make use of the law of total probability.

(ii) Let D be the event "A occurs before B." Then, we may write that

$$P[D \mid C \text{ occurs on the first repetition}] = P[D].$$

Question no. 11

Transistors are drawn at random and with replacement from a box containing a very large number of transistors, some of which are defectless and others are defective, and are tested one at a time. We continue until either a defective transistor has been obtained or three transistors in all have been tested. Describe the sample space Ω for this random experiment.

Question no. 12

Let A and B be events such that $P[A] = 1/3$ and $P[B' \mid A] = 5/7$. Calculate $P[B]$ if B is a subset of A.

Question no. 13

In a certain university, the proportion of full, associate, and assistant professors, and of lecturers is 30%, 40%, 20%, and 10% respectively, of which 60%, 70%, 90%, and 40% hold a PhD. What is the probability that a person taken at random among those teaching at this university holds a PhD?

Question no. 14

All the items in stock in a certain store bear a code made up of five letters. If the same letter is never used more than once in a given code, how many different codes can there be?

Question no. 15

A fair die is rolled twice, independently. Consider the events

A = "the first number that shows up is a 6;"

B = "the sum of the two numbers obtained is equal to 7;"

C = "the sum of the two numbers obtained is equal to 7 or 11."

(a) Calculate $P[B \mid C]$.

(b) Calculate $P[A \mid B]$.

(c) Are A and B independent events?

Question no. 16

A commuter has two vehicles, one being a compact car and the other one a minivan. Three times out of four, he uses the compact car to go to work and the remainder of the time he uses the minivan. When he uses the compact car (resp., the minivan), he gets home before 5:30 p.m. 75% (resp., 60%) of the time. However, the minivan has air conditioning. Calculate the probability that

(a) he gets home before 5:30 p.m. on a given day;

(b) he used the compact car if he did not get home before 5:30 p.m.;

(c) he uses the minivan and he gets home after 5:30 p.m.;

(d) he gets home before 5:30 p.m. on two (independent) consecutive days and he does not use the same vehicle on these two days.

Question no. 17

Rain is forecast half the time in a certain region during a given time period. We estimate that the weather forecasts are accurate two times out of three. Mr. X goes out every day and he really fears being caught in the rain without an umbrella. Consequently, he always carries his umbrella if rain is forecast. Moreover, he even carries his umbrella one time out of three if rain is not forecast. Calculate the probability that it is raining and Mr. X does not have his umbrella.

Question no. 18

A fair die is rolled three times, independently. Let F be the event "the first number obtained is smaller than the second one, which is itself smaller than the third one." Calculate $P[F]$.

Question no. 19

We consider the set of all families having exactly two children. We suppose that each child has a 50–50 chance of being a boy. Let the events be

A_1 = "both sexes are represented among the children;"

A_2 = "at most one child is a girl."

(a) Are A_1 and A_2' incompatible events?

(b) Are A_1 and A_2' independent events?

(c) We also suppose that the probability that the third child of an arbitrary family is a boy is equal to 11/20 if the first two children are boys, to 2/5 if the first two children are girls, and to 1/2 in the other cases. Knowing that the third child of a given family is a boy, what is the probability that the first two are also boys?

Exercises

Question no. 1

We study the traffic (in one direction) on two roads, 1 and 2, which merge to form road 3 (see Figure 2.12). Roads 1 and 2 have the same capacity (number of lanes) and road 3 has a greater capacity than road 1 and road 2. During rush hours, the probability that the traffic is congested on road 1 (resp., road 2) is equal to 0.1 (resp., 0.3). Moreover, given that traffic is congested on road 2, it is also congested on road 1 one time out of three. We define the events

$$A, B, C = \text{"traffic is congested on roads 1, 2, 3, respectively."}$$

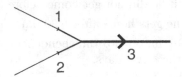

Fig. 2.12. Figure for Exercise no. 1.

(a) Calculate the probability that traffic is congested
  (i) on roads 1 and 2;
  (ii) on road 2, given that it is congested on road 1;
  (iii) on road 1 only;
  (iv) on road 2 only;
  (v) on road 1 or on road 2;
  (vi) neither on road 1 nor on road 2.

(b) On road 3, traffic is congested with probability

  1   if it is congested on roads 1 and 2;
  0.15 if it is congested on road 2 only;
  0.1 if it is neither congested on road 1 nor on road 2.

Calculate the probability that traffic is congested
  (i) on road 3;
  (ii) on road 1, given that it is congested on road 3.

Question no. 2

We roll a die and then we toss a coin. If we obtain "tails," then we roll the die a second time. Suppose that the die and the coin are fair. What is the probability of

(a) obtaining "heads" or a 6 on the first roll of the die;

(b) obtaining no 6s;

(c) obtaining exactly one 6;

(d) having obtained "heads," given that we obtained exactly one 6.

Question no. 3 (see Example 2.4.1)

A device is composed of two components, A and B, subject to random failures. The components are connected in parallel and, consequently, the device is down only if both components are down. The two components are not independent. We estimate that the probability of

a failure of component A is equal to 0.2;

a failure of component B is equal to 0.8 if component A is down;

a failure of component B is equal to 0.4 if component A is active.

(a) Calculate the probability of a failure
  (i) of component A if component B is down;
  (ii) of exactly one component.

(b) In order to increase the reliability of the device, a third component, C, is added in such a way that components A, B, and C are connected in parallel. The probability that component C breaks down is equal to 0.2, independently of the state (up or down) of components A and B. Given that the device is active, what is the probability that component C is down?

Question no. 4

In a factory producing electronic parts, the quality control is ensured through three tests as follows:

- each component is subjected to test no. 1;
- if a component passes test no. 1, then it is subjected to test no. 2;
- if a component passes test no. 2, then it is subjected to test no. 3;
- as soon as a component fails a test, it is returned for repair.

We define the events

$$A_i = \text{"the component fails test no. } i, \text{ for } i = 1, 2, 3.\text{"}$$

From past experience, we estimate that

$$P[A_1] = 0.1, \quad P[A_2 \mid A_1'] = 0.05 \quad \text{and} \quad P[A_3 \mid A_1' \cap A_2'] = 0.02.$$

The elementary outcomes of the sample space Ω are: $\omega_1 = A_1$, $\omega_2 = A_1' \cap A_2$, $\omega_3 = A_1' \cap A_2' \cap A_3$, and $\omega_4 = A_1' \cap A_2' \cap A_3'$.

(a) Calculate the probability of each elementary outcome.

(b) Let R be the event "the component must be repaired."

   (i) Express R in terms of A_1, A_2, A_3.

   (ii) Calculate the probability of R.

   (iii) Calculate $P[A_1' \cap A_2 \mid R]$.

(c) We test three components and we define the events

$$R_k = \text{``the } k\text{th component must be repaired, for } k = 1, 2, 3\text{''} \text{ and}$$

$$B = \text{``at least one of the three components passes all three tests.''}$$

We assume that the events R_k are independent.

   (i) Express B in terms of R_1, R_2, R_3.

   (ii) Calculate $P[B]$.

Question no. 5

Let A, B, and C be events such that $P[A] = 1/2$, $P[B] = 1/3$, $P[C] = 1/4$, and $P[A \cap C] = 1/12$. Furthermore, A and B are incompatible. Calculate $P[A \mid B \cup C]$.

Question no. 6

In a group of 20,000 men and 10,000 women, 6% of men and 3% of women suffer from a certain disease. What is the probability that a member of this group suffering from the disease in question is a man?

Question no. 7

Two tokens are taken at random and without replacement from an urn containing 10 tokens, numbered from 1 to 10. What is the probability that the larger of the two numbers obtained is 3?

Question no. 8

We consider the system in Figure 2.13. All components fail independently of each other. During a certain time period, the type A components fail with probability 0.3 and component B (resp., C) fails with probability 0.01 (resp., 0.1). Calculate the probability that the system is not down at the end of this period.

Question no. 9

A sample of size 20 is drawn (without replacement) from a lot of infinite size containing 2% defective items. Calculate the probability of obtaining at least one defective item in the sample.

Question no. 10

A lot contains 10 items, of which one is defective. The items are examined one by one, without replacement, until the defective item has been found. What is the probability that this defective item will be (a) the second item examined? (b) the ninth item examined?

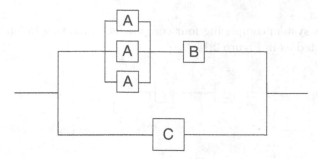

Fig. 2.13. Figure for Exercise no. 8.

Question no. 11

A bag holds two coins: a fair one and one with which we always get "heads." A coin is drawn at random and is tossed. Knowing that "heads" was obtained, calculate

(a) the probability that the fair coin was drawn;

(b) the probability of obtaining "heads" on a second toss of the same coin.

Question no. 12

The diagnosis of a physician in regard to one of her patients is unsure. She hesitates between three possible diseases. From past experience, we were able to construct the following tables:

S_i	S_1	S_2	S_3	S_4
$P[D_1 \mid S_i]$	0.2	0.1	0.6	0.4

S_i	S_1	S_2	S_3	S_4
$P[D_2 \mid S_i]$	0.2	0.5	0.5	0.3

S_i	S_1	S_2	S_3	S_4
$P[D_3 \mid S_i]$	0.6	0.3	0.1	0.2

where the D_is represent the diseases and the S_is are the symptoms. In addition, we assume that the four symptoms are incompatible, exhaustive, and equiprobable.

(a) Independently of the symptom present in the patient, what is the probability that he or she suffers from the first disease?

(b) What is the probability that the patient suffers from the second disease and presents symptom S_1?

(c) Given that the patient suffers from the third disease, what is the probability that he or she presents symptom S_2?

(d) We consider two independent patients. What is the probability that they do not suffer from the same disease, if we assume that the three diseases are incompatible?

Question no. 13

We consider a system comprising four components operating independently of each other and connected as in Figure 2.14.

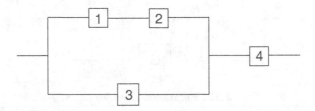

Fig. 2.14. Figure for Exercise no. 13.

The reliability of each component is supposed constant, over a certain time period, and is given by the following table:

Component	1	2	3	4
Reliability	0.9	0.95	0.95	0.99

(a) What is the probability that the system operates at the end of this time period?

(b) What is the probability that component no. 3 is down and the system still operates?

(c) What is the probability that at least one of the four components is down?

(d) Given that the system is down, what is the probability that it will resume operating if we replace component no. 1 by an identical (nondefective) component?

Question no. 14

A box contains 8 brand A and 12 brand B transistors. Two transistors are drawn at random and without replacement. What is the probability that they are both of the same brand?

Question no. 15

What is the reliability of the system shown in Figure 2.15 if the four components operate independently of each other and all have a reliability equal to 0.9 at a given time instant?

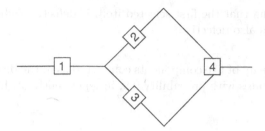

Fig. 2.15. Figure for Exercise no. 15.

Question no. 16
   Let A_1 and A_2 be two events such that $P[A_1] = 1/4$, $P[A_1 \cap A_2] = 3/16$, and $P[A_2 \mid A_1'] = 1/8$. Calculate $P[A_2']$.

Question no. 17
   A fair coin is tossed until either "heads" is obtained or the total number of tosses is equal to 3. Given that the random experiment ended with "heads," what is the probability that the coin was tossed only once?

Question no. 18
   In a room, there are four 18-year-old male students, six 18-year-old female students, six 19-year-old male students, and x 19-year-old female students. What must be the value of x, if we want age and sex to be independent when a student is taken at random in the room?

Question no. 19
   Stores S_1, S_2, and S_3 of the same company have, respectively, 50, 70, and 100 employees, of which 50%, 60%, and 75% are women. A person working for this company is taken at random. If the employee selected is a woman, what is the probability that she works in store S_3?

Question no. 20
   Harmful nitrogen oxides constitute 20%, in terms of weight, of all pollutants present in the air in a certain metropolitan area. Emissions from car exhausts are responsible for 70% of these nitrogen oxides, but for only 10% of all the other air pollutants. What percentage of the total pollution for which emissions from car exhausts are responsible are harmful nitrogen oxides?

Question no. 21
   Three machines, M_1, M_2, and M_3, produce, respectively, 1%, 3%, and 5% defective items. Moreover, machine M_1 produces twice as many items on an arbitrary day as machine M_2, which itself produces three times as many items as machine M_3. An item is taken at random among those manufactured on a given day, then a second item is taken at random among those manufactured by the machine that produced the first

selected item. Knowing that the first selected item is defective, what is the probability that the second one is also defective?

Question no. 22

A machine is made up of five components connected as in the diagram of Figure 2.16. Each component operates with probability 0.9, independently of the other components.

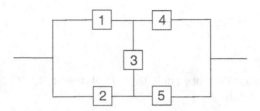

Fig. 2.16. Figure for Exercise no. 22.

(a) Knowing that component no. 1 is down, what is the probability that the machine operates?

(b) Knowing that component no. 1 is down and that the machine still operates, what is the probability that component no. 3 is active?

Question no. 23

Before being declared to conform to the technical norms, devices must pass two quality control tests. According to the data gathered so far, 75% of the devices tested in the course of a given week passed the first test. The devices are subjected to the second test, whether they pass the first test or not. We found that 80% of the devices that passed the second test had also passed the first one. Furthermore, 20% of those that failed the second test had passed the first one.

(a) What is the probability that a given device passed the second test?

(b) Find the probability that, for a given device, the second test contradicts the first one.

(c) Calculate the probability that a given device failed the second test, knowing that it passed the first one.

Question no. 24

In a certain workshop, the probability that a part manufactured by an arbitrary machine is nondefective, that is, conforms to the technical norms, is equal to 0.9. The quality control engineer proposes to adopt a procedure that classifies as nondefective with probability 0.95 the parts indeed conforming to the norms, and with only probability 0.15 those not conforming to these norms. It is decided that every part will be subjected to this quality control procedure twice, independently.

(a) What is the probability that a part having passed the procedure twice does indeed conform to the norms?

(b) Suppose that if a part fails the first control test, then it is withdrawn immediately. Let B_j be the event "a given part passed (if the case may be) the jth control test," for $j = 1, 2$. Calculate (i) $P[B_2]$ and (ii) $P[B_1' \cap B_2']$.

Question no. 25

We have 20 type I components, of which 5 are defective, and 30 type II components, of which 15 are defective.

(a) We wish to construct a system comprising 10 type I components and 5 type II components connected in series. What is the probability that the system will operate if the components are taken at random?

(b) How many different systems comprising four components connected in series, of which at least two are of type I, can be constructed, if the order of the components is taken into account?

Remarks. (i) We suppose that we can differentiate two components of the same type.

(ii) When a system is made up of components connected *in series*, then it operates if and only if every component operates.

Question no. 26

A system is made up of n components, including components A and B.

(a) Show that if the components are connected in series, then the probability that there are exactly r components between A and B is given by

$$\frac{2(n - r - 1)}{(n - 1)n} \quad \text{for } r \in \{0, 1, \ldots, n - 2\}.$$

(b) Calculate the probability that there are exactly r components between A and B if the components are connected in circle.

(c) Suppose that $n = 5$ and that the components are connected in series. Calculate the probability of operation of the subsystem constituted of components A, B and the r components placed between them if the components operate independently of each other and all have a reliability of 0.95.

Question no. 27

A man owns a car and a motorcycle. Half the time, he uses his motorcycle to go to work. One-third of the time, he drives his car to work and, the remainder of the time, he uses public transportation. He gets home before 5:30 p.m. 75% of the time when he uses his motorcycle. This percentage is equal to 60% when he drives his car and to 2% when he uses public transportation. Calculate the probability that

(a) he used public transportation if he got home after 5:30 p.m. on a given day;

(b) he got home before 5:30 p.m. on two consecutive (independent) days and he used public transportation on exactly one of these two days.

Question no. 28

In a certain airport, a shuttle coming from the city center stops at each of the four terminals to let passengers get off. Suppose that the probability that a given passenger gets off at a particular terminal is equal to $1/4$. If there are 20 passengers using the shuttle and if they occupy seats numbered from 1 to 20, what is the probability that the passengers sitting in seats nos. 1 to 4 all get off

(a) at the same stop?

(b) at different stops?

Question no. 29

A square grid consists of 289 points. A particle is at the center of the grid. Every second, it moves at random to one of the four nearest points from the one it occupies. When the particle arrives at the boundary of the grid, it is absorbed.

(a) What is the probability that the particle is absorbed after eight seconds?

(b) Let A_i be the event "the particle is at the center of the grid after i seconds." Calculate $P[A_4]$ (knowing that A_0 is certain, by assumption).

Question no. 30

Five married couples bought 10 tickets for a concert. In how many ways can they sit (in the same row) if

(a) the five men want to sit together?

(b) the two spouses in each couple want to sit together?

Multiple choice questions

Question no. 1

Two weeks prior to the most recent general election, a poll conducted among 1000 voters revealed that 48% of them intended to vote for the party in power. A survey made after the election, among the same sample of voters, showed that 90% of the persons who indeed voted for the party in power intended to vote for this party, and 10% of those who voted for another party intended (two weeks prior to the election) to vote for the party in power. Let the events be

A = "a voter, taken at random in the sample, intended to vote for the party in power;"

B = "a voter, taken at random in the sample, voted for the party in power."

(A) From the statement of the problem, we can write that $P[A] = 0.48$ and that

(a) $P[A \cap B] = 0.9$; $P[A \cap B'] = 0.1$
(b) $P[B \mid A] = 0.9$; $P[B' \mid A] = 0.1$
(c) $P[A \mid B] = 0.9$; $P[A \mid B'] = 0.1$
(d) $P[A' \cap B] = 0.9$; $P[A \cap B'] = 0.1$
(e) $P[A \mid B] = 0.9$; $P[B' \mid A] = 0.1$

(B) The probability of event B is given by

(a) 0.45   (b) 0.475   (c) 0.48   (d) 0.485   (e) 0.50   (f) 0.515

(C) Are events A and B' incompatible?

(a) yes   (b) no   (c) we cannot conclude from the information provided

(D) Are events A and B' independent?

(a) yes   (b) no   (c) we cannot conclude from the information provided

(E) Do events A and B' form a partition of the sample space Ω?

(a) yes   (b) no   (c) we cannot conclude from the information provided

(F) Let E be "a voter, taken at random among the 1000 voters polled, did not vote as he intended to two weeks prior to the election (in regard to the party in power)." We can write that

(a) $P[E] = P[A \mid B'] + P[A' \mid B]$
(b) $P[E] = P[A \cap B] + P[A' \cap B']$
(c) $P[E] = P[B' \mid A] + P[B \mid A']$
(d) $P[E] = P[A \cap B'] + P[A' \cap B']$
(e) $P[E] = P[A \cap B'] + P[A' \cap B]$

Question no. 2

Let A and B be two events such that

$$P[A \cap B] = P[A' \cap B] = P[A \cap B'] = p.$$

Calculate $P[A \cup B]$.

(a) p (b) $2p$ (c) $3p$ (d) $3p^2$ (e) p^3

Question no. 3

We have nine electronic components, of which one is defective. Five components are taken at random to construct a system in series. What is the probability that the system does not operate?

(a) 1/3   (b) 4/9   (c) 1/2   (d) 5/9   (e) 2/3

Question no. 4

Two dice are rolled simultaneously. If a sum of 7 or 11 is obtained, then a coin is tossed. How many elementary outcomes [of the form (die1, die2) or (die1, die2, coin)] are there in the sample space Ω?

(a) 28   (b) 30   (c) 36   (d) 44   (e) 72

Question no. 5

Let A and B be two independent events such that $P[A] < P[B]$, $P[A \cap B] = 6/25$, and $P[A \mid B] + P[B \mid A] = 1$. Calculate $P[A]$.

(a) 1/25   (b) 1/5   (c) 6/25   (d) 2/5   (e) 3/5

Question no. 6

In a certain lottery, 4 balls are drawn at random and without replacement among 25 balls numbered from 1 to 25. The player wins the grand prize if the 4 balls that she selected are drawn in the order indicated on her ticket. What is the probability of winning the grand prize?

(a) $\dfrac{1}{12,650}$ (b) $\dfrac{24}{390,625}$ (c) $\dfrac{1}{303,600}$ (d) $\dfrac{1}{390,625}$ (e) $\dfrac{1}{6,375,600}$

Question no. 7

New license plates are made up of three letters followed by three digits. If we suppose that the letters I and O are not used and that no plates bear the digits 000, how many different plates can there be?

(a) $24^3 \times 9^3$ (b) $(26 \times 25 \times 24)(10 \times 9 \times 8)$ (c) $24^3 \times (10 \times 9 \times 8)$
(d) $24^3 \times 999$ (e) $25^3 \times 9^3$

Question no. 8

Let $P[A \mid B] = 1/2$, $P[B'] = 1/3$, and $P[A \cap B'] = 1/4$. Calculate $P[A]$.

(a) 1/4   (b) 1/3   (c) 5/12   (d) 1/2   (e) 7/12

Question no. 9

In the lottery known as 6/49, first 6 balls are drawn at random and without replacement among 49 balls numbered from 1 to 49. Next, a seventh ball (the *bonus number*) is drawn at random among the 43 remaining balls. A woman selected what she thinks would be the six winning numbers for the next draw. What is the probability that this woman actually did not select any of the seven balls that will be drawn (including the bonus number)?

(a) $\dfrac{\binom{42}{6}}{\binom{49}{6}}$ (b) $\dfrac{\binom{42}{7}}{\binom{49}{6}}$ (c) $\dfrac{\binom{42}{6}}{\binom{49}{7}}$ (d) $\dfrac{\binom{43}{6}}{\binom{49}{7}}$ (e) $\dfrac{\binom{42}{7}}{\binom{49}{7}}$

Question no. 10

In a quality control procedure, every electronic component manufactured is subjected to (at most) three tests. After the first test, an arbitrary component is classified as either "good," "average," or "defective," and likewise after the second test. Finally, after the last test, the components are classified as either "good" or "defective." As soon as a component is classified as defective after a test, it is returned to the factory for repair. The following random experiment is performed: a component is taken at random and the result of each test it is subjected to is recorded. How many elementary outcomes are there in the sample space Ω?

(a) 3   (b) 8   (c) 11   (d) 18   (e) 21

Question no. 11

Let $P[A] = 1/3$, $P[B] = 1/2$, $P[C] = 1/4$, $P[A \mid B] = 1/2$, $P[B \mid A] = 3/4$, $P[A \mid C] = 1/3$, $P[C \mid A] = 1/4$, and $P[B \cap C] = 0$. Calculate the probability $P[A \mid B \cup C]$.

(a) 0   (b) 1/3   (c) 4/9   (d) 5/6   (e) 1

Question no. 12

A fair die is rolled twice (independently). Consider the events

A = "the two numbers obtained are different;"

B = "the first number obtained is a 6;"

C = "the two numbers obtained are even."

Which pairs of events are the only ones comprised of independent events?

(a) no pairs   (b) (A, B) (c) (A, B) and (B, C) (d) (A, B) and (A, C)
(e) the three pairs

Question no. 13

A man plays a series of games for which the probability of winning a given game, from the second one, is equal to 3/4 if he won the previous game and to 1/4 otherwise. Calculate the probability that he wins games nos. 2 and 3 consecutively if the probability that he wins the first game is equal to 1/2.

(a) 3/16   (b) 1/4   (c) 3/8   (d) 9/16   (e) 5/8

Question no. 14

A box contains two coins, one of them being fair but the other one having two "heads." A coin is taken at random and is tossed twice, independently. Calculate the probability that the fair coin was selected if "heads" was obtained twice.

(a) 1/5   (b) 1/4   (c) 1/3   (d) 1/2   (e) 3/5

3

Random variables

The elements of a sample space may take diverse forms: real numbers, but also brands of components, colors, "good," or "defective," and so on. Because it is easier to work with real numbers, in this chapter we transform all the elementary outcomes into numerical values, by means of *random variables*. We consider the most important particular cases and define the main functions that characterize random variables.

3.1 Introduction

Definition 3.1.1. *A* **random variable** *is a real-valued function defined on a sample space.*

Example 3.1.1. (i) Suppose that a coin is tossed. The function X that associates the number 1 with the outcome "heads" and the number 0 with the outcome "tails" is a random variable.

(ii) Suppose now that a die is rolled. The function X that associates with each elementary outcome the number obtained (so that X is the *identity function* in this case) is also a random variable.

Example 3.1.2. Consider the random experiment that consists in observing the time T that a person must wait in line to use an automatic teller machine. The function T is a random variable.

3.1.1 Discrete case

Definition 3.1.2. *A random variable is said to be of* **discrete** *type if the number of different values it can take is finite or countably infinite.*

M. Lefebvre, *Basic Probability Theory with Applications*, Springer Undergraduate Texts in Mathematics and Technology, DOI: 10.1007/978-0-387-74995-2_3,

Definition 3.1.3. *The function p_X that associates with each possible value of the (discrete) random variable X the probability of this value is called the* **probability (mass)** **function** *of X.*

Let $\{x_1, x_2, \ldots\}$ be the set of possible values of the discrete random variable X. The function p_X has the following properties:

(i) $p_X(x_k) \geq 0$ for all x_k;

(ii) $\sum_{k=1}^{\infty} p_X(x_k) = 1$.

Example 3.1.1 (continued). (i) If the coin is fair (or unbiased, or well-balanced), we may write that

x	0	1
$p_X(x)$	1/2	1/2

(ii) Similarly, if the die is fair, then we have the following table:

x	1	2	3	4	5	6
$p_X(x)$	1/6	1/6	1/6	1/6	1/6	1/6

Definition 3.1.4. *The function F_X that associates with each real number x the probability $P[X \leq x]$ that the random variable X takes on a value smaller than or equal to this number is called the* **distribution function** *of X. We have:*

$$F_X(x) = \sum_{x_k \leq x} p_X(x_k).$$

Remark. The function F_X is *nondecreasing* and *right-continuous*.

Example 3.1.1 (continued). (i) In the case of the coin, we easily find that (if $P[\{\text{heads}\}] = 1/2$)

x	0	1
$F_X(x)$	1/2	1

Remark. More completely, we may write that

$$F_X(x) = \begin{cases} 0 & \text{if } x < 0, \\ 1/2 & \text{if } 0 \leq x < 1, \\ 1 & \text{if } x \geq 1, \end{cases}$$

where x is an arbitrary real number.

(ii) If the die is well-balanced, then we deduce from the function $p_X(x)$ the following table:

x	1	2	3	4	5	6
$F_X(x)$	1/6	1/3	1/2	2/3	5/6	1

(see Figure 3.1).

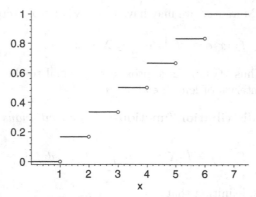

Fig. 3.1. Distribution function of the random variable in Example 3.1.1 (ii).

3.1.2 Continuous case

Definition 3.1.5. *A random variable that may take an uncountably infinite number of values is said to be of* **continuous type**.

Example 3.1.2 (continued). Because the set of possible values of the random variable T in Example 3.1.2 is the interval $(0, \infty)$, T is a *continuous* random variable.

Remark. We assume in Example 3.1.2 that the person cannot arrive and use the ATM immediately, otherwise T would be a random variable of *mixed type*, that is, a variable that is discrete and continuous at the same time. We do not insist on this type of random variable in this textbook.

Definition 3.1.6. *The* **(probability) density function** *of a continuous random variable X is a function f_X defined for all $x \in \mathbb{R}$ and having the following properties:*
(i) $f_X(x) \geq 0$ for any real number x;
(ii) if A is any subset of $\mathbb{R}$, then

$$P[X \in A] = \int_A f_X(x)\, dx.$$

Remarks. (i) Because X is a real-valued function, so that it must assume some value in the interval $(-\infty, \infty)$, we can write that

$$1 = P[X \in (-\infty, \infty)] = \int_{-\infty}^{\infty} f_X(x)\, dx.$$

(ii) The density function is different from the probability function p_X of a discrete random variable. Indeed, $f_X(x)$ does *not* give the probability that the random variable

X takes on the value x. Moreover, we may have that $f_X(x) > 1$. Actually, we may write that

$$f_X(x)\epsilon \simeq P\left[x - \frac{\epsilon}{2} \leq X \leq x + \frac{\epsilon}{2}\right],$$

where $\epsilon > 0$ is small. Thus, $f_X(x)\,\epsilon$ is approximately equal to the probability that X takes on a value in an interval of length ϵ about x.

Definition 3.1.7. *The* **distribution function** F_X *of a continuous random variable X is defined by*

$$F_X(x) = P[X \leq x] = \int_{-\infty}^{x} f_X(u)\,du.$$

We deduce from this definition that

$$P[X = x] = P[x \leq X \leq x] = P[X \leq x] - P[X < x]$$
$$= \int_{-\infty}^{x} f_X(u)\,du - \int_{-\infty}^{x^-} f_X(u)\,du = 0$$

for any real number x, where x^- means that the range of the integral is the open interval $(-\infty, x)$. That is, *before* performing the random experiment, the probability of obtaining a particular value of a *continuous* random variable is equal to zero. Therefore, if we take a point at random in the interval $[0, 1]$, we may assert that the point that we will obtain did not have, *a priori*, any chance of being selected!

We also deduce from the previous definition that

$$\frac{d}{dx}F_X(x) = f_X(x) \tag{3.1}$$

for any x where $F_X(x)$ is differentiable.

Remarks. (i) If X is a continuous random variable, then its distribution function F_X is also continuous. However, a continuous function is not necessarily differentiable at all points. Furthermore, the density function of X may be discontinuous, as in the next example. Actually, f_X is a piecewise continuous function, that is, a function having at most a finite number of *jump discontinuities* (see p. 2). We say that f_X has a jump discontinuity at x_0 if both $\lim_{x\downarrow x_0} f_X(x)$ and $\lim_{x\uparrow x_0} f_X(x)$ exist, but are different.

(ii) Every random variable X has a distribution function F_X. To simplify the presentation, we could theoretically define the density function f_X as the derivative of F_X, whether X is a discrete, continuous, or mixed type random variable. We mentioned in the previous remark that when F_X is a continuous function, its derivative is a piecewise continuous function. However, in the case of a discrete random variable, the distribution function F_X is a *step* or *staircase function*, as in Figure 3.1. The derivative of a step function is equal to zero everywhere, except at the jump points x_k, $k = 1, 2, \ldots$. We can write that

$$\frac{d}{dx}F_X(x) = \sum_{k=1}^{\infty} P[X = x_k]\delta(x - x_k),$$

where $P[X = x_k] = \lim_{x \downarrow x_k} F_X(x) - \lim_{x \uparrow x_x} F_X(x)$, and $\delta(\cdot)$ is the Dirac delta function (see p. 4). Similarly, the density function of a mixed type random variable involves the Dirac delta function. To avoid using this *generalized function*, the vast majority of authors prefer to consider the discrete and continuous random variables (and vectors) separately. In the discrete case, we define the probability mass function $p_X(x_k) = P[X = x_k]$, as we did above, rather than the density function.

Example 3.1.3. Suppose that the waiting time (in minutes) to be served at a counter in a bank is a continuous random variable X having the density function (see Figure 3.2)

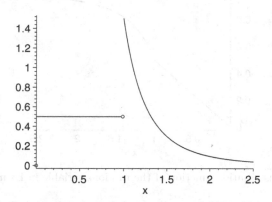

Fig. 3.2. Density function of the random variable in Example 3.1.3.

$$f_X(x) = \begin{cases} 0 & \text{if } x < 0, \\ 1/2 & \text{if } 0 \le x < 1, \\ 3/(2x^4) & \text{if } x \ge 1. \end{cases}$$

Note that the function f_X is a valid density function, because $f_X(x) \ge 0$ for all x and

$$\int_0^{\infty} f_X(x)\, dx = \int_0^1 \frac{1}{2}\, dx + \int_1^{\infty} \frac{3}{2x^4}\, dx = \frac{1}{2} - \frac{1}{2x^3}\Big|_1^{\infty} = \frac{1}{2} + \frac{1}{2} = 1.$$

Calculate (a) the distribution function of X and (b) the conditional probability $P[X > 2 \mid X > 1]$.

Solution. (a) By definition,

$$F_X(x) = \begin{cases} \displaystyle\int_{-\infty}^{x} 0\, du = 0 & \text{if } x < 0, \\[2ex] \displaystyle\int_{-\infty}^{0} 0\, du + \int_{0}^{x} \frac{1}{2}\, du = \frac{x}{2} & \text{if } 0 \le x < 1, \\[2ex] \displaystyle\int_{-\infty}^{0} 0\, du + \int_{0}^{1} \frac{1}{2}\, du + \int_{1}^{x} \frac{3}{2u^4}\, du = 1 - \frac{1}{2x^3} & \text{if } x \ge 1 \end{cases}$$

(see Figure 3.3).

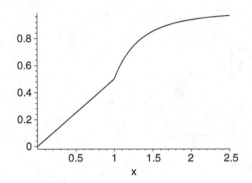

Fig. 3.3. Distribution function of the random variable in Example 3.1.3.

(b) We seek

$$\begin{aligned} P[X > 2 \mid X > 1] &= \frac{P[\{X > 2\} \cap \{X > 1\}]}{P[X > 1]} = \frac{P[X > 2]}{P[X > 1]} \\[1ex] &= \frac{1 - P[X \le 2]}{1 - P[X \le 1]} = \frac{1 - F_X(2)}{1 - F_X(1)} = \frac{1 - \frac{15}{16}}{1 - \frac{1}{2}} = \frac{1}{8}. \end{aligned}$$

Remark. Because X is a continuous random variable, $P[X < x] = P[X \le x] = F_X(x)$ for any real number x. It follows that $P[X \ge 2 \mid X \ge 1] = 1/8$ as well. In general, we have:

$$P[a \le X \le b] = P[a < X \le b] = P[a \le X < b] = P[a < X < b]$$

if X is a *continuous* random variable.

3.2 Important discrete random variables

3.2.1 Binomial distribution

Suppose that we perform repetitions of a certain random experiment and that we divide the set of possible outcomes into two mutually exclusive and exhaustive subsets: $B_1 \cap B_2 = \emptyset$ and $B_1 \cup B_2 = \Omega$. That is, B_1 and B_2 constitute a partition of the sample space Ω (see p. 35). If the elementary outcome that occurs belongs to B_1, then we say that the experiment resulted in a *success*; in the opposite case, it resulted in a *failure*.

Definition 3.2.1. *Let X be the (discrete) random variable that counts the number of successes obtained in n repetitions of a random experiment, where n is fixed. If*
(i) the probability p of success is constant *for the n trials and*
(ii) the trials are independent,
then we say that X has (or follows) a **binomial distribution** *with parameters n and p. We write that $X \sim B(n,p)$.*

Remark. A *parameter* is a symbol that appears in the definition of a random variable and that can take different values. For example, in the case of the binomial distribution, n can take on the values $1, 2, \ldots$, and p all the values in the interval $(0,1)$. In practice, the parameter p is generally unknown.

Now, suppose that, when we performed the n trials, we first obtained x consecutive successes S and then $n - x$ consecutive failures F. By independence, we may write that the probability of this elementary outcome is

$$P[\underbrace{SS\ldots S}_{x \text{ times}}\,\underbrace{FF\ldots F}_{(n-x) \text{ times}}] = \{P[S]\}^x \{P[F]\}^{n-x} = p^x(1-p)^{n-x}.$$

Hence, given that we can place the x successes among the n trials in $\binom{n}{x}$ different ways, we deduce that the probability function of the random variable $X \sim B(n,p)$ is given by

$$p_X(x) = \binom{n}{x} p^x q^{n-x} \quad \text{for } x = 0, 1, \ldots, n,$$

where $q := 1 - p$.

Remarks. (i) We have that $p_X(x) \geq 0$ for all x and, by Newton's *binomial formula*,

$$\sum_{x=0}^{n} p_X(x) = \sum_{x=0}^{n} \binom{n}{x} p^x q^{n-x} = (p+q)^n = 1,$$

as should be.

(ii) The distribution function of X is

$$F_X(k) = \sum_{x=0}^{k} \binom{n}{x} p^x q^{n-x} \quad \text{for } k = 0, 1, \dots, n.$$

There is no simple formula (without a summation symbol) for F_X. To evaluate this function, we can use a pocket calculator or a statistical software package. Some values of the distribution function F_X are given in Table B.1, page 276.

(iii) The shape and the position of the probability function p_X depend on the parameters n and p (see Figure 3.4).

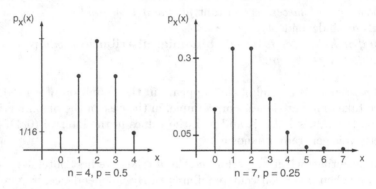

Fig. 3.4. Probability functions of binomial random variables.

(iv) If X has a binomial distribution with parameters $n = 1$ and p, we also say that X follows a *Bernoulli distribution* with parameter p. We thus have:

$$p_X(x) = p^x(1-p)^{1-x} \quad \text{if } x = 0, 1,$$
$$= \begin{cases} 1 - p & \text{if } x = 0, \\ p & \text{if } x = 1. \end{cases}$$

Moreover, we can show that if the random variables $X_1, \dots, X_n$ are *independent* (see Section 4.1) and if they all have a Bernoulli distribution with parameter p, then

$$X := \sum_{i=1}^{n} X_i \sim B(n, p).$$

Finally, we say that a binomial random variable counts the number of successes in n *Bernoulli trials*, that is, in n *independent* trials for which the probability p of success is the *same* from trial to trial.

Example 3.2.1. In an airport, five radars are in operation and each radar has a 0.9 probability of detecting an arriving airplane. The radars operate independently of each other.

(a) Calculate the probability that an arriving airplane will be detected by at least four radars.

(b) Knowing that at least three radars detected a given airplane, what is the probability that the five radars detected this airplane?

(c) What is the smallest number n of radars that must be installed if we want an arriving airplane to be detected by at least one radar with probability 0.9999?

Solution. Let X be the number of radars that detect the airplane.

(a) We have that $X \sim B(n = 5, p = 0.9)$. We seek

$$P[X \geq 4] = \binom{5}{4}(0.9)^4(0.1) + (0.9)^5 = (0.9)^4[5 \times (0.1) + 0.9] \simeq 0.9185.$$

Remark. Let Y be the number of radars that do *not* detect the arriving airplane. We find in Table B.1, page 276, that

$$P[X \geq 4] = P[Y \leq 1] \simeq 0.9185.$$

(b) We want

$$P[X = 5 \mid X \geq 3] = \frac{P[\{X = 5\} \cap \{X \geq 3\}]}{P[X \geq 3]} = \frac{P[X = 5]}{P[X \geq 3]}$$
$$\overset{(a)}{=} \frac{P[Y = 0]}{P[Y \leq 2]} \overset{\text{Tab. B.1}}{\simeq} \frac{0.5905}{0.9914} \simeq 0.596.$$

(c) We now have that $X \sim B(n, p = 0.9)$ and we seek (the smallest) n such that $P[X \geq 1] = 0.9999$. We have:

$$P[X \geq 1] = 1 - P[X = 0] = 1 - \binom{n}{0}(0.9)^0(0.1)^{n-0} = 1 - (0.1)^n$$
$$= 0.9999 \quad \Longleftrightarrow \quad (0.1)^n = 0.0001 = (0.1)^4.$$

Thus, we may write that $n_{\min} = 4$.

Remark. Note that, n being a positive *integer*, we cannot, in general, find a value of n for which the probability requested is *exactly* equal to a given number p. We must rather find the smallest n for which the probability of the event in question is greater than or equal to p. For instance, here if we had required the probability of detecting the airplane to be (at least) 0.9995, then the answer would have been the same: $n_{\min} = 4$.

3.2.2 Geometric and negative binomial distributions

Definition 3.2.2. *Let X be the random variable that counts the number of Bernoulli trials needed to obtain a first success. We say that X has a **geometric distribution** with parameter p, where p is the probability of success on any trial. We write that $X \sim Geo(p)$ [or $Geom(p)$].*

We have:

$$p_X(x) = P[\underbrace{FF\ldots F}_{(x-1)\ \text{times}}\ S] \overset{\text{ind.}}{=} \{P[F]\}^{x-1}P[S] = q^{x-1}p$$

for $x = 1, 2, \ldots$. We observe that the function p_X is strictly decreasing (see Figure 3.5).

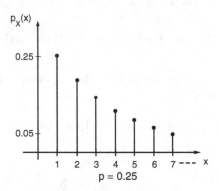

Fig. 3.5. Probability function of a geometric random variable.

Remarks. (i) We have that $p_X(x) \geq 0$ for all x and

$$\sum_{x=1}^{\infty} p_X(x) = p\sum_{x=1}^{\infty} q^{x-1} = p\frac{1}{1-q} = 1,$$

so that the function p_X is a valid probability function.

(ii) The distribution function of X is given by

$$F_X(x) = \sum_{k=1}^{x} p_X(k) = p\sum_{k=1}^{x} q^{k-1} = p\frac{1-q^x}{1-q} = 1 - q^x$$

for $x = 1, 2, \ldots$. Note that we then deduce that $P[X > x] = q^x$.

(iii) Making use of the formula $P[X > x] = q^x$, we can show that

$$P[X > x + y \mid X > x] = P[X > y] \quad \text{for any } x, y \in \{0, 1, 2, \ldots\}.$$

This property is known as the *memoryless* property of the geometric distribution.

Remark. The geometric distribution is sometimes defined as the number of Bernoulli trials performed *before* the first success occurs.

Definition 3.2.3. *Let X be the random variable that counts the number of Bernoulli trials performed until the rth success occurs, where $r = 1, 2, \ldots$. We say that X has a* **negative binomial distribution** *with parameters r and p. We write that $X \sim NB(r, p)$.*

Note that the geometric distribution is the particular case of the negative binomial distribution obtained with $r = 1$. We get the probability function of X as follows:

$$p_X(x) = P[\ \underbrace{F \ldots F}_{(x-r)\ \text{times}}\ \underbrace{S \ldots S}_{r\ \text{times}}] + P[\ \underbrace{F \ldots F}_{(x-r-1)\ \text{times}}\ SF\ \underbrace{S \ldots S}_{(r-1)\ \text{times}}\ ]$$

$$+ \cdots + P[\ \underbrace{S \ldots S}_{(r-1)\ \text{times}}\ \underbrace{F \ldots F}_{(x-r)\ \text{times}}\ S]$$

$$= \binom{x-1}{r-1} p^r (1-p)^{x-r} \quad \text{for } x = r, r+1, \ldots . \tag{3.2}$$

by independence and incompatibility, because there are $\binom{x-1}{r-1}$ different ways of placing the $r-1$ successes among the first $x-1$ trials (the xth trial being necessarily a success).

Remarks. (i) The negative binomial distribution is also known as the *Pascal distribution*.

(ii) As in the case of the binomial distribution, the shape and the position of the function p_X vary according to the values taken by the parameters r and p.

(iii) Notice the difference between the binomial and the *negative* binomial distributions: a binomial random variable counts the number of successes in a *fixed* number (n) of *trials*, whereas in the case of the negative binomial distribution, the random variable denotes the number of trials required to obtain a *fixed* number (r) of *successes*. If $X \sim NB(r, p)$, then we can write that

$$P[X = x] = P[B(x - 1, p) = r - 1]\, p.$$

Moreover, we can show the following relation between the two distributions:

$$P[NB(r, p) \leq x] = P[B(r + x, p) \geq r].$$

Example 3.2.2. A man shoots at a target until he has hit it twice. Suppose that the probability that a given shot hits the target is equal to 0.8. What is the probability that the man must shoot exactly four times?

Solution. Let X be the number of shots needed to end the random experiment. Then, if we assume that the shots are independent, we may write that $X \sim \text{NB}(r = 2, p = 0.8)$. We seek

$$P[X = 4] \equiv p_X(4) = \binom{3}{1}(0.8)^2(1 - 0.8)^2 = 3 \times (0.64)(0.04) = 0.0768.$$

Remark. If the man stops shooting as soon as he hits the target, then $X \sim \text{Geo}(p = 0.8)$ and

$$P[X = 4] = (1 - 0.8)^3(0.8) = 0.0064.$$

3.2.3 Hypergeometric distribution

Suppose that we perform n repetitions of a random experiment, but that the probability of success varies from one repetition to another. For example, we take *without* replacement n objects in a lot of *finite* size N, and we count the number of defective objects obtained. In this case, we cannot use the binomial distribution. Suppose that d objects, among the N, are defective (or possess a certain characteristic). Let X be the number of defective objects obtained among the n drawn. We have:

$$p_X(x) = \frac{\binom{d}{x} \cdot \binom{N-d}{n-x}}{\binom{N}{n}} \quad \text{for } x = 0, 1, \ldots, n. \tag{3.3}$$

Indeed, there are $\binom{N}{n}$ different (equiprobable) *samples* of size n and, among them, there are $\binom{d}{x} \cdot \binom{N-d}{n-x}$ with exactly x defective and $n - x$ nondefective objects.

Remark. We have that $\binom{n}{k} = 0$ if $k < 0$ or $k > n$.

Definition 3.2.4. *We say that the random variable X whose probability function is given by Formula (3.3) follows a* **hypergeometric distribution** *with parameters N, n, and d. We write that $X \sim \text{Hyp}(N, n, d)$.*

We must have that $N \in \{1, 2, \ldots\}$, $n \in \{(0), 1, 2, \ldots, N\}$, and $d \in \{(0), 1, \ldots, N\}$. Moreover, if the size N of the lot is large in comparison to the size n of the sample, then the fact of taking the objects without replacement will not influence much the probability of getting a defective object from draw to draw. That is, it is almost as if the objects were taken *with* replacement. Now, in that case, we may write that $X \sim \text{B}(n, p = d/N)$. Hence, we deduce that the binomial distribution can be used to approximate the hypergeometric distribution. To be more precise, we can show that if d and N tend to infinity in such a way that d/N converges to p, but n is kept constant, then the distribution of $X \sim \text{Hyp}(N, n, d)$ tends to that of a $\text{B}(n, p)$ random variable. In practice, the approximation obtained should be good when $n/N < 0.1$.

Remark. The quantity n/N is called the *sampling fraction*.

Example 3.2.3. Lots containing 25 devices each are subjected to the following *sampling plan*: a sample of 5 devices is taken at random and without replacement, and the lot is accepted if and only if the sample contains less than 3 defective devices. Calculate, supposing that there are exactly 4 defective devices in a particular lot,
(a) the probability that this lot is accepted;
(b) an approximation of the probability calculated in (a) with the help of a binomial distribution.

Solution. Let X be the number of defective devices in the sample. We have that $X \sim$ Hyp($N = 25, n = 5, d = 4$).
(a) Let F be the event "the lot is accepted." We seek

$$P[F] = P[X \le 2] = \sum_{x=0}^{2} \frac{\binom{4}{x} \cdot \binom{21}{5-x}}{\binom{25}{5}}$$
$$\simeq 0.3830 + 0.4506 + 0.1502 \simeq 0.984.$$

(b) We can write that

$$P[X \le 2] \simeq P[Y \le 2], \quad \text{where } Y \sim \text{B}(n = 5, p = 4/25)$$
$$= \sum_{y=0}^{2} \binom{5}{y} (4/25)^{y} (21/25)^{5-y} \simeq 0.4182 + 0.3983 + 0.1517 \simeq 0.968.$$

Note that here we have that $n/N = 5/25 = 0.2$, which is greater than 0.1. Therefore, we did not expect the approximation to be very good. If we replace $N = 25$ by $N = 100$, so that $n/N = 0.05$, we obtain that

$$P[\text{Hyp}(N = 100, n = 5, d = 4) \le 2] \simeq 0.9998$$

and

$$P[\text{B}(n = 5, p = 4/100) \le 2] \simeq 0.9994,$$

which is a better approximation. Finally, if we keep the same ratio $d/N = 4/25 = 0.16$, but if we assume that $N = 100$ and $d = 16$, then we calculate

$$P[\text{Hyp}(N = 100, n = 5, d = 16) \le 2] \simeq 0.9720,$$

and with $N = 200$ and $d = 32$, we find that

$$P[\text{Hyp}(N = 200, n = 5, d = 32) \le 2] \simeq 0.9700.$$

Notice that the quality of the approximation $P[X \le 2] \simeq 0.968$ obtained with a binomial distribution increases with increasing N, even though d/N is always equal to 0.16.

3.2.4 Poisson distribution and process

Let X be a random variable having a binomial distribution with parameters n and p. We can show that if n tends to infinity and p decreases to 0, in such a way that the product np remains equal to the constant λ, then the probability function of X converges to the function $p_X(x)$ given by

$$p_X(x) = \frac{e^{-\lambda}\lambda^x}{x!} \quad \text{for } x = 0, 1, \ldots . \tag{3.4}$$

Definition 3.2.5. *We say that the discrete random variable X whose probability function is given by Formula (3.4) has a **Poisson distribution** with parameter $\lambda > 0$. We write that $X \sim Poi(\lambda)$.*

Remarks. (i) Making use of the formula

$$e^x = 1 + x + \frac{x^2}{2!} + \cdots,$$

we easily show that the function defined in (3.4) is a valid probability function.

(ii) The Greek letter α is also often used for the parameter of the Poisson distribution. In *statistics*, we write θ to designate an arbitrary parameter of a random variable.

(iii) The shape of the probability function p_X depends on the value of the parameter λ.

(iv) To evaluate the distribution function of a Poisson random variable, we can use a pocket calculator or a statistical software package. Table B.2, page 278, gives many values of this function.

(v) We deduce from what precedes that we can use a Poisson distribution with parameter $\lambda = np$ to approximate the binomial distribution with parameters n and p. In general, the *Poisson approximation* should be good if $n > 20$ and $p < 0.05$. If the value of p is greater than $1/2$, then we must consider the number of failures (with probability $1 - p < 1/2$) rather than the number of successes before performing the approximation by the Poisson distribution.

Example 3.2.4. A new type of brakes is being studied. The company manufacturing these brakes claims that they could last at least 100,000 km for 90% of the vehicles that will use them. A laboratory simulated the driving of 100 cars using these brakes. Let X be the number of cars whose brakes will not last 100,000 km.

(a) What distribution does X follow?

(b) We will doubt the claimed percentage if the brakes must be changed on 17 cars or more before 100,000 km. What is, approximately, the probability of observing this event if, in fact, the claimed percentage is exact?

Solution. (a) By definition, if we assume that the cars are independent, we may write that $X \sim \mathrm{B}(n = 100, p = 0.1)$.

(b) We want $P[X \geq 17]$. We have that $P[X \geq 17] \simeq P[Y \geq 17]$, where $Y \sim \mathrm{Poi}(\lambda = 100(0.1) = 10)$. We then find in Table B.2, page 278, that

$$P[Y \geq 17] = 1 - P[Y \leq 16] \simeq 1 - 0.9730 = 0.0270.$$

Remark. Here, we have that $n = 100$ and $p = 0.1$. The value of n is very large, which is preferable, but that of p is slightly too large to expect a good approximation. Actually, we find that $P[X \geq 17] \simeq 0.021$.

Poisson process

Suppose that the random variable $N(t)$ denotes the number of events that will occur in the interval $[0, t]$. For instance, we can be interested in the number of failures of a machine, or in the number of customers, or in the number of telephone calls in the interval $[0, t]$. If we make the following assumptions:

(i) $N(0) = 0$;

(ii) the value of $N(t_4) - N(t_3)$ is independent of the value taken by $N(t_2) - N(t_1)$ if $0 \leq t_1 < t_2 \leq t_3 < t_4$;

(iii) $N(t + s) - N(t) \sim \mathrm{Poi}(\lambda s)$, for $s, t \geq 0$,

then the set $\{N(t), t \geq 0\}$ of random variables is called a *Poisson process* with *rate* $\lambda > 0$.

Remarks. (i) Condition (ii) above means that what happens in two disjoint intervals is *independent*. Furthermore, condition (iii) implies that the distribution of the number of events in an arbitrary interval depends only on the length of this interval. We say that the Poisson process has *independent* and *stationary increments*, respectively.

(ii) Conditions (i) and (iii) imply that $N(t) \equiv N(t + 0) - N(0) \sim \mathrm{Poi}(\lambda t)$.

(iii) The Poisson process is a very important particular case of what is known as a *stochastic* or *random process*. A stochastic process is a set $\{X(t), t \in T\}$ of random variables $X(t)$ (see Chapter 6). The set T is a subset of $\mathbb{R}$. In the case of the Poisson process, we take $T = [0, \infty)$. For every particular value t_0 of t, we get a random variable $N(t_0)$ having a Poisson distribution with parameter λt_0. It is important to distinguish between the random *variable* $N(t)$ and the random *process* $\{N(t), t \geq 0\}$.

(iv) The Poisson process is a particular *continuous-time Markov chain*, and is used abundantly in *communication theory* and in *queueing theory*, which is the subject of Chapter 6.

Example 3.2.5. Telephone calls arrive at an exchange according to a Poisson process with rate $\lambda = 2$ per minute (i.e., calls arrive at the average rate of 2 per minute, according to a Poisson distribution). Calculate the probability that exactly 2 calls will be received during each of the first 5 minutes of a given hour. What is the probability that exactly 10 calls will be received during the first 5 minutes of the hour in question?

Solution. Let $N(1)$ be the number of calls received during a 1-minute period. We can write that $N(1) \sim \text{Poi}(2 \cdot 1)$. We first calculate

$$P[N(1) = 2] = P[\text{Poi}(2) = 2] = \frac{e^{-2}2^2}{2!} = 2e^{-2}.$$

Next, let M be the number of minutes, among the 5 minutes considered, during which exactly 2 calls will be received. By independence, we may write that M follows a binomial distribution with parameters $n = 5$ and $p = 2e^{-2}$. We seek

$$P[M = 5] = \binom{5}{5}(2e^{-2})^5(1 - 2e^{-2})^{5-5} = 32e^{-10} \simeq 0.00145.$$

To obtain the probability that exactly 10 calls will be received during the first 5 minutes of the hour considered, we calculate

$$P[N(5) = 10] = P[\text{Poi}(2 \cdot 5) = 10] = \frac{e^{-10}10^{10}}{10!} \simeq 0.1251.$$

3.3 Important continuous random variables

3.3.1 Normal distribution

Definition 3.3.1. *Let X be a continuous random variable that can take any real value. If its density function is given by*

$$f_X(x) = \frac{1}{\sqrt{2\pi}\sigma} \exp\left\{-\frac{(x-\mu)^2}{2\sigma^2}\right\} \quad \text{for } -\infty < x < \infty,$$

*then we say that X has a **normal** (or **Gaussian**) distribution with parameters μ and σ^2, where $\mu \in \mathbb{R}$ and $\sigma > 0$. We write that $X \sim N(\mu, \sigma^2)$.*

Remark. The parameter μ is actually equal to the *mean* of X, and σ is the *standard deviation* of X (see Section 3.5). Furthermore, the standard deviation of a random variable is the square root of the *variance* of this variable. Therefore, in the case of the normal distribution, σ^2 is its variance.

The normal distribution is the most important continuous distribution, largely because of the *central limit theorem*, which is stated in Chapter 4. Moreover, all normal distributions have the same general shape, namely that of a *bell* (see Figure 3.6).

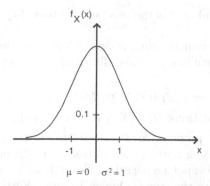

Fig. 3.6. Density function of a normal random variable.

The functions $f_X(x)$ are symmetrical with respect to the parameter μ. That is,

$$f_X(\mu - x) = f_X(\mu + x) \quad \text{for all } x \in \mathbb{R}.$$

The points $\mu - \sigma$ and $\mu + \sigma$ are those where the function f_X changes its direction of concavity. Finally, the larger σ is, the more flattened the curve is. Conversely, if σ is small, then the curve is concentrated around the mean μ.

Now, let $X \sim N(\mu, \sigma^2)$. We can show (see Section 3.4) that if we define the random variable $Z = (X - \mu)/\sigma$, then $Z \sim N(0,1)$. The notation Z is used, in general, for the $N(0,1)$ distribution. Its density function is often denoted by $\phi(z)$.

Remark. The $N(0,1)$ distribution is called the *standard* or *unit* normal distribution.

The main values of the distribution function of the $N(0,1)$ distribution, denoted by $\Phi(z)$, are presented in Table B.3, page 279. With the help of this table, we can calculate the probability $P[a \leq X \leq b]$ for any normal distribution. The table gives the value of $\Phi(z)$ for positive z. By symmetry, we may write that $\Phi(-z) = 1 - \Phi(z)$.

If we look for the number a for which $P[X \leq a] = p \geq 1/2$, we first find the number z in Table B.3 that corresponds to this probability p (sometimes we must interpolate in the table), and then we set

$$a = \mu + z \cdot \sigma.$$

If $p < 1/2$, the formula becomes (by symmetry)

$$a = \mu - z \cdot \sigma.$$

Finally, the numbers b that correspond to the main values of the probability $p :=$ $P[X > b]$, for instance, $p = 0.05$, $p = 0.01$, and so on, are given in Table B.4, page 280. Note that these numbers can be written as follows:

$$b = \Phi^{-1}(1 - p) \equiv Q^{-1}(p),$$

where $Q(z) := 1 - \Phi(z)$ and Q^{-1} is the *inverse function* of Q.

Example 3.3.1. Suppose that the compressive strength X (in pounds per square inch) of a certain type of concrete has a normal distribution with parameters $\mu = 4200$ and $\sigma^2 = (400)^2$.

(a) Calculate the probability $P[3000 \leq X \leq 4500]$.

(b) Solve the following equations: (i) $P[X \leq a] = 0.95$; (ii) $P[X \geq b] = 0.9$.

Remark. In fact, the model proposed for the compressive strength of the concrete *cannot* be the *exact* model, because a normal distribution can take on any real value, whereas the compressive strength cannot be negative. Nevertheless, the model in question may be a good approximation to the true (unknown) model. Furthermore, we find that the probability that a random variable having a $N(\mu, \sigma^2)$ distribution will take on a value in the interval $[\mu - 3\sigma, \mu + 3\sigma]$ is greater than 99.7%.

Solution. (a) We have:

$$
\begin{aligned}
P[3000 \leq X \leq 4500] &= P\left[\frac{3000 - 4200}{400} \leq \frac{X - 4200}{400} \leq \frac{4500 - 4200}{400} \right] \\
&= P[-3 \leq Z \leq 0.75] = \Phi(0.75) - \Phi(-3) \\
&\stackrel{\text{Tab. B.3}}{\simeq} 0.7734 - 0.0013 = 0.7721.
\end{aligned}
$$

(b) (i) We find in Table B.3, page 279, that $P[Z \leq 1.645] \simeq 0.95$ (see also Table B.4, page 280). Hence, we may write that

$$
a \simeq 4200 + (1.645)(400) = 4858.
$$

ii) We have that $P[X \geq b] = 0.9 \Leftrightarrow P[X < b] = 0.1$. Next, we may write that

$$
\begin{aligned}
P[X < b] &= 0.1 \Longleftrightarrow P\left[Z < \frac{b - 4200}{400} \right] = 0.1 \Longleftrightarrow \Phi\left(\frac{b - 4200}{400} \right) = 0.1 \\
&\Longleftrightarrow Q\left(\frac{b - 4200}{400} \right) = 0.9.
\end{aligned}
$$

Finally, we find in Table B.4 that the value that corresponds to $Q^{-1}(0.1)$ is approximately equal to 1.282. Because, by symmetry, $Q^{-1}(0.9) = -Q^{-1}(0.1)$, it follows that

$$
b \simeq 4200 + (-1.282)(400) = 3687.2.
$$

Making use of the *central limit theorem* (see Chapter 4), we can show that, if n is large enough, we can use a normal distribution to approximate the binomial distribution with parameters n and p. Let $X \sim B(n, p)$. The *de Moivre–Laplace approximation* is the following:

$$P[X = x] \simeq f_Y(x) \quad \text{for } x = 0, 1, \ldots, n,$$

where $Y \sim N(np, npq)$.

Remarks. (i) Thus, we replace a *probability* at a point by the value of a *density function* evaluated at this point. Recall that $f_Y(x)$ is *not* the probability that Y takes on the value x. Indeed, because Y is a continuous random variable, we know that $P[Y = x] = 0$ for all $x \in \mathbb{R}$.

(ii) The *mean* of the binomial distribution is given by np, and its *variance* by $np(1-p)$ (see Section 3.5). It is therefore logical to choose $\mu_Y = np$ and $\sigma_Y^2 = npq$.

When we want to evaluate (approximately) a probability of the form $P[a \leq X \leq b]$ with the help of a normal distribution, we use the following formula:

$$P[a \leq X \leq b] \simeq P\left[Z \leq \frac{b + 0.5 - np}{\sqrt{npq}}\right] - P\left[Z \leq \frac{a - 0.5 - np}{\sqrt{npq}}\right], \tag{3.5}$$

where a and b are integers.

Remarks. (i) To approximate the probability $P[a < X < b]$, we must write that

$$P[a < X < b] = P[a + 1 \leq X \leq b - 1] \tag{3.6}$$

$$\simeq P\left[Z \leq \frac{b - 0.5 - np}{\sqrt{npq}}\right] - P\left[Z \leq \frac{a + 0.5 - np}{\sqrt{npq}}\right].$$

(ii) The term 0.5 in Formula (3.5) [and (3.6)] is a *continuity correction* factor that most authors recommend using, because we replace a *discrete* distribution by a *continuous* distribution.

(iii) The approximation obtained should be good if $np \geq 5$ when $p \leq 0.5$, or if $n(1-p) \geq 5$ when $p \geq 0.5$. The normal distribution being symmetrical, it is easier to approximate the distribution of $X \sim B(n, p)$ with $p \simeq 1/2$ than to approximate the distribution of a binomial random variable having a very small or very large parameter p. Actually, in that case, we should use the *Poisson approximation* that we saw in the preceding section.

Example 3.3.2. A manufacturing process produces 10% defective items. A sample of 200 items is drawn at random. Let X be the number of defective items in the sample. Use a normal distribution to calculate (approximately) the probability $P[X = 20]$.

Solution. We may assume (see the first remark below) that the random variable X has a $B(n = 200, p = 0.1)$ distribution. So, we have that $np = 20 > 5$. We set

$$P[X = 20] \simeq f_Y(20), \quad \text{where } Y \sim N(20, 18)$$

$$= \frac{1}{\sqrt{2\pi}\sqrt{18}} \exp\left\{-\frac{(20 - 20)^2}{2 \cdot 18}\right\} \simeq 0.0904.$$

We can also proceed as follows:

$$P[X = 20] = P[20 \leq X \leq 20]$$
$$\simeq P\left[Z \leq \frac{20 + 0.5 - 20}{\sqrt{18}}\right] - P\left[Z \leq \frac{20 - 0.5 - 20}{\sqrt{18}}\right]$$
$$\simeq \Phi(0.12) - \Phi(-0.12) \stackrel{\text{Tab. B.3}}{\simeq} 2(0.5478) - 1 = 0.0956.$$

Remarks. (i) In the statement of the problem, it is not specified whether the sample is taken *with* or *without* replacement. At any rate, in order to use the hypergeometric distribution, we need the value of the size N of the lot, which is not specified either. Therefore, we must assume that all the items in the sample have a 0.1 probability of being defective, independently from one item to the other.

(ii) The exact value, obtained by using the binomial distribution, is

$$P[X = 20] = \binom{200}{20}(0.1)^{20}(0.9)^{180} \simeq 0.0936.$$

(iii) The Poisson approximation should work well in this example, because n is very large and p is relatively small. We find that

$$P[X = 20] \simeq P[\text{Poi}(\lambda = 20) = 20] = e^{-20}\frac{20^{20}}{20!} \simeq 0.0888.$$

Thus, the *normal approximation* is actually better in this particular example.

3.3.2 Gamma distribution

Definition 3.3.2. *The* **gamma function**, *denoted by* Γ, *is defined by*

$$\Gamma(u) = \int_0^\infty x^{u-1}e^{-x}\,dx \quad \text{for } u > 0. \tag{3.7}$$

We can show that $\Gamma(u) = (u-1)\Gamma(u-1)$ if $u > 1$. Because we find directly that $\Gamma(1) = 1$, we may write that

$$\Gamma(u) = (u-1)! \quad \text{if } u \in \{1, 2, \ldots\}.$$

Moreover, we have that $\Gamma(1/2) = \sqrt{\pi}$.

Definition 3.3.3. *Let X be a continuous random variable whose density function is given by*

$$f_X(x) = \frac{\lambda}{\Gamma(\alpha)}(\lambda x)^{\alpha-1}e^{-\lambda x} \quad \text{for } x > 0.$$

We say that X has a **gamma distribution** with parameters $\alpha > 0$ and $\lambda > 0$. We write that $X \sim G(\alpha, \lambda)$.

Remark. The parameter λ is a *scale* parameter, whereas α is a *shape* parameter. The shape of the density function f_X changes a lot with α, when α is relatively small (see Figure 3.7). When α becomes large, $f_X(x)$ tends to a normal density, which is a consequence of the *central limit theorem* because, when α is an integer, the random variable X can be represented as a sum of α random variables (see the next remark). When $0 < \alpha < 1$, the function $f_X(x)$ tends to infinity as x decreases to 0.

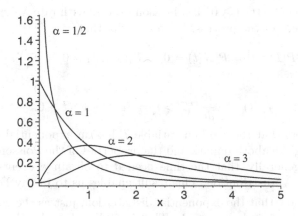

Fig. 3.7. Density functions of various random variables having a gamma distribution with $\lambda = 1$.

In general, we cannot give a simple formula, that is, without an integration sign in it, for the distribution function of a random variable having a gamma distribution. However, if the parameter α is a (positive) integer n, we can show that

$$P[\mathrm{G}(n, \lambda) \leq x] = P[\mathrm{Poi}(\lambda x) \geq n] = 1 - P[\mathrm{Poi}(\lambda x) \leq n - 1].$$

Note that this formula enables us to express the distribution function of a *continuous* random variable $X \sim \mathrm{G}(n, \lambda)$ in terms of that of a *discrete* random variable $Y \sim \mathrm{Poi}(\lambda x)$:

$$F_X(x) = 1 - F_Y(n - 1).$$

Particular cases

(i) If α is a natural number, then we also say that X follows an *Erlang distribution*, which is important in *queueing theory*.

(ii) If $\alpha = n/2$, where $n \in \{1, 2, \ldots\}$, and $\lambda = 1/2$, then the gamma distribution is also known as the *chi-square distribution* with n *degrees of freedom*. We write that $X \sim \chi_n^2$. This distribution is very useful in *statistics*.

(iii) If $\alpha = 1$, then the density function f_X becomes

$$f_X(x) = \lambda e^{-\lambda x} \quad \text{for } x > 0.$$

We say that X has an *exponential distribution* with parameter $\lambda > 0$. We write that $X \sim \text{Exp}(\lambda)$.

Remark. If $X_1, \ldots, X_n$ are *independent* random variables and if $X_i \sim \text{Exp}(\lambda)$, for $i = 1, \ldots, n$, then $Y := \sum_{i=1}^{n} X_i$ follows a gamma distribution with parameters n and λ (see Chapter 4).

Suppose now that $\{N(t), t \geq 0\}$ is a Poisson process with rate λ. Let T be the arrival time of the first event of the process. We may write that

$$P[T > t] = P[N(t) = 0] = P[\text{Poi}(\lambda t) = 0] = e^{-\lambda t}.$$

It follows that

$$f_T(t) = -\frac{d}{dt} P[T > t] = \lambda e^{-\lambda t} \quad \text{for } t > 0.$$

Thus, we may assert that the random variable T has an exponential distribution with parameter λ. Using the above remark and the properties of the Poisson process, we may also assert, more generally, that the time needed to obtain n events (from any time instant) has a $G(n, \lambda)$ distribution. This result enables us to justify Formula (3.3.2).

Remark. We can show that the exponential distribution, just as the geometric distribution, possesses the *memoryless property*. That is, if $X \sim \text{Exp}(\lambda)$, then

$$P[X > t + s \mid X > t] = P[X > s] \quad \text{for } s, t \geq 0.$$

Actually, only the geometric and exponential distributions possess this memoryless property. Furthermore, in the case of the geometric distribution, the property is only valid for s and $t \in \{0, 1, 2, \ldots\}$.

Example 3.3.3. The lifetime (in years) of a radio has an exponential distribution with parameter $\lambda = 1/10$. If we buy a five-year-old radio, what is the probability that it will work for less than 10 additional years?

Solution. Let X be the total lifetime of the radio. We have that $X \sim \text{Exp}(\lambda = 1/10)$. We seek

$$P[X < 15 \mid X > 5] = 1 - P[X \geq 15 \mid X > 5] = 1 - P[X > 10] = P[X \leq 10]$$

$$= \int_0^{10} \frac{1}{10} e^{-x/10} \, dx = -e^{-x/10} \Big|_0^{10} = 1 - e^{-1} \simeq 0.6321.$$

Because of its memoryless property, the exponential distribution is widely used in *reliability*. This property implies that the *failure rate* of a device is constant over time. The exponential distribution appears in the theory of *stochastic processes* and in *queueing theory* as well.

An extension of the exponential distribution to the entire real line is obtained by defining

$$f_X(x) = \frac{\lambda}{2} e^{-\lambda|x|} \quad \text{for } -\infty < x < \infty,$$

where λ is a positive constant. We say that the random variable X has a *double exponential distribution*, or a *Laplace distribution*, with parameter λ.

3.3.3 Weibull distribution

Definition 3.3.4. *Let X be a continuous random variable whose density function is of the form*

$$f_X(x) = \lambda \beta x^{\beta-1} \exp\left\{-\lambda x^\beta\right\} \quad \text{for } x > 0.$$

We say that X follows a **Weibull distribution** *with parameters $\lambda > 0$ and $\beta > 0$. We write that $X \sim W(\lambda, \beta)$.*

The Weibull distribution generalizes the exponential distribution, which is obtained by taking $\beta = 1$. It is important in reliability. Like the gamma distribution, it can be used in numerous applications because of the various shapes taken by its density function depending on the values given to the parameter β. It is also one of the distributions known as *extreme value distributions*. These distributions are used to model phenomena that occur very rarely, such as extremely cold or hot temperatures, exceptional floods of rivers, and so on.

Example 3.3.4. Let T denote the temperature (in degrees Celsius) in a certain city during the month of July. Suppose that

$$X := T - 30 \mid \{T > 30\} \sim W(0.8, 0.5).$$

That is, given that the temperature is above 30 degrees, it has a Weibull distribution with parameters $\lambda = 0.8$ and $\beta = 0.5$. Thus,

$$f_X(x) = 0.4 x^{-1/2} \exp\left\{-0.8 x^{1/2}\right\} \quad \text{for } x > 0$$

(see Figure 3.8). Notice that the function $f_X(x)$ diverges when x decreases to 0.

Using the results in Section 3.5, we find that the *average* temperature in this city (in July), when it exceeds 30°C, is equal to 33.125°C. We obtain the same value if we suppose that X is exponentially distributed with parameter $\lambda = 1/3.125 = 0.32$ instead. However, as can be seen in Figure 3.9, the Weibull distribution W(0.8, 0.5) goes to zero more slowly than the Exp(0.32) distribution does. Therefore, *extreme* temperatures (above approximately 42°C in this particular example) are more likely.

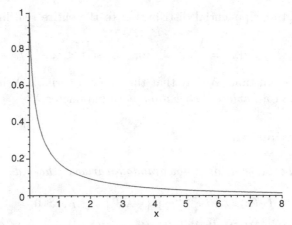

Fig. 3.8. Probability density function of a W(0.8, 0.5) random variable.

We have:

$$P[T > 35 \mid T > 30] = P[X > 5] = \int_{5}^{\infty} 0.4x^{-1/2} \exp\left\{-0.8x^{1/2}\right\} dx \simeq 0.1672.$$

3.3.4 Beta distribution

Definition 3.3.5. *Let X be a continuous random variable whose density function is given by*

$$f_X(x) = \frac{\Gamma(\alpha + \beta)}{\Gamma(\alpha)\Gamma(\beta)} x^{\alpha-1}(1-x)^{\beta-1} \quad \text{for } 0 < x < 1,$$

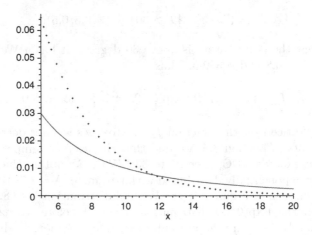

Fig. 3.9. Probability density functions of a W(0.8, 0.5) (continuous line) and an Exp(0.32) (broken line) random variables in the interval [5, 20].

where $\alpha > 0$ and $\beta > 0$. We say that X has a **beta distribution** *with parameters α and β. We write that $X \sim Be(\alpha, \beta)$.*

If $X \sim Be(\alpha, \beta)$ and $Y := a + (b - a)X$, where $a < b$, Y is said to have a *generalized beta* distribution.

Particular case

Let $X \sim Be(\alpha, \beta)$. If $\alpha = \beta = 1$, then we have:

$$f_X(x) = 1 \quad \text{for } 0 < x < 1.$$

We say that the continuous random variable X follows a *uniform distribution* on the interval $(0, 1)$. We write that $X \sim U(0, 1)$. In general, we have that $X \sim U(a, b)$ if (see Figure 3.10)

$$f_X(x) = \frac{1}{b - a} \quad \text{for } a < x < b.$$

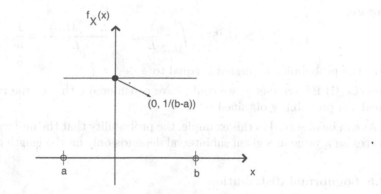

Fig. 3.10. Probability density function of a uniform random variable on the interval (a, b).

Remarks. (i) This density function is obtained, for example, when a point is taken *at random* in the interval (a, b). Because the probability that the selected point is *close* to x, where $a < x < b$, is the same for all x, the function $f_X(x)$ must be *constant* in the interval (a, b). Note that a random variable having a uniform distribution on the interval (a, b) also has a generalized beta distribution with parameters $\alpha = \beta = 1$.

(ii) We can show that if exactly one event of a Poisson process took place in the interval $(0, t]$, then the time instant T_1 at which this event occurred has a uniform distribution on $(0, t]$. That is,

$$T_1 \mid \{N(t) = 1\} \sim U(0, t].$$

Example 3.3.5. A point is taken at random on a line segment of length L. What is the probability that the length of the shorter segment divided by the length of the longer one is smaller than $1/4$?

Solution. Suppose, without loss of generality, that the segment starts at 0. Let X be the point selected. Then, $X \sim U[0, L]$.
(i) If $X \in [0, L/2]$, then we must have:

$$\frac{X}{L - X} < \frac{1}{4} \quad \Longleftrightarrow \quad 4X < L - X \quad \Longleftrightarrow \quad X < \frac{L}{5}.$$

(ii) If $X \in (L/2, L]$, then

$$\frac{L - X}{X} < \frac{1}{4} \quad \Longleftrightarrow \quad 4L - 4X < X \quad \Longleftrightarrow \quad X > \frac{4L}{5}.$$

We have:

$$P[X < L/5] = \int_0^{L/5} \frac{1}{L - 0} \, dx = \frac{L/5}{L} = \frac{1}{5}.$$

Likewise,

$$P[X > 4L/5] = \int_{4L/5}^{L} \frac{1}{L} \, dx = \frac{L - (4L/5)}{L} = \frac{1}{5}.$$

Thus, the probability requested is equal to $\frac{1}{5} + \frac{1}{5} = \frac{2}{5}$.

Remarks. (i) By symmetry, we could have considered either of the two cases and multiplied the probability obtained by 2.

(ii) As can be observed in this example, the probability that the uniform random variable X takes on a value in a given subinterval depends only on the length of this subinterval.

3.3.5 Lognormal distribution

Definition 3.3.6. *Let X be a continuous random variable taking only positive values. If $Y := \ln X$ follows a $N(\mu, \sigma^2)$ distribution, then we say that X has a **lognormal distribution** with parameters μ and σ^2. We write that $X \sim LN(\mu, \sigma^2)$. The density function of X is given by*

$$f_X(x) = \frac{1}{\sqrt{2\pi}\sigma x} \exp\left\{-\frac{(\ln x - \mu)^2}{2\sigma^2}\right\} \quad \text{for } x > 0,$$

where $\mu \in \mathbb{R}$ and $\sigma > 0$.

Remark. In many situations, the lognormal distribution may be a more realistic model than the normal distribution, because it is always positive. For instance, the weight of manufactured items could have a lognormal distribution.

Example 3.3.6. Let $X \sim \mathrm{LN}(5,4)$. Calculate $P[X \leq 100]$.

Solution. We have:

$$P[X \leq 100] = P[\ln X \leq \ln 100] = P[Y \leq \ln 100], \quad \text{where } Y \sim N(5,4)$$
$$= P\left[Z \leq \frac{\ln 100 - 5}{2}\right] \simeq \varPhi(-0.2) \overset{\text{Tab. B.3}}{\simeq} 0.4207.$$

3.4 Functions of random variables

Because a random variable is a real-valued *function* and the *composition* of two functions is another function, we can assert that if X is a random variable, then $Y := g(X)$, where g is a real-valued function defined on the real line, is a random variable as well. In this section, we show how to obtain the probability function or the density function of Y.

3.4.1 Discrete case

Because a *function* g associates a single real number with each possible value of the random variable X, we can assert that if X is a variable of discrete type, then $Y = g(X)$ will also be a discrete random variable, whatever the function g is. Indeed, Y cannot take on more different values than X. To obtain the probability function of Y, we apply the transformation g to each possible value of X and we add the probabilities of all values x of the random variable X that correspond to the same y.

Example 3.4.1. Let X be a discrete random variable whose probability function is given by

x	-1	0	1
$p_X(x)$	$1/4$	$1/4$	$1/2$

We define $Y = 2X$. Because the function $g : x \to 2x$ is *bijective* [i.e., to a given $y = g(x) = 2x$, there corresponds one and only one x and vice versa], the number of possible values of the random variable Y will be the same as the number of possible values of X. We find that

y	-2	0	2	Σ
$p_Y(y)$	$1/4$	$1/4$	$1/2$	1

Now, let $W = X^2$. Because to two values of X, namely -1 and 1, there corresponds the same value $w = 1$, we must add $p_X(-1)$ and $p_X(1)$ to obtain $p_W(1)$. We thus have:

w	0	1	Σ
$p_W(w)$	$1/4$	$3/4$	1

In the case when the random variable X can take a (countably) infinite number of values, we apply the transformation to an arbitrary value of X, and we try to find a general formula for the function $p_Y(y)$.

Example 3.4.2. Let $X \sim \text{Geo}(p)$ and $Y = X^2$. Because $X \in \{1, 2, \ldots\}$, the quadratic function is (here) a bijective transformation and we easily calculate

$$p_Y(y) = p_X(\sqrt{y}) = q^{\sqrt{y}-1}p \quad \text{for } y = 1, 4, 9, \ldots .$$

3.4.2 Continuous case

The *composition* of two continuous functions is another continuous function. Consequently, if X is a continuous random variable and if g is a continuous function, then $Y := g(X)$ is a continuous random variable as well. In this book, we consider only the case when the function g is *bijective*. In that case, the inverse function $g^{-1}(y)$ exists and we can use the following proposition to obtain the density function of the new random variable Y.

Proposition 3.4.1. *Suppose that the equation* $y = g(x)$ *has a unique solution:* $x = g^{-1}(y)$. *Then, the density function of* Y *is given by the formula*

$$f_Y(y) = f_X[g^{-1}(y)] \left| \frac{d}{dy} g^{-1}(y) \right|.$$

Example 3.4.3. We can use the above proposition to prove the result stated in Section 3.3: if X has a $\text{N}(\mu, \sigma^2)$ distribution, then $Z := (X - \mu)/\sigma$ follows a standard normal distribution. Indeed, we have:

$$z = g(x) = (x - \mu)/\sigma \quad \Longleftrightarrow \quad g^{-1}(z) = \mu + \sigma z.$$

It follows that

$$f_Z(z) = f_X(\mu + \sigma z) \left| \frac{d}{dz}(\mu + \sigma z) \right| = \frac{1}{\sqrt{2\pi}\sigma} \exp(-z^2/2)|\sigma| = \phi(z)$$

for $-\infty < z < \infty$, because $\sigma > 0$. Similarly, we can prove that if $Y := aX + b$, then Y has a normal distribution with parameters $a\mu + b$ and $a^2\sigma^2$.

Example 3.4.4. Let $X \sim \text{U}(0, 1)$ and $Y = -\theta \ln X$, where $\theta > 0$. First, note that the possible values of the random variable Y are those located in the interval $(0, \infty)$. Next, we have that $g^{-1}(y) = e^{-y/\theta}$, so that

$$f_Y(y) = f_X\left(e^{-y/\theta}\right) \left| \frac{d}{dy} e^{-y/\theta} \right| = 1 \cdot \left| -\frac{1}{\theta} e^{-y/\theta} \right| = \frac{1}{\theta} e^{-y/\theta}$$

for $0 < y < \infty$. Thus, we can assert that Y has an exponential distribution with parameter $1/\theta$. We make use of this result in *simulation*. Indeed, if we wish to generate an observation of a random variable following an exponential distribution with parameter $\lambda = 2$ (for instance), it suffices to generate an observation x of a $U(0,1)$ distribution, and then to apply the transformation $y = -(1/2) \ln x$. Therefore, it is not necessary to write a special computer program to generate observations of exponential distributions. Computer languages in general, and even many pocket calculators, allow us to generate *pseudo-random* observations of uniform distributions.

3.5 Characteristics of random variables

In this section, we present some numerical quantities that enable us to characterize a random variable X. All the quantities are obtained by computing the *mathematical expectation* of various functions of X.

Definition 3.5.1. *We define the* **mathematical expectation** *of a function $g(X)$ of a random variable X by*

$$
E[g(X)] = \begin{cases} \displaystyle\sum_{i=1}^{\infty} g(x_i) p_X(x_i) & \text{if } X \text{ is discrete,} \\[2mm] \displaystyle\int_{-\infty}^{\infty} g(x) f_X(x)\, dx & \text{if } X \text{ is continuous.} \end{cases} \tag{3.8}
$$

Properties. (i) $E[c] = c$, for any constant c.

(ii) $E[c_1 g(X) + c_0] = c_1 E[g(X)] + c_0$, for all constants c_1 and c_0.

Remarks. (i) E is therefore a *linear operator*.

(ii) The mathematical expectation may be infinite and it may even not exist.

(iii) If X is a random variable of *mixed type*, that is, a random variable that is discrete and continuous at the same time, then $E[g(X)]$ can be computed by decomposing the problem into two parts. For example, suppose that we toss a coin for which the probability of getting "tails" is $3/8$. If we get "tails," then the random variable X takes on the value 1; otherwise, X is a number taken at random in the interval $[2, 4]$. We can obtain $E[g(X)]$ as follows:

$$
E[g(X)] = 1 \times \frac{3}{8} + \left(\int_2^4 g(x) \cdot \frac{1}{2}\, dx \right) \times \frac{5}{8}.
$$

Definition 3.5.2. *The **mean** (or the **expected value**) of a random variable X is given by*

$$\mu_X \equiv E[X] = \begin{cases} \displaystyle\sum_{i=1}^{\infty} x_i p_X(x_i) & \text{if } X \text{ is discrete}, \\[2em] \displaystyle\int_{-\infty}^{\infty} x f_X(x)\, dx & \text{if } X \text{ is continuous}. \end{cases} \tag{3.9}$$

Example 3.5.1. Let $X \sim \text{Poi}(\lambda)$. We have:

$$\mu_X = \sum_{x=0}^{\infty} x e^{-\lambda} \frac{\lambda^x}{x!} = e^{-\lambda} \sum_{x=1}^{\infty} \frac{\lambda^x}{(x-1)!} = e^{-\lambda} \lambda \sum_{x=1}^{\infty} \frac{\lambda^{x-1}}{(x-1)!}$$

$$= e^{-\lambda} \lambda \left\{ 1 + \frac{\lambda}{1!} + \frac{\lambda^2}{2!} + \ldots \right\} = e^{-\lambda} \lambda e^{\lambda} = \lambda.$$

Example 3.5.2. Let $X \sim \text{Exp}(\lambda)$. We calculate

$$\mu_X = \int_0^{\infty} x \lambda e^{-\lambda x}\, dx = \lambda \cdot \frac{1}{\lambda^2} = \frac{1}{\lambda}$$

because, in general, if $a > 0$, then we have:

$$\int_0^{\infty} x^n e^{-ax}\, dx \stackrel{y=ax}{=} \int_0^{\infty} \left(\frac{y}{a} \right)^n e^{-y} \frac{dy}{a} \tag{3.10}$$

$$= \frac{1}{a^{n+1}} \int_0^{\infty} y^n e^{-y}\, dy = \frac{\Gamma(n+1)}{a^{n+1}} = \frac{n!}{a^{n+1}}.$$

Remark. We could also have integrated the function $x\lambda e^{-\lambda x}$ by parts.

Definition 3.5.3. *A **median** of a random variable X is a value x_m ($\equiv \tilde{x}$) for which*

$$P[X \leq x_m] \geq \frac{1}{2} \quad \text{and} \quad P[X \geq x_m] \geq \frac{1}{2}.$$

Remarks. (i) When X is a discrete random variable, the median is not necessarily unique. It is *not* unique if there exists a real number a such that $F_X(a) = 1/2$. For example, let $X \sim \text{B}(n = 2, p = 1/2)$. We have that $F_X(1) = P[X = 0] + P[X = 1] = 1/4 + 1/2 = 3/4$. In this case, the number $x_m = 1$ satisfies the above definition, because $P[X \leq 1] = 3/4 \geq 1/2$ and $P[X \geq 1] = 3/4 \geq 1/2$ as well. Furthermore, $x_m = 1$ is the only number for which both inequalities are satisfied at the same time. On the other hand, let

x	1	2	3
$p_X(x)$	1/4	1/4	1/2

The real number 2 satisfies the definition of the median, but so does the number 2.5 (for instance). In fact, every number in the interval $[2,3]$ is a median of X. Remark that if we change the probabilities as follows: $P[X = 1] = 1/4$, $P[X = 2] = 1/8$, and $P[X = 3] = 5/8$, then the median $x_m = 3$ is unique. Likewise, if $P[X = 1] = 1/4$, $P[X = 2] = 3/8$, and $P[X = 3] = 3/8$, then $x_m = 2$ is the unique median of X.

(ii) When X is a continuous random variable taking all its values in a single (finite or infinite) interval, the median *is* unique and can be defined as follows:

$$P[X \le x_m] = 1/2 \quad (\Longrightarrow P[X \ge x_m] = 1/2).$$

Example 3.5.1 (continued). Suppose that $X \sim$ Poi(2). We find in Table B.2, page 278, that $P[X \le 1] \simeq 0.4060$ and $P[X \le 2] \simeq 0.6767$. There is no number x_m such that $P[X \le x_m] = P[X \ge x_m] = 1/2$. However, we have:

$$P[X \le 2] \simeq 0.6767 \ge 1/2 \quad \text{and} \quad P[X \ge 2] \simeq 1 - 0.4060 \ge 1/2.$$

Moreover, 2 is the only number for which *both* inequalities are satisfied. Hence, $x_m = 2$ is *the* median of X.

Example 3.5.2 (continued). If $X \sim$ Exp(λ), then we have:

$$P[X \le x_m] = \int_0^{x_m} \lambda e^{-\lambda x}\, dx = -e^{-\lambda x}\Big|_0^{x_m} = 1 - e^{-\lambda x_m}.$$

It follows that

$$P[X \le x_m] = 1/2 \quad \Longleftrightarrow \quad 1 - e^{-\lambda x_m} = 1/2 \quad \Longleftrightarrow \quad x_m = \frac{\ln 2}{\lambda}.$$

We can check that we indeed have:

$$P\left[X \ge \frac{\ln 2}{\lambda}\right] = \int_{\frac{\ln 2}{\lambda}}^{\infty} \lambda e^{-\lambda x}\, dx = -e^{-\lambda x}\Big|_{\frac{\ln 2}{\lambda}}^{\infty} = \exp\left\{-\lambda\frac{\ln 2}{\lambda}\right\} = 1/2.$$

Thus, *the* median is given by $x_m = (\ln 2)/\lambda$.

The median is useful when the random variable X may take on very large values (in absolute value) as compared to the others. Indeed, the median is less influenced by these extreme values than the mean μ_X is.

The median, in the continuous case, is a particular case of the notion of *quantile*.

Definition 3.5.4. *Let X be a continuous random variable whose set of possible values is an arbitrary interval (a, b). The number x_p is called the $100(1 - p)$th* **quantile** *of X if*

$$P[X \leq x_p] = 1 - p,$$

where $0 < p < 1$.

If $100p$ is an integer, then x_p is also called the $100(1 - p)$th *percentile* of X. The median of a continuous random variable is therefore the 50th percentile of X. The 25th percentile is also known as the *first quartile*, the 50th percentile is the *second quartile*, and the 75th percentile is the *third quartile*. Finally, the difference between the third and the first quartile is called the *interquartile range*.

Definition 3.5.5. *A* **mode** *of a random variable X is any value x that corresponds to a local maximum for $p_X(x)$ or $f_X(x)$.*

Remark. The mode is thus not necessarily unique. A distribution having a single mode is said to be *unimodal*.

Example 3.5.1 (continued). Let $X \sim \text{Poi}(2)$. From Table B.2, page 278, we obtain that $P[X = 0] \simeq 0.135$, $P[X = 1] \simeq 0.271$, $P[X = 2] \simeq 0.271$, $P[X = 3] \simeq 0.180$, and so on. Hence, X has two modes: at $x = 1$ and at $x = 2$.

Remark. We indeed have:

$$P[X = 1] = e^{-2}\frac{2^1}{1!} = e^{-2}\frac{2^2}{2!} = P[X = 2] \quad (\simeq 0.271).$$

That is, the two probabilities are *exactly* equal.

Example 3.5.2 (continued). Let $X \sim \text{Exp}(\lambda)$. Then, X being a continuous random variable, we can use differential calculus to find its mode(s). We have:

$$\frac{d}{dx}f_X(x) = \frac{d}{dx}\lambda e^{-\lambda x} = -\lambda^2 e^{-\lambda x} \neq 0$$

for all $x \in (0, \infty)$. However, because $f_X(x)$ is a strictly decreasing function, we can assert that *the* mode of X is at $x = 0^+$ [$f_X(x)$ tends to a minimum as $x \to \infty$].

The various quantities defined above are *measures of central position*. We continue by defining *measures of dispersion*.

Definition 3.5.6. *The* **range** *of a random variable is the difference between the largest and the smallest value that this variable can take.*

For example, the range of a random variable $X \sim \text{B}(n, p)$ is equal to $n - 0 = n$. Likewise, if $X \sim \text{Exp}(\lambda)$, then its range is $\infty - 0 = \infty$.

Definition 3.5.7. *The* **variance** *of a random variable X is defined by*

$$
\sigma_X^2 \equiv \mathrm{VAR}[X] = \begin{cases} \displaystyle\sum_{i=1}^{\infty} (x_i - \mu_X)^2 p_X(x_i) & \text{if } X \text{ is discrete,} \\[4mm] \displaystyle\int_{-\infty}^{\infty} (x - \mu_X)^2 f_X(x)\, dx & \text{if } X \text{ is continuous.} \end{cases}
$$

Remarks. (i) We deduce at once from the definition that the variance of X is always nonnegative (it can be infinite). Actually, it is strictly positive, except if the random variable X is a constant, which is a *degenerate* random variable. Finally, the larger the variance is, the more spread out is the distribution of the random variable around its mean.

(ii) We also define the *standard deviation* of a random variable X by

$$
\mathrm{STD}[X] = \sqrt{\mathrm{VAR}[X]} \equiv \sigma_X.
$$

We often prefer to work with the standard deviation rather than with the variance of a random variable, because it is easier to interpret. Indeed, the standard deviation is expressed in the same units of measure as X, whereas the units of the variance are the squared units of X.

Now, we may write that

$$
\mathrm{VAR}[X] \equiv E[(X - E[X])^2],
$$

and we can show that

$$
\mathrm{VAR}[X] = E[X^2] - (E[X])^2. \tag{3.11}
$$

Example 3.5.1 (continued). When $X \sim \mathrm{Poi}(\lambda)$, we calculate

$$
E[X^2] = \sum_{x=0}^{\infty} x^2 e^{-\lambda} \frac{\lambda^x}{x!} = e^{-\lambda}\lambda \sum_{x=1}^{\infty} x \frac{\lambda^{x-1}}{(x-1)!} = e^{-\lambda}\lambda \sum_{x=1}^{\infty} \frac{d}{d\lambda} \frac{\lambda^x}{(x-1)!}
$$

$$
= e^{-\lambda}\lambda \frac{d}{d\lambda} \sum_{x=1}^{\infty} \frac{\lambda^x}{(x-1)!} = e^{-\lambda}\lambda \frac{d}{d\lambda}\left(\lambda e^{\lambda}\right)
$$

$$
= e^{-\lambda}\lambda(e^{\lambda} + \lambda e^{\lambda}) = \lambda + \lambda^2.
$$

Then, using Formula (3.11) with $E[X] = \lambda$, we find that

$$
\mathrm{VAR}[X] = (\lambda + \lambda^2) - (\lambda)^2 = \lambda.
$$

Thus, in the case of the Poisson distribution, its parameter λ is both its mean and its variance.

Example 3.5.2 (continued). We already found that if $X \sim \text{Exp}(\lambda)$, then $E[X] = 1/\lambda$. We now calculate [see (3.10)]

$$E[X^2] = \int_0^\infty x^2 \lambda e^{-\lambda x}\, dx = \lambda \frac{2!}{\lambda^{2+1}} = \frac{2}{\lambda^2}.$$

It follows that

$$\text{VAR}[X] = \frac{2}{\lambda^2} - \left(\frac{1}{\lambda}\right)^2 = \frac{1}{\lambda^2}.$$

Note that, in the case of the exponential distribution, its mean and its standard deviation are equal.

Table 3.1, page 89, gives the mean and the variance of the various probability distributions found in Sections 3.2 and 3.3.

The main properties of the *mathematical operator* VAR are the following:
(i) $\text{VAR}[c] = 0$, for any constant c.
(ii) $\text{VAR}[c_1 g(X) + c_0] = c_1^2\, \text{VAR}[g(X)]$, for any constants c_1 and c_0. Thus, the operator VAR is *not* linear.

The mean and the variance of a random variable are particular cases of the quantities known as the *moments* of this variable.

Definition 3.5.8. *The* **moment of order** k *or* **kth-order** *(or simply kth)* **moment** *of a random variable* X *about a point* **a** *is defined by*

$$E[(X - a)^k] = \begin{cases} \displaystyle\sum_{i=1}^\infty (x_i - a)^k p_X(x_i) & \text{if } X \text{ is discrete,} \\[2em] \displaystyle\int_{-\infty}^\infty (x - a)^k f_X(x)\, dx & \text{if } X \text{ is continuous.} \end{cases}$$

Particular cases
(i) The kth-order *moment about the origin*, or *noncentral moment*, of X is

$$E[X^k] \equiv \mu_k' = \sum_{i=1}^\infty x_i^k p_X(x_i) \quad \text{or} \quad \int_{-\infty}^\infty x^k f_X(x)\, dx$$

for $k = 0, 1, \ldots$. We have that $\mu_0' = 1$ and $\mu_1' = \mu_X = E[X]$.
(ii) The kth-order *moment about the mean*, or *central moment*, of X is

$$E[(X - \mu_X)^k] \equiv \mu_k = \sum_{i=1}^\infty (x_i - \mu_X)^k p_X(x_i) \quad \text{or} \quad \int_{-\infty}^\infty (x - \mu_X)^k f_X(x)\, dx$$

for $k = 0, 1, \ldots$. We have that $\mu_0 = 1$, $\mu_1 = 0$, and $\mu_2 = \sigma_X^2$.

Table 3.1. Means and variances of the probability distributions of Sections 3.2 and 3.3

Distribution	Parameters	Mean	Variance
Bernoulli	p	p	pq
Binomial	n and p	np	npq
Hypergeometric	N, n, and d	$n \cdot \dfrac{d}{N}$	$n \cdot \dfrac{d}{N} \cdot \left(1 - \dfrac{d}{N}\right) \cdot \left(\dfrac{N-n}{N-1}\right)$
Geometric	p	$\dfrac{1}{p}$	$\dfrac{q}{p^2}$
Pascal	r and p	$\dfrac{r}{p}$	$\dfrac{rq}{p^2}$
Poisson	λ	λ	λ
Uniform	$[a,b]$	$\dfrac{a+b}{2}$	$\dfrac{(b-a)^2}{12}$
Exponential	λ	$\dfrac{1}{\lambda}$	$\dfrac{1}{\lambda^2}$
Laplace	λ	0	$\dfrac{2}{\lambda^2}$
Gamma	α and λ	$\dfrac{\alpha}{\lambda}$	$\dfrac{\alpha}{\lambda^2}$
Weibull	λ and β	$\dfrac{\Gamma(1+\beta^{-1})}{\lambda^{1/\beta}}$	$\dfrac{\Gamma(1+2\beta^{-1}) - \Gamma^2(1+\beta^{-1})}{\lambda^{2/\beta}}$
Normal	μ and σ^2	μ	σ^2
Beta	α and β	$\dfrac{\alpha}{\alpha+\beta}$	$\dfrac{\alpha\beta}{(\alpha+\beta+1)(\alpha+\beta)^2}$
Lognormal	μ and σ^2	$e^{\mu+\frac{1}{2}\sigma^2}$	$e^{2\mu+\sigma^2}(e^{\sigma^2}-1)$

Example 3.5.2 (continued). In the case when $X \sim \text{Exp}(\lambda)$, we calculate [see (3.10)]

$$E[X^k] = \int_0^\infty x^k \lambda e^{-\lambda x}\, dx = \lambda \frac{k!}{\lambda^{k+1}} = \frac{k!}{\lambda^k}.$$

Remark. We can show, making use of *Newton's binomial formula*, that

$$\mu_k \equiv E[(X - \mu_X)^k] = \sum_{i=0}^{k}(-1)^i \binom{k}{i}\mu'_{k-i}\mu_X^i.$$

This formula enables us to check that $\text{VAR}[X] = E[X^2] - (E[X])^2$, which may be rewritten as follows:

$$\mu_2 = \mu'_2 - \mu_X^2 = \mu'_2 - (\mu'_1)^2.$$

Two other quantities that are used to characterize the distribution of a random variable X are the *skewness* and *kurtosis* coefficients. These two coefficients are defined in terms of the moments of X.

First, the quantity $\mu_3 \equiv E[(X - \mu_X)^3]$ is used to measure the degree of *asymmetry* of probability distributions. If the distribution of X is symmetrical with respect to its mean μ_X, then $\mu_3 = 0$ (if we assume that μ_3 exists). If $\mu_3 > 0$ (resp., < 0), then the distribution is said to be *right-skewed* (resp., *left-skewed*).

Remark. Actually, if the distribution is symmetrical with respect to its mean and if all its central moments exist, then we may write that $\mu_{2k+1} = 0$, for $k = 0, 1, \ldots$.

Definition 3.5.9. *The **skewness** (coefficient) of a random variable X is defined by*

$$\beta_1 = \frac{\mu_3^2}{\sigma^6}.$$

Remarks. (i) The coefficient β_1 is a unitless quantity.

(ii) Some authors prefer to work with the coefficient $\gamma_1 := \sqrt{\beta_1}$.

Example 3.5.2 (continued). We can write that

$$\mu_3 \equiv E[(X - \mu_X)^3] = E[X^3 - 3\mu_X X^2 + 3\mu_X^2 X - \mu_X^3].$$

As shown in Chapter 4, the mathematical expectation of a *linear combination* of random variables can be obtained by replacing each variable by its mean (by linearity of the expectation operator). It follows that

$$\mu_3 = E[X^3] - 3\mu_X E[X^2] + 3\mu_X^2 E[X] - \mu_X^3.$$

Making use of the formula $E[X^k] = k!/\lambda^k$, we obtain:

$$\mu_3 = \frac{3!}{\lambda^3} - \frac{3}{\lambda}\frac{2!}{\lambda^2} + \frac{3}{\lambda^2}\frac{1!}{\lambda} - \frac{1}{\lambda^3} = \frac{2}{\lambda^3}.$$

So, we find that

$$\beta_1 = \frac{(2/\lambda^3)^2}{(1/\lambda^2)^3} = 4.$$

Thus, all the exponential distributions have the same skewness β_1, whatever the value of the parameter λ is, which reflects the fact that they all have the same shape (see Figure 3.11).

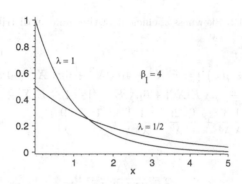

Fig. 3.11. Skewness coefficient of exponential distributions.

Example 3.5.3. Let $X \sim U(a, b)$. The density function $f_X(x)$ is constant, therefore it is symmetrical with respect to $\mu_X = (a + b)/2$. Given that all the moments of the random variable X exist (because X is bounded by a and b), it follows that $\beta_1 = 0$ (see Figure 3.12).

Definition 3.5.10. *The **kurtosis** (coefficient) of a random variable X is the unitless quantity*

$$\beta_2 = \frac{\mu_4}{\sigma^4}.$$

Remarks. (i) When the distribution of X is symmetrical, β_2 measures the relative *thickness* of the tails of the distribution with respect to its central part.

(ii) As in the case of the coefficient β_1, some authors use a different coefficient: $\gamma_2 := \beta_2 - 3$. Because $\beta_2 = 3$ if $X \sim N(\mu, \sigma^2)$, the quantity γ_2 is chosen so that the kurtosis of all normal distributions is equal to zero.

Example 3.5.2 (continued). Making use once again of the formula $E[X^k] = k!/\lambda^k$, we may write that

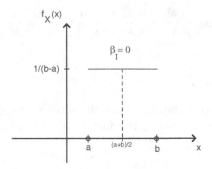

Fig. 3.12. Skewness coefficient of the uniform distribution.

$$\mu_4 \equiv E[(X - \mu_X)^4] = E[X^4 - 4\mu_X X^3 + 6\mu_X^2 X^2 - 4\mu_X^3 X + \mu_X^4]$$
$$= E[X^4] - 4\mu_X E[X^3] + 6\mu_X^2 E[X^2] - 4\mu_X^3 E[X] + \mu_X^4$$
$$= \frac{4!}{\lambda^4} - \frac{4}{\lambda} \frac{3!}{\lambda^3} + \frac{6}{\lambda^2} \frac{2!}{\lambda^2} - \frac{4}{\lambda^3} \frac{1}{\lambda} + \frac{1}{\lambda^4} = \frac{9}{\lambda^4}.$$

Therefore, we have:

$$\beta_2 = \frac{9/\lambda^4}{(1/\lambda^2)^2} = 9,$$

which is independent of the parameter λ.

Example 3.5.4. As we mentioned above, we can show that if $X \sim N(\mu, \sigma^2)$, then $\beta_2 = 3$. A random variable whose density function looks a lot like that of a normal distribution is the *Student distribution* with n degrees of freedom, which is important in statistics. Its kurtosis is given by

$$\beta_2 = \frac{6}{n-4} + 3 \quad \text{if } n > 4.$$

Note that the coefficient β_2 is larger than that of normal distributions, reflecting the fact that the density function $f_X(x)$ tends less rapidly to 0 as x tends to $\pm\infty$ than it does in the case of normal distributions. However, note also that β_2 decreases to 3 as n tends to infinity.

Example 3.5.5. Let $X \sim B(1, p)$. That is, X has a Bernoulli distribution with parameter p. We have:

$$\mu_k' \equiv E[X^k] = \sum_{x=0}^{1} x^k p_X(x) = 0^k q + 1^k p = p$$

for $k = 1, 2, \ldots$. In particular, $E[X] = \mu_1' = p$, so that

$$\mu_k = E[(X - p)^k] = \sum_{x=0}^{1}(x - p)^k p_X(x) = (-p)^k q + (1 - p)^k p.$$

It follows that $\mathrm{VAR}[X] = \mu_2 = p^2 q + q^2 p = pq(p + q) = pq$. We also have:

$$\mu_3 = -p^3 q + q^3 p = pq(-p^2 + q^2)$$

and

$$\mu_4 = p^4 q + q^4 p = pq(p^3 + q^3).$$

Then, we calculate

$$\beta_1 \equiv \frac{\mu_3^2}{\sigma^6} = \frac{p^2 q^2(-p^2 + q^2)^2}{(pq)^3} = \frac{p^3}{q} + \frac{q^3}{p} - 2pq$$

and

$$\beta_2 \equiv \frac{\mu_4}{\sigma^4} = \frac{pq(p^3 + q^3)}{(pq)^2} = \frac{p^2}{q} + \frac{q^2}{p}.$$

Remark. The coefficients β_1 and β_2 are often used to compare an arbitrary distribution with a normal distribution, for which $\beta_1 = 0$ (by symmetry) and $\beta_2 = 3$. For example, if X follows a chi-square distribution with n degrees of freedom (see the gamma distribution above), then $\beta_1 = 8/n$ and

$$\beta_2 = 3\left(\frac{4}{n} + 1\right).$$

Note that β_1 decreases to 0 and β_2 decreases to 3 as n tends to infinity. Actually, the density function of X tends to that of a normal distribution, which is a consequence of the *central limit theorem* (see Chapter 4).

We end this section by giving a proposition that enables us to obtain a bound for a certain probability, when the mean and the variance of the random variable involved are (known and) finite.

Proposition 3.5.1. (Bienaymé–Chebyshev inequality) *For an arbitrary constant $a > 0$, we have:*

$$P[\mu_X - a\sigma_X \le X \le \mu_X + a\sigma_X] > 1 - \frac{1}{a^2}$$

for any random variable X whose variance $\mathrm{VAR}[X] \equiv \sigma_X^2$ is finite.

Remarks. (i) For the variance of X to be finite, its mean $E[X]$ must also be finite.

(ii) Generally, we say that the mean (resp., the variance) of a random variable X does not exist if $E[X] = \pm\infty$ (resp., $\mathrm{VAR}[X] = \infty$). This is the reason why in many books the validity condition for the Bienaymé–Chebyshev inequality is that the variance of X must exist. We can, however, distinguish between the case when the mean (or the variance) of X is infinite and that when it does not exist. For instance, the mean of a *Cauchy distribution* (see p. 8) does not exist, because we find that $E[X] = \infty - \infty$, which is not defined. Then, its variance does not exist either.

Example 3.5.6. If X is a random variable for which $E[X] = 0$ and $\mathrm{VAR}[X] = 1$, then we may write that

$$P[-3 \le X \le 3] > 1 - \frac{1}{9} = 0.\bar{8}.$$

In the case of a $N(0,1)$ distribution, this probability is actually greater than 99.7% (as mentioned above). If X follows a uniform distribution on the interval $[-\sqrt{3}, \sqrt{3}]$, so that its mean is zero and its variance is equal to 1, then the probability in question is equal to 1 (because $[-\sqrt{3}, \sqrt{3}] \subset [-3, 3]$).

3.6 Exercises for Chapter 3

Solved exercises

Question no. 1
    Let

$$p_X\left(x\right) = \begin{cases} a/8 \text{ if } x = -1, \\ a/4 \text{ if } x = 0, \\ a/8 \text{ if } x = 1, \end{cases}$$

where $a > 0$. Find the constant a.

Question no. 2
    Let

$$f_X\left(x\right) = \frac{3}{4}\left(1 - x^2\right) \quad \text{if } -1 < x < 1.$$

Calculate $F_X\left(0\right)$.

Question no. 3
    Calculate the standard deviation of X if

$$p_X\left(x\right) = \frac{1}{3} \quad \text{for } x = 1, 2 \text{ or } 3.$$

Question no. 4
    Suppose that

$$f_X(x) = 2x \quad \text{if } 0 < x < 1.$$

Calculate $E[X^{1/2}]$.

Question no. 5
    Calculate the 25th percentile of the random variable X for which

$$F_X(x) = \frac{x^2}{4} \quad \text{if } 0 < x < 2.$$

Question no. 6

Let

$$f_X(x) = \frac{1}{2} \quad \text{if } 0 < x < 2.$$

We define $Y = X + 1$. Calculate $f_Y(y)$.

Question no. 7

Two items are taken at random and without replacement from a box containing 5 brand A and 10 brand B items. Let X be the number of brand A items among the two selected. What is the probability distribution of X and its parameters?

Question no. 8

Suppose that $X \sim B(n = 5, p = 0.25)$. What is the mode of X, that is, the most probable value of X?

Question no. 9

Ten percent of the articles produced by a certain machine are defective. If 10 (independent) articles fabricated by this machine are taken at random, what is the probability that exactly two of them are defective?

Question no. 10

Calculate $P[X \geq 1 \mid X \leq 1]$ if $X \sim \text{Poi}(\lambda = 5)$.

Question no. 11

Failures occur according to a Poisson process with rate $\lambda = 2$ per day. Calculate the probability that, in the course of two consecutive days, exactly one failure (in all) will occur.

Question no. 12

In a certain lake, there are 200 type I and 50 type II fish. We draw, without replacement, five fish from the lake. Use a binomial distribution to calculate approximately the probability that we get no type II fish.

Question no. 13

Let $X \sim B(n = 50, p = 0.01)$. Use a Poisson distribution to calculate approximately $P[X \geq 4]$.

Question no. 14

A fair coin is tossed until "heads" is obtained. What is the probability that the random experiment will end on the fifth toss?

Question no. 15

The lifetime X of a radio has an exponential distribution with mean equal to ten years. What is the probability that a ten-year-old radio will still work after ten additional years?

Question no. 16
   Suppose that $X \sim \text{Exp}(\lambda)$. Find the value of λ such that $P[X > 1] = 2P[X > 2]$.

Question no. 17
   Let $X \sim \text{G}(\alpha = 2, \lambda = 1)$. What other probability distribution can be used to calculate *exactly* $P[X < 4]$? Give also the parameter(s) of this distribution.

Question no. 18
   The customers of a salesman arrive according to a Poisson process, at the (average) rate of two customers per hour. What is the distribution of the time X needed until ten customers have arrived? Give also the parameter(s) of this distribution.

Question no. 19
   Calculate $P[|X| < 1/2]$ if $X \sim \text{N}(0, 1)$.

Question no. 20
   Calculate the 10th percentile of $X \sim \text{N}(1, 2)$.

Question no. 21
   Devices are made up of five independent components. A given device operates if at least four of its five components are active. Each component operates with probability 0.95. We receive a very large batch of these devices. We inspect the devices, taken at random and with replacement, one at a time until a first device that does not operate has been obtained.

(a) What is the probability that a device taken at random operates?

(b) What is the expected value of the number of devices that will have to be inspected?

Question no. 22
   City buses pass by a certain street corner, between 7:00 a.m. and 7:30 p.m., according to a Poisson process at the (average) rate of four per hour.

(a) What is the probability that at least 30 minutes elapse between the first and the third bus?

(b) What is the variance of the waiting time between the first and the third bus?

(c) Given that a woman has been waiting for 5 minutes, what is the probability that she will have to wait 15 more minutes?

Question no. 23
   Suppose that the length X (in meters) of an arbitrary parking place follows a $\text{N}(\mu, 0.01\mu^2)$ distribution.

(a) A man owns a luxury car whose length is 15% greater than the average length of a parking place. What proportion of free parking places can he use?

(b) Suppose that $\mu = 4$. What should be the length of a car if we want its owner to be able to use 90% of the free parking places?

Question no. 24

A student gets the correct answers, on average, to half of the probability problems she attempts to solve. In an exam, there are ten independent questions. What is the probability that she can solve more than half?

Question no. 25

Let X be a random variable having a binomial distribution with parameters $n = 100$ and $p = 0.1$. Use a Poisson distribution to calculate approximately $P[X = 15]$.

Question no. 26

Let

$$f_X(x) = \sqrt{\frac{2}{\pi} - x^2} \quad \text{for } -\sqrt{\frac{2}{\pi}} \le x \le \sqrt{\frac{2}{\pi}}.$$

Calculate $P[X < 0]$.

Question no. 27

The results of an intelligence quotient (IQ) test for the pupils of a certain elementary school showed that the IQ of these pupils follows (approximately) a normal distribution with parameters $\mu = 100$ and $\sigma^2 = 225$. What total percentage of pupils have an IQ smaller than 91 or greater than 130?

Question no. 28

A certain assembly requires 59 nondefective transistors. We have at our disposal 60 transistors taken at random from those fabricated by a machine that is known to produce 5% defective transistors.

(a) Calculate the probability that the assembly can be made.

(b) Obtain an approximate value of the probability in (a) with the help of a Poisson distribution.

(c) Suppose that, in fact, we have a very large number of transistors at our disposal, of which 5% are defective. What is the probability that we will have to take exactly 60 transistors at random to get 59 nondefective ones?

Question no. 29

Let X be the delivery time (in days) for a certain product. We know that X is a continuous random variable whose mean is 7 and whose standard deviation is equal to 1. Determine a time interval for which, whatever the distribution of X is, we can assert that the delivery times will be in this interval with a probability of at least 0.9.

Question no. 30

The entropy H of a continuous random variable X is defined by $H = E[-\ln f_X(X)]$, where f_X is the density function of X and ln denotes the natural logarithm. Calculate the entropy of a Gaussian random variable with zero mean and variance $\sigma^2 = 2$.

Question no. 31

The number N of devices that a technician must try to repair during the course of an arbitrary workday is a random variable having a geometric distribution with parameter $p = 1/8$. We estimate the probability that he manages to repair a given device to be equal to 0.95, independently from one device to another.

(a) What is the probability that the technician manages to repair exactly five devices, before his second failure, during a given workday, if we assume that he will receive at least seven out-of-order devices in the course of this particular workday?

(b) If, in the course of a given workday, the technician received exactly ten devices for repair, what is the probability that he managed to repair exactly eight of those?

(c) Use a Poisson distribution to calculate approximately the probability in part (b).

(d) Suppose that exactly eight of the ten devices in part (b) have indeed been repaired. If we take three devices at random and without replacement among the ten that the technician had to repair, what is the probability that the two devices he could not repair are among those?

Question no. 32

The number X of raisins in an arbitrary cookie has a Poisson distribution with parameter λ. What value of λ must be chosen if we want the probability that at most 2 cookies, in a bag of 20, contain no raisins to be 0.925?

Question no. 33

The storage tank of a gas station is filled once a week. Suppose that the weekly demand X (in thousands of liters) is a random variable following an exponential distribution with parameter $\lambda = 1/10$. What must the capacity of the storage tank be if we want the probability of exhausting the weekly supply to be 0.01?

Question no. 34

We are interested in the lifetime X (in years) of a machine. From past experience, we estimate the probability that a machine of this type lasts for more than nine years to be 0.1.

(a) We propose the following model for the density function of X:

$$f_X(x) = \frac{a}{(x+1)^b} \quad \text{for } x \geq 0,$$

where $a > 0$ and $b > 1$. Find the constants a and b.

(b) If we propose a normal distribution with mean $\mu = 7$ for X, what must the value of the parameter σ be?

(c) We consider ten machines of this type, which are assumed to be independent. Calculate the probability that eight or nine of these machines last for less than nine years.

Exercises

Question no. 1

A continuous random variable has the following density function:

$$f_X(x) = \begin{cases} cxe^{-x/2} & \text{if } x \geq 0, \\ 0 & \text{if } x < 0. \end{cases}$$

(a) Calculate the constant c.

(b) Obtain (integrating by parts) the distribution function $F_X(x)$.

(c) Find the mean of X.

(d) Calculate the standard deviation of X.

(e) Show that the median of X is located between 3 and 4.

Indication. See (3.10).

Question no. 2

A merchant receives a batch of 100 electrical devices. To save time, she decides to use the following sampling plan: she takes two devices, at random and without replacement, and she decides to accept the whole batch if the two devices selected are nondefective. Let X be the random variable denoting the number of defective devices in the sample.

(a) Give the probability distribution of X as well as the parameters of this distribution.

(b) If the batch contains exactly two defective devices, calculate the probability that it is accepted.

(c) Approximate the probability computed in part (b) with the help of a binomial distribution.

(d) Approximate the probability calculated in part (c) by using a Poisson distribution.

Remark. Give your answers with four decimals.

Question no. 3

In a dart game, the player aims at a circular target having a radius of 25 centimeters. Let X be the distance (in centimeters) between the dart's impact point and the center of the target. Suppose that

$$P[X \leq x] = \begin{cases} c\pi x^2 & \text{if } 0 \leq x \leq 25, \\ 1 & \text{if } x > 25, \end{cases}$$

where c is a constant.

(a) Calculate

    (i) the constant c;

    (ii) the density function, $f_X(x)$, of X;

    (iii) the mean of X;

(iv) the probability $P[X \leq 10 \mid X \geq 5]$.

(b) It costs $1 to throw a dart and the player wins

$$
\begin{cases}
\$10 \text{ if } X \leq r, \\
\$1 \text{ if } r < X \leq 2r, \\
\$0 \text{ if } 2r < X \leq 25.
\end{cases}
$$

For what value of r is the average gain of the player equal to $0.25?

Question no. 4

Telephone calls arrive at an exchange according to a Poisson process with rate λ per minute. We know, from past experience, that the probability of receiving exactly one call during a one-minute period is three times that of receiving no calls during the same time period. For each of the following questions, give the probability distribution of the random variable and calculate, if the case may be, the requested probability.

(a) Let X be the number of calls received over a one-minute period. What is the probability $P[2 \leq X \leq 4]$?

(b) Let Y be the number of calls received over a three-minute period. Calculate the probability $P[Y \geq 4]$.

(c) Let W_1 be the waiting time (in minutes) until the first call, from time $t = 0$. Calculate $P[W_1 \leq 1]$.

(d) Let W_2 be the waiting time (in minutes) between the first and the second call. Calculate $P[W_2 > 1]$.

(e) Let W be the waiting time until the second call, from time $t = 0$. Give the probability distribution of W as well as its parameters.

(f) We consider 100 consecutive one-minute periods and we denote by U the number of periods during which no calls were received. Calculate $P[U \leq 1]$.

Question no. 5

We have ten (independent) machines at our disposal, each producing 2% defective items.

(a) How many items will be fabricated by the first machine, on average, *before* it produces a first defective item?

(b) We take at random one item fabricated by each machine. What is the probability that at most two items among the ten selected are defective?

(c) Redo part (b), using a Poisson approximation.

(d) How many items fabricated by the first machine must be taken, at a minimum, in order that the probability of obtaining at least one defective item be greater than $1/2$ (assuming that the items are independent of one another)?

Question no. 6

We are interested in the proportion θ of defectives in a batch of manufactured articles. We decide to draw, at random and with replacement, a sample of 20 articles from the batch.

(a) We denote by X the number of defective articles in the sample.

(i) Give the probability function $p_X(x)$.

(ii) Give the probability function $p_X(x)$ if the draws are made without replacement and if there are 1000 articles in the batch.

(b) If the draws are made with replacement and if $\theta = 0.25$, calculate

(i) $P[X = 10]$;

(ii) $P[X \geq 10]$, by using a Poisson approximation.

Question no. 7

Let X be a random variable having the density function

$$f_X(x) = \begin{cases} c(1 - x^2) & \text{if } -1 \leq x \leq 1, \\ 0 & \text{if } |x| > 1, \end{cases}$$

where c is a positive constant. Calculate (a) the constant c; (b) the mean of X; (c) the variance of X; (d) the distribution function $F_X(x)$.

Question no. 8

The average number of faulty articles produced by a certain manufacturing process is equal to six per 25-minute period, according to a Poisson process. We consider a given production hour divided into 12 five-minute periods. Let

X be the number of faulty articles produced over a five-minute period;

Y be the number of five-minute periods needed to obtain a first period during which no faulty articles are produced;

Z be the number of periods, among the 12, during which no faulty articles are produced.

(a) Give the distribution of X, Y, and Z as well as their parameter(s).

(b) During which period, on average, will no faulty articles be produced for the first time?

(c) What is the probability that, during exactly 2 of the 12 periods, will no faulty articles be produced?

(d) What is the probability that exactly two faulty articles have been produced during a given five-minute period, given that at most four faulty articles have been produced during this time period?

Question no. 9
   Calculate the variance of $\sqrt{X}$ if

$$p_X(x) = \begin{cases} 1/4 \text{ if } x = 0, \\ 1/2 \text{ if } x = 1, \\ 1/4 \text{ if } x = 2. \end{cases}$$

Question no. 10
   Calculate the 30th percentile of the continuous random variable X whose density function is

$$f_X(x) = \begin{cases} x \text{ if } 0 \le x \le \sqrt{2}, \\ 0 \text{ elsewhere.} \end{cases}$$

Question no. 11
   A 300-page book contains 200 typos. Calculate, using a Poisson distribution, the probability that a particular page contains at least two typos.

Question no. 12
   Based on past data, we estimate that 85% of the articles produced by a certain machine are defective. If the machine produces 20 articles per hour, what is the probability that 8 or 9 articles fabricated over a 30-minute period are defective?

Question no. 13
   Calculate $P[X \ge 8]$ if

$$f_X(x) = \begin{cases} \frac{1}{96}x^3 e^{-x/2} \text{ if } x \ge 0, \\ 0 \qquad \text{elsewhere.} \end{cases}$$

Question no. 14
   The lifetime of a certain electronic component follows an exponential distribution with mean equal to five years. Knowing that a given component is one year old, what is the probability that it will fail during its fourth year of operation?

Question no. 15
   A security system is composed of ten components operating independently of one another. For the system to be operational, at least five components must be active. To check whether the system is operational, we periodically inspect four of its components taken at random (and without replacement). The system is deemed operational if at least three of the four components inspected are active. If, actually, only four of the ten components are active, what is the probability that the system is deemed operational?

Question no. 16
   Calculate the 25th percentile of a continuous random variable X whose density function is

$$f_X(x) = \begin{cases} xe^{-x^2/2} \text{ if } x \ge 0, \\ 0 \quad \text{if } x < 0. \end{cases}$$

Question no. 17

Calculate the probability of obtaining exactly three "tails" in 15 (independent) tosses of a coin for which the probability of getting "tails" is 0.4.

Question no. 18

A sample of four parts is drawn without replacement from a lot of ten parts, of which one is defective. Calculate the probability that the defective part is included in the sample.

Question no. 19

Customers arrive at a counter, according to a Poisson process, at the average rate of five per minute. What is the probability that the number of customers is greater than or equal to ten in a given three-minute period?

Question no. 20

The arrivals of customers at a counter constitute a Poisson process with rate $\lambda = 1$ per two-minute time period. Calculate the probability that the waiting time until the next customer (from any time instant) is smaller than ten minutes.

Question no. 21

Let X be a random variable having a $N(10, 2)$ distribution. Find its 90th percentile.

Question no. 22

Let

$$
F_X(x) = \begin{cases} 0 & \text{if } x < 0, \\ x/2 & \text{if } 0 \le x \le 1, \\ x/6 + 1/3 & \text{if } 1 < x < 4, \\ 1 & \text{if } x \ge 4 \end{cases}
$$

be the distribution function of the continuous random variable X.

(a) Calculate the density function of X.

(b) What is the 75th percentile of X?

(c) Calculate the expected value of X.

(d) Calculate $E[1/X]$.

(e) We define

$$
Y = \begin{cases} -1 \text{ if } X \le 1, \\ 1 \text{ if } X > 1. \end{cases}
$$

(i) Find $F_Y(0)$.

(ii) Calculate the variance of Y.

Question no. 23

A box contains 100 brand A and 50 brand B transistors.

(a) Transistors are drawn one by one, at random and with replacement, until a first brand B transistor has been obtained. What is the probability that nine or ten transistors will have to be drawn?

(b) What is the minimum number of transistors that must be drawn, at random and with replacement, if we want the probability of obtaining only brand A transistors to be smaller than 1/3?

Question no. 24

Parts are fabricated in series. To perform a quality control check, every hour we draw, at random and without replacement, 10 parts from a box containing 25. The fabrication process is deemed under (statistical) control if at most one of the inspected parts is defective.

(a) If all the inspected boxes contain exactly two defective parts, what is the probability that the fabrication process is deemed under control at least seven times during the course of an eight-hour workday?

(b) Use a Poisson distribution to evaluate approximately the probability calculated in part (a).

(c) Knowing that, on the last quality control check performed in part (a), the fabrication process was deemed under control, what is the probability that the corresponding sample of 10 parts contained no defectives?

Question no. 25

Let X be a random variable whose probability function is given by

x	-1	0	3
$p_X(x)$	0.5	0.2	0.3

(a) Calculate the standard deviation of X.

(b) Calculate the mathematical expectation of X^3.

(c) Find the distribution function of X.

(d) We define $Y = X^2 + X + 1$. Find $p_Y(y)$.

Question no. 26

In a particular factory, there were 25 industrial accidents in 2005. Every year, the factory closes for summer holidays for two weeks in July. Answer the following questions, assuming that the industrial accidents occur according to a Poisson process.

(a) What is the probability that exactly one of the 25 accidents occurred during the first two weeks of 2005?

(b) If the average rate of industrial accidents remains the same in 2006, what is the probability that there will be exactly one accident during the first two weeks of that year?

Question no. 27

In a certain lottery, four balls are drawn at random and without replacement among 20 balls numbered from 1 to 20. The player wins a prize if the combination that he has chosen comprises at least two winning numbers. A man decides to buy one ticket per week until he has won a prize. What is the probability that he will have to buy less than ten tickets?

Question no. 28

The density function of the random variable X is given by

$$f_X(x) = \begin{cases} 6x\,(1-x) & \text{if } 0 \le x \le 1, \\ 0 & \text{elsewhere.} \end{cases}$$

(a) Calculate the mathematical expectation of $1/X$.

(b) Obtain the distribution function of X.

(c) We define

$$Y = \begin{cases} 2 \text{ if } X \ge 1/4, \\ 0 \text{ if } X < 1/4. \end{cases}$$

Calculate $E[Y^k]$, where k is a natural number.

(d) Let $Z = X^2$. Find the density function of Z.

Question no. 29

The concentration X of reactant in a chemical reaction is a random variable whose density function is

$$f_X(x) = \begin{cases} 2\,(1-x) & \text{if } 0 < x < 1, \\ 0 & \text{elsewhere.} \end{cases}$$

The amount Y (in grams) of final product is given by $Y = 3X$.

(a) What is the probability that the concentration of reactant is equal to 1/2? Justify.

(b) Calculate the variance of Y.

(c) Obtain the density function of Y.

(d) What is the minimum amount of final product that, in 95% of the cases, will not be exceeded?

Question no. 30

An insurance company employs 20 salespersons. Each salesperson works at the office or on the road. We estimate that a given salesperson is at the office at 2:30 p.m., on any workday, with probability 0.2, independently of the other workdays and of the other salespersons.

(a) The company wants to install a minimum number of desks, so that an arbitrary salesperson finds a free desk in at least 90% of the cases. Find this minimum number of desks.

(b) Calculate the minimum number of desks in part (a) by using a Poisson approximation.

(c) A woman telephoned the office at 2:30 p.m. on the last two workdays in order to talk to a particular salesperson. Given that she did not manage to talk to the salesperson in question, what is the probability that she will have to phone at least two more times, assuming that she always phones at 2:30 p.m.?

Question no. 31

Calculate $\mathrm{VAR}[e^X]$ if X is a random variable whose probability function is given by

$$p_X(x) = \begin{cases} 1/4 \text{ if } x = 0, \\ 1/4 \text{ if } x = 1, \\ 1/2 \text{ if } x = 4, \\ 0 \ \text{ otherwise.} \end{cases}$$

Question no. 32

The density function of the random variable X is

$$f_X(x) = \begin{cases} -x \text{ if } -1 \leq x \leq 0, \\ x \text{ if } 0 < x \leq 1, \\ 0 \text{ elsewhere.} \end{cases}$$

Calculate $F_X(1/2)$.

Question no. 33

Let

$$f_X(x) = \begin{cases} 1/e \text{ if } 0 < x < e, \\ 0 \ \text{ elsewhere.} \end{cases}$$

Calculate $f_Y(y)$, where $Y := -2\ln X$.

Question no. 34

A lot contains 20 items, of which two are defective. Three items are drawn at random and with replacement. Given that at least one defective item was obtained, what is the probability that three defectives were obtained?

Question no. 35

Calls arrive at an exchange according to a Poisson process, at the average rate of two per minute. What is the probability that, during at least one of the first five minutes of a given hour, no calls arrive?

Question no. 36

A box contains 20 granite-type and 5 basalt-type rocks. Ten rocks are taken at random and without replacement. Use a binomial distribution to calculate approximately the probability of obtaining the 5 basalt-type rocks in the sample.

Question no. 37

A fair coin is tossed until "heads" has been obtained ten times. What is the variance of the number of tosses needed to end the random experiment?

Question no. 38

Let X be a random variable having an exponential distribution with parameter λ. We define $Y = \text{int}(X)+1$, where $\text{int}(X)$ designates the *integer part* of X. Calculate $F_Y(y)$.

Question no. 39

Let
$$f_X(x) = \begin{cases} 4x^2 e^{-2x} & \text{if } x \geq 0, \\ 0 & \text{if } x < 0. \end{cases}$$

Calculate the variance of X.

Question no. 40

Suppose that $X \sim N(1, \sigma^2)$. Find σ if $P[-1 < X < 3] = 1/2$.

Question no. 41

The distribution function of the discrete random variable X is given by
$$F_X(x) = \begin{cases} 0 & \text{if } x < 0, \\ 1/2 & \text{if } 0 \leq x < 1, \\ 1 & \text{if } x \geq 1. \end{cases}$$

Calculate (a) $p_X(x)$; (b) $E[\cos(\pi X)]$.

Question no. 42

At least half of the engines of an airplane must operate to enable it to fly. If each engine operates, independently of the others, with probability 0.6, is an airplane having four engines more reliable than a two-engine airplane? Justify.

Question no. 43

We define $Y = |X|$, where X is a continuous random variable whose density function is
$$f_X(x) = \begin{cases} 3/4 & \text{if } -1 \leq x \leq 0, \\ 1/4 & \text{if } 1 \leq x \leq 2, \\ 0 & \text{elsewhere.} \end{cases}$$

What is the 95th percentile of Y?

Question no. 44

The probability that a part produced by a certain machine conforms to the technical specifications is equal to 0.95, independently from one part to the other. We collect parts produced by this machine until we have obtained *one* part that conforms to the technical specifications. This random experiment is repeated on 15 consecutive (independent) days. Let X be the number of days, among the 15 days considered, during which we had to collect at least two parts to get one part conforming to the technical specifications.

(a) What is the mean value of X?

(b) Use a Poisson distribution to calculate approximately the conditional probability $P[X = 2 \mid X \geq 1]$.

Question no. 45
Ten samples of size 10 are drawn at random and without replacement from identical lots containing 100 articles, of which two are defective. A given lot is accepted if at most one defective article is found in the corresponding sample. What is the probability that less than nine of the ten lots are accepted?

Question no. 46
The number X of particles emitted by a certain radioactive source during a one-hour period is a random variable following a Poisson distribution with parameter $\lambda = \ln 5$. Furthermore, we assume that the emissions of particles are independent from hour to hour.

(a) (i) Calculate the probability that during at least 30 hours, among the 168 hours of a given week, no particles are emitted.

(ii) Use a Poisson distribution to calculate approximately the probability in part (i).

(b) Calculate the probability that the fourth hour, during which no particles are emitted, takes place over the course of the first day of the week considered in part (a).

Question no. 47
The duration X (in hours) of major power failures, in a given region, follows approximately a normal distribution with mean $\mu = 2$ and standard deviation $\sigma = 0.75$. Find the duration x_0 for which the probability that an arbitrary major power failure lasts at least 30 minutes more than x_0 is equal to 0.06.

Question no. 48
A continuous random variable X has the following density function:

$$f_X(x) = \begin{cases} \dfrac{x}{k} e^{-x^2/2k} & \text{if } x > 0, \\ 0 & \text{elsewhere,} \end{cases}$$

where $k > 0$ is a constant.

(a) Calculate the mean and the variance of X.

(b) What is the effect of the constant k on the shape of the function f_X?

Remark. You can calculate (using a mathematical software package, if possible) the coefficients β_1 and β_2 to answer this question.

Question no. 49
In a particular region, the daily temperature X (in degrees Celsius) during the month of September has a normal distribution with parameters $\mu = 15$ and $\sigma^2 = 25$.

(a) Let Y be the random variable designating the temperature, given that it is above 17 degrees Celsius. That is, $Y := X \mid \{X > 17\}$. Calculate the density function of Y.

(b) Calculate the (exact) probability that during the month of September the temperature exceeds 17 degrees Celsius on exactly ten days.

Question no. 50

The amount X of rain (in millimeters) that falls over a 24-hour period, in a given region, is a random variable such that

$$F_X(x) = \begin{cases} 0 & \text{if } x < 0, \\ 3/4 & \text{if } x = 0, \\ 1 - \frac{1}{4}e^{-x^2} & \text{if } x > 0. \end{cases}$$

Calculate (a) the expected value of X; (b) the mathematical expectation and the variance of the random variable $Y := e^{X^2/2}$.

Reminder. The variable X is an example of what is known as a random variable of *mixed type*, because it can take on the value 0 with a positive probability, but all the positive real numbers have a zero probability of occurring (as in the case of a continuous random variable). To answer the previous questions, one must make use of the formulas for discrete and continuous random variables at the same time (see p. 83).

Question no. 51

The number of typos in a 500-page book has a Poisson distribution with parameter $\lambda = 2$ per page, independently from one page to the other.

(a) What is the probability that more than ten pages will have to be taken, at random and with replacement, to obtain three pages containing at least two typos each?

(b) Suppose that there are actually 20 pages, among the 500, that contain exactly five typos each.

(i) If 100 pages are taken, at random and without replacement, what is the probability that less than five pages contain exactly five typos each?

(ii) We consider 50 identical copies of this book. If the random experiment in part (i) is repeated for each of these books, what is the probability that, for exactly 30 of the 50 copies, less than five pages with exactly five typos each are obtained?

Question no. 52

A manufacturer sells an article at a fixed price s. He reimburses the purchase price to every customer who discovers that the weight of the article is smaller than a given weight w_0 and he recuperates the article, whose value of the reusable raw material is r ($< s$). The weight W follows approximately a normal distribution with mean μ and variance σ^2. An appropriate setting enables one to fix μ to any desirable value, but it is not possible to fix the value of σ. The cost price C is a function of the weight of the article: $C = \alpha + \beta W$, where α and β are positive constants.

(a) Give an expression for the profit Z in terms of W.

(b) We can show that the average profit, $z(\mu)$, is given by

$$z(\mu) = s - \alpha - \beta\mu - (s - r)P[W < w_0].$$

Find the value μ_0 of μ that maximizes $z(\mu)$.

Question no. 53
In a collection of 20 rocks, 10 are of basalt type and 10 are of granite type. Five rocks are taken at random and without replacement to perform chemical analyses. Let X be the number of basalt-type rocks in the sample.

(a) Give the probability distribution of X as well as its parameters.

(b) Calculate the probability that the sample contains only rocks of the same type.

Multiple choice questions

Question no. 1
Let
$$f_X(x) = \begin{cases} 1/2 & \text{if } 0 < x < 1, \\ 1/(2x) & \text{if } 1 \le x < e. \end{cases}$$

Calculate $P[X < 2 \mid X > 1]$.

(a) $\dfrac{\ln 2}{4}$ (b) $\dfrac{\ln 2}{2}$ (c) $\dfrac{2\ln 2}{3}$ (d) $\ln 2$ (e) 1

Question no. 2
Suppose that $X \sim U(0,1)$. Find $E[(X - E[X])^3]$.

(a) 0   (b) 1/4   (c) 1/3   (d) 1/2   (e) 2/3

Question no. 3
Let $X \sim B(n = 2, p = 0.5)$. Calculate $P[X \ge 1 \mid X \le 1]$.

(a) 0   (b) 1/4   (c) 1/2   (d) 2/3   (e) 1

Question no. 4
Find $E[X^2]$ if $p_X(0) = e^{-\lambda}$ and

$$p_X(x) = \frac{e^{-\lambda} \lambda^{|x|}}{2 \; |x|!} \quad \text{for } x = \ldots, -2, -1, 1, 2, \ldots,$$

where $\lambda > 0$.

(a) λ (b) $\lambda^2 + \lambda$ (c) $\lambda^2 - \lambda$ (d) λ^2 (e) $2\lambda^2$

Question no. 5
   Let
$$f_X(x) = \begin{cases} e^x/2 & \text{if } x < 0, \\ e^{-x}/2 & \text{if } x \geq 0. \end{cases}$$
Calculate the variance of X.
(a) 1   (b) 3/2   (c) 2   (d) 3   (e) 4

Question no. 6
   Suppose that
$$F_X(x) = \begin{cases} 0 & \text{if } x < -1, \\ 1/4 & \text{if } -1 \leq x < 0, \\ 3/4 & \text{if } 0 \leq x < 2, \\ 1 & \text{if } x \geq 2. \end{cases}$$
Calculate $p_X(0) + p_X(1)$.
(a) 0   (b) 1/4   (c) 1/2   (d) 3/4   (e) 1

Question no. 7
   Let $f_X(x) = 2xe^{-x^2}$, for $x > 0$. We set $Y = \ln X$. Find $f_Y(y)$, for any $y \in \mathbb{R}$.
(a) $2e^{2y}e^{-e^{2y}}$ (b) $2e^{2y}e^{-y^2}$ (c) $2e^y e^{-e^{2y}}$ (d) $2e^{-2y}$ (e) e^{-y}

Question no. 8
   Calculate $P[X < 5]$ if $X \sim G(\alpha = 5, \lambda = 1/2)$.
(a) 0.109   (b) 0.243   (c) 0.5   (d) 0.757   (e) 0.891

Question no. 9
   Suppose that X is a discrete random variable whose set of possible values is $\{0, 1, 2, \ldots\}$. Calculate $P[X = 0]$ if $E[t^X] = e^{\lambda(t-1)}$, where t is a real constant.
(a) 0   (b) 1/4   (c) 1/2   (d) $e^{-2\lambda}$ (e) $e^{-\lambda}$

Question no. 10
   Calculate $P[X^2 < 4]$ if $X \sim \text{Exp}(\lambda = 2)$.
(a) $1 - e^{-4}$ (b) $2(1 - e^{-4})$ (c) $\frac{1}{2}e^{-4}$ (d) e^{-4} (e) $2e^{-4}$

Question no. 11
   Calculate $P[0 < X < 2]$ if
$$F_X(x) = \begin{cases} 0 & \text{if } x < 0, \\ 1/2 & \text{if } x = 0, \\ 1 - \frac{1}{2}e^{-x} & \text{if } x > 0. \end{cases}$$
(a) $\dfrac{e^{-1} - e^{-3}}{2}$ (b) $\dfrac{1}{2} - \dfrac{e^{-2}}{2}$ (c) $\dfrac{1}{2}$ (d) $1 - \dfrac{e^{-2}}{2}$ (e) $1 - \dfrac{e^{-3}}{2}$

Remark. The random variable X is of *mixed type* (see p. 83 and Exercise no. 50, p. 109). Note that the function $F_X(x)$ is discontinuous at the point $x = 0$.

Question no. 12
We define $Y = |X|$, where X is a continuous random variable whose density function is

$$f_X(x) = \begin{cases} 1/2 \text{ if } -1 \leq x \leq 1, \\ 0 \text{ elsewhere.} \end{cases}$$

Find $f_Y(y)$.

(a) $\dfrac{1}{2}$ if $-1 \leq y \leq 1$ (b) $\dfrac{y+1}{2}$ if $-1 \leq y \leq 1$ (c) y if $0 \leq y \leq 1$
(d) $2y$ if $0 \leq y \leq 1$ (e) 1 if $0 \leq y \leq 1$

Question no. 13
Let

x	1	4	9
$p_X(x)$	1/4	1/4	1/2

We have that $E[\sqrt{X}] = 9/4$ and $E[X] = 23/4$. Calculate the standard deviation of $\sqrt{X} + 4$.

(a) $\dfrac{\sqrt{11}}{4}$ (b) $\dfrac{\sqrt{11}}{4} + 2$ (c) $\dfrac{\sqrt{11}}{4} + 4$ (d) $\dfrac{11}{6}$ (e) $\dfrac{11}{6} + 2$

Question no. 14
Suppose that

$$f_X(x) = \begin{cases} 1/e \text{ if } 0 < x < e, \\ 0 \text{ elsewhere.} \end{cases}$$

Find a function $g(x)$ such that if $Y := g(X)$, then $f_Y(y) = e^{-y-1}$, for $y > 1$.
(a) e^{-x} (b) e^{x-1} (c) e^x (d) $-\ln x$ (e) $\ln x$

Question no. 15
Calculate the third-order central moment of the discrete random variable X whose probability function is

x	-1	0	1
$p_X(x)$	1/8	1/2	3/8

(a) $-3/32$ (b) 0 (c) $1/64$ (d) $3/32$ (e) $1/4$

Question no. 16
Let X be a continuous random variable defined on the interval (a, b). What is the density function of the random variable $Y := F_X(X)$?

(a) It is not defined   (b) 0   (c) 1 if $0 \leq y \leq 1$ (d) $f_X(y)$
(e) $2y$ if $0 \leq y \leq 1$

Question no. 17

The rate of suicides in a certain city is equal to four per month, according to a Poisson distribution, independently from one month to the other. Calculate the probability that during at least one month over the course of a given year there will be at least eight suicides.

(a) 0.0520   (b) 0.1263   (c) 0.4731   (d) 0.5269   (e) 0.8737

Question no. 18

A telephone survey has been conducted to determine public opinion on the construction of a new nuclear plant. There are 150,000 subscribers whose phone numbers are published in the telephone directory of a certain city and we assume that 90,000 among them would express a negative opinion if asked. Let X be the number of negative opinions obtained in 15 calls made at random (among the 150,000 listed numbers). Calculate approximately $P[X = 9]$ if we also assume that nobody was contacted more than once.

(a) 0.1666   (b) 0.1766   (c) 0.1866   (d) 0.1966   (e) 0.2066

Question no. 19

A multiple choice examination comprises 30 questions. For each question, five answers are proposed. Every correct answer is worth two points and for every wrong answer 1/2 point is deducted. Suppose that a student has already answered 20 questions. Then, she decides to select the letter a for each of the remaining 10 questions, without even reading these questions. If the correct answers are distributed at random among the letters a, b, c, d and e, what is her expected total mark (on 60), assuming that she has four chances out of five of having the correct answer to any of the first 20 (independent) questions she has already done?

(a) 26   (b) 28   (c) 30   (d) 32   (e) 36

Question no. 20

Let $p_X(x) = (3/4)^{x-1}(1/4)$, for $x = 1, 2, \ldots$. Calculate the expected value of the discrete random variable X, given that X is greater than 2.

(a) 4   (b) 5   (c) 6   (d) 7   (e) 8

4

Random vectors

The notion of random variables can be generalized to the case of two (or more) dimensions. In this textbook, we consider in detail only two-dimensional *random vectors*. However, the extension of the various definitions to the multidimensional case is immediate. In this chapter, we also state the most important theorem in probability theory, namely the *central limit theorem*.

4.1 Discrete random vectors

The *joint probability function*

$$p_{X,Y}(x_j, y_k) := P\left[\{X = x_j\} \cap \{Y = y_k\}\right] \equiv P[X = x_j, Y = y_k]$$

of the pair of discrete random variables (X, Y), whose possible values are a (finite or countably infinite) set of points (x_j, y_k) in the plane, has the following properties:

(i) $p_{X,Y}(x_j, y_k) \geq 0 \ \forall (x_j, y_k)$;

(ii) $\sum_{j=1}^{\infty} \sum_{k=1}^{\infty} p_{X,Y}(x_j, y_k) = 1$.

The *joint distribution function* $F_{X,Y}$ is defined by

$$F_{X,Y}(x, y) = P[\{X \leq x\} \cap \{Y \leq y\}] = \sum_{x_j \leq x} \sum_{y_k \leq y} p_{X,Y}(x_j, y_k).$$

Example 4.1.1. Consider the joint probability function $p_{X,Y}$ given by the following table:

M. Lefebvre, *Basic Probability Theory with Applications*, Springer Undergraduate Texts in Mathematics and Technology, DOI: 10.1007/978-0-387-74995-2_4,

$y\backslash x$	-1	0	1
0	1/16	1/16	1/16
1	1/16	1/16	2/16
2	2/16	1/16	6/16

We can check that the function $p_{X,Y}$ possesses the two properties of joint probability functions stated above. Furthermore, given that only the points $(-1,0)$ and $(0,0)$ are such that $x_j \leq 0$ *and* $y_k \leq 1/2$, we may write that

$$F_{X,Y}(0,1/2) = p_{X,Y}(-1,0) + p_{X,Y}(0,0) = \frac{1}{8}.$$

When the function $p_{X,Y}$ is summed over all possible values of Y (resp., X), the resulting function is called the *marginal probability function* of X (resp., Y). That is,

$$p_X(x_j) = \sum_{k=1}^{\infty} p_{X,Y}(x_j,y_k) \quad \text{and} \quad p_Y(y_k) = \sum_{j=1}^{\infty} p_{X,Y}(x_j,y_k).$$

Example 4.1.1 (continued). We find that

x	-1	0	1	Σ
$p_X(x)$	1/4	3/16	9/16	1

and

y	0	1	2	Σ
$p_Y(y)$	3/16	1/4	9/16	1

Definition 4.1.1. *Two discrete random variables, X and Y, are said to be* **independent** *if and only if*

$$p_{X,Y}(x_j,y_k) = p_X(x_j)p_Y(y_k) \quad \text{for any point } (x_j,y_k). \tag{4.1}$$

Example 4.1.1 (continued). We have that $p_{X,Y}(-1,0) = 1/16$, $p_X(-1) = 1/4$ and $p_Y(0) = 3/16$. Because $1/16 \neq (1/4)(3/16)$, X and Y are *not* independent random variables.

Finally, let A_X be an event defined in terms of the random variable X. For instance, $A_X = \{X \geq 0\}$. We define the *conditional probability function* of Y, given the event A_X, by

$$p_Y(y \mid A_X) \equiv P[Y = y \mid A_X] = \frac{P[\{Y = y\} \cap A_X]}{P[A_X]} \quad \text{if } P[A_X] > 0.$$

Likewise, we define

$$p_X(x \mid A_Y) \equiv P[X = x \mid A_Y] = \frac{P[\{X = x\} \cap A_Y]}{P[A_Y]} \quad \text{if } P[A_Y] > 0.$$

Remark. If X and Y are independent discrete random variables, then we may write that

$$p_Y(y \mid A_X) \equiv p_Y(y) \quad \text{and} \quad p_X(x \mid A_Y) \equiv p_X(x).$$

Example 4.1.1 (continued). Let $A_X = \{X = 1\}$. We have:

$$p_Y(y \mid X = 1) = \frac{P[\{Y = y\} \cap \{X = 1\}]}{P[X = 1]} = \frac{p_{X,Y}(1, y)}{p_X(1)}$$

$$= \frac{16}{9} p_{X,Y}(1, y) = \begin{cases} 1/9 \text{ if } y = 0, \\ 2/9 \text{ if } y = 1, \\ 2/3 \text{ if } y = 2. \end{cases}$$

Example 4.1.2. A box contains one brand A, two brand B, and three brand C transistors. Two transistors are drawn at random and with replacement. Let X (resp., Y) be the number of brand A (resp., brand B) transistors among the two selected at random.

(a) Calculate the joint probability function $p_{X,Y}(x, y)$.

(b) Find the marginal probability functions.

(c) Are the random variables X and Y independent? Justify.

(d) Calculate the probability $P[X = Y]$.

Solution. (a) The possible values of the pair (X, Y) are: $(0,0)$, $(0,1)$, $(1,0)$, $(1,1)$, $(0,2)$, and $(2,0)$. Because the transistors are taken *with* replacement, so that the draws are independent, we obtain the following table:

$y \backslash x$	0	1	2
0	1/4	1/6	1/36
1	1/3	1/9	0
2	1/9	0	0

For instance, we have:

$$p_{X,Y}(0,0) = (1/2)(1/2) = 1/4$$

(by independence of the draws) and

$$p_{X,Y}(1,0) \stackrel{\text{inc.}}{=} P[A_1 \cap C_2] + P[C_1 \cap A_2] \stackrel{\text{ind.,sym.}}{=} 2(1/6)(1/2) = 1/6,$$

where A_k is the event "a brand A transistor is obtained on the kth draw," and so on. We can check that the sum of all the fractions in the table is equal to 1.

Remark. In fact, the random variables X and Y follow binomial distributions with parameters $n = 2$ and $p = 1/6$, and with parameters $n = 2$ and $p = 1/3$, respectively.

(b) From the table in part (a), we find that

x	0	1	2	Σ
$p_X(x)$	25/36	5/18	1/36	1

and

y	0	1	2	Σ
$p_Y(y)$	4/9	4/9	1/9	1

(c) The random variables X and Y are *not* independent, because, for instance,

$$p_{X,Y}(2,2) = 0 \neq p_X(2)p_Y(2) = (1/36)(1/9).$$

Remark. The random variables X and Y could not be independent, because the relation $0 \leq X + Y \leq 2$ must be satisfied.

(d) We calculate

$$P[X = Y] = p_{X,Y}(0,0) + p_{X,Y}(1,1) + p_{X,Y}(2,2) = \frac{1}{4} + \frac{1}{9} + 0 = \frac{13}{36}.$$

4.2 Continuous random vectors

Let X and Y be two continuous random variables. The generalization of the notion of density function to the two-dimensional case is the *joint density function* $f_{X,Y}$ of the pair (X,Y). This function is such that

$$P[x < X \leq x + \epsilon, y < Y \leq y + \delta] = \int_y^{y+\delta} \int_x^{x+\epsilon} f_{X,Y}(u,v)\,du\,dv$$

and has the following properties:

(i) $f_{X,Y}(x,y) \geq 0$ for any point (x,y);

(ii) $\int_{-\infty}^{\infty} \int_{-\infty}^{\infty} f_{X,Y}(x,y)\,dx\,dy = 1$.

Remark. In n dimensions, a continuous random vector possesses a joint density function defined on $\mathbb{R}^n$ (or on an uncountably infinite subset of $\mathbb{R}^n$). This function is nonnegative and its integral on $\mathbb{R}^n$ is equal to 1.

The *joint distribution function* is defined by

$$F_{X,Y}(x,y) = P[X \leq x, Y \leq y] = \int_{-\infty}^{y} \int_{-\infty}^{x} f_{X,Y}(u,v)\,du\,dv.$$

Example 4.2.1. Consider the function $f_{X,Y}$ defined by

$$f_{X,Y}(x,y) = cxye^{-x^2-y^2} \quad \text{for } x \geq 0, y \geq 0,$$

where $c > 0$ is a constant. We have that (i) $f_{X,Y}(x,y) \geq 0$ for any point (x,y) with $x \geq 0$ and $y \geq 0$ [$f_{X,Y}(x,y) = 0$ elsewhere] and (ii)

$$\int_0^\infty \int_0^\infty cxye^{-x^2-y^2}\,dx\,dy = c \int_0^\infty xe^{-x^2}\,dx \int_0^\infty ye^{-y^2}\,dy$$

$$= c \left[-\frac{e^{-x^2}}{2} \Big|_0^\infty \right] \left[-\frac{e^{-y^2}}{2} \Big|_0^\infty \right] = c(1/2)(1/2) = c/4.$$

So, this function is a valid joint density function if and only if the constant c is equal to 4.

The joint distribution function of the pair (X,Y) is given by

$$F_{X,Y}(x,y) = \int_0^y \int_0^x 4uve^{-u^2-v^2}\,du\,dv$$

$$= \int_0^x 2ue^{-u^2}\,du \int_0^y 2ve^{-v^2}\,dv$$

$$= \left[-e^{-u^2} \Big|_0^x \right] \left[-e^{-v^2} \Big|_0^y \right] = (1 - e^{-x^2})(1 - e^{-y^2})$$

for $x \geq 0$ and $y \geq 0$.

Remark. We have that $F_{X,Y}(x,y) = 0$ if $x < 0$ or $y < 0$.

The *marginal density functions* of X and Y are defined by

$$f_X(x) = \int_{-\infty}^\infty f_{X,Y}(x,y)\,dy \quad \text{and} \quad f_Y(y) = \int_{-\infty}^\infty f_{X,Y}(x,y)\,dx.$$

Remark. We can easily generalize the previous definitions to the case of three or more random variables. For instance, the joint density function of the random vector (X,Y,Z) is a nonnegative function $f_{X,Y,Z}(x,y,z)$ such that

$$\int_{-\infty}^\infty \int_{-\infty}^\infty \int_{-\infty}^\infty f_{X,Y,Z}(x,y,z)\,dx\,dy\,dz = 1.$$

Moreover, the joint density function of the random vector (X,Y) is obtained as follows:

$$f_{X,Y}(x,y) = \int_{-\infty}^\infty f_{X,Y,Z}(x,y,z)\,dz.$$

Finally, the marginal density function of the random variable X is given by

$$f_X(x) = \int_{-\infty}^{\infty} \int_{-\infty}^{\infty} f_{X,Y,Z}(x,y,z)\, dy\, dz.$$

Definition 4.2.1. *The continuous random variables X and Y are said to be* **independent** *if and only if*

$$f_{X,Y}(x,y) = f_X(x) f_Y(y) \quad \text{for any point } (x,y). \tag{4.2}$$

Example 4.2.1 (continued). We have:

$$f_X(x) = \int_0^{\infty} 4xy e^{-x^2-y^2}\, dy = 2xe^{-x^2} \int_0^{\infty} 2y e^{-y^2}\, dy$$
$$= 2xe^{-x^2}\left[-e^{-y^2} \Big|_0^{\infty} \right] = 2xe^{-x^2} \quad \text{for } x \geq 0.$$

Then, by symmetry, we may write that

$$f_Y(y) = 2y e^{-y^2} \quad \text{for } y \geq 0.$$

Furthermore, because

$$f_X(x) f_Y(y) = 4xy e^{-x^2-y^2} = f_{X,Y}(x,y)$$

for any point (x,y) (with $x \geq 0$ and $y \geq 0$), the random variables X and Y are independent.

Finally, let A_Y be an event depending only on Y. For example, $A_Y = \{0 \leq Y \leq 1\}$. The *conditional density function* of X, given that A_Y occurred, is given by

$$f_X(x \mid A_Y) = \frac{\int_{A_Y} f_{X,Y}(x,y) dy}{P[A_Y]} \quad \text{if } P[A_Y] > 0.$$

If A_Y is an event of the form $\{Y = y\}$, we can show that

$$f_X(x \mid Y = y) = \frac{f_{X,Y}(x,y)}{f_Y(y)} \quad \text{if } f_Y(y) > 0.$$

That is, the conditional density function $f_X(x \mid Y = y)$ is obtained by dividing the joint density function of (X,Y), evaluated at the point (x,y), by the marginal density function of Y evaluated at the point y.

Remarks. (i) If X and Y are two independent continuous random variables, then we have:

$$f_X(x \mid A_Y) \equiv f_X(x) \quad \text{and} \quad f_Y(y \mid A_X) \equiv f_Y(y).$$

(ii) In general, if X is a continuous random variable, then we can write that

$$P[Y \in A_Y] = \int_{-\infty}^{\infty} P[Y \in A_Y \mid X = x] f_X(x) dx, \tag{4.3}$$

where A_Y is an event that involves only the random variable Y. In the case when X is discrete, we have:

$$P[Y \in A_Y] = \sum_{k=1}^{\infty} P[Y \in A_Y \mid X = x_k] p_X(x_k). \tag{4.4}$$

These formulas are extensions of the *law of total probability* from Chapter 2. We also have, for instance:

$$P[Y > X] = \int_{-\infty}^{\infty} P[Y > X \mid X = x] f_X(x) dx$$

$$= \int_{-\infty}^{\infty} P[Y > x \mid X = x] f_X(x) dx, \tag{4.5}$$

and so on.

Definition 4.2.2. *The* **conditional expectation** *of the random variable Y, given that $X = x$, is defined by*

$$E[Y \mid X = x] = \begin{cases} \sum_{j=1}^{\infty} y_j p_Y(y_j \mid X = x) & \text{if } (X, Y) \text{ is discrete,} \\ \int_{-\infty}^{\infty} y f_Y(y \mid X = x) dy & \text{if } (X, Y) \text{ is continuous.} \end{cases}$$

Remarks. (i) We can show that

$$E[Y] = E[E[Y \mid X]] := \begin{cases} \sum_{k=1}^{\infty} E[Y \mid X = x_k] p_X(x_k) & \text{if } X \text{ is discrete,} \\ \int_{-\infty}^{\infty} E[Y \mid X = x] f_X(x) dx & \text{if } X \text{ is continuous.} \end{cases}$$

(ii) In general,

$$E[g(Y)] = E[E[g(Y) \mid X]]$$

for any function $g(\cdot)$. It follows that

$$\text{VAR}[Y] = E[E[Y^2 \mid X]] - \{E[E[Y \mid X]]\}^2 .$$

Example 4.2.2. Let

$$f_{X,Y}(x,y) = \begin{cases} k(x^2 + y^2) & \text{for } 0 \le x \le a,\, 0 \le y \le b, \\ 0 & \text{elsewhere.} \end{cases}$$

(a) Calculate the constant k.

(b) Find the marginal density functions of X and Y. Are X and Y independent?

(c) Obtain the conditional density functions $f_X(x \mid Y = y)$, $f_Y(y \mid X = x)$, and $f_Y(y \mid X < a/2)$.

(d) Find the distribution function $F_{X,Y}(x,y)$.

Solution. (a) We have:

$$1 = \int_0^b \int_0^a k(x^2 + y^2)\,dx\,dy = k \int_0^b \left(\frac{x^3}{3} + xy^2 \right) \Big|_0^a dy$$

$$= k \int_0^b \left(\frac{a^3}{3} + ay^2 \right) dy = k \left[\frac{a^3 y}{3} + \frac{ay^3}{3} \right] \Big|_0^b.$$

Thus, k must be given by

$$k = \left[\frac{a^3 b}{3} + \frac{ab^3}{3} \right]^{-1} = \frac{3}{(ab)(a^2 + b^2)}.$$

(b) We can write that

$$f_X(x) = \int_0^b k(x^2 + y^2)\,dy = k \left\{ yx^2 + \frac{y^3}{3} \Big|_0^b \right\}$$

$$= k \left(bx^2 + \frac{b^3}{3} \right) \quad \text{for } 0 \le x \le a,$$

where k has been calculated in part (a). Similarly, we find that

$$f_Y(y) = k \left(ay^2 + \frac{a^3}{3} \right) \quad \text{for } 0 \le y \le b.$$

Now, we have:

$$f_X(0)f_Y(0) = k \left(\frac{b^3}{3} \right) k \left(\frac{a^3}{3} \right) = k^2 \frac{a^3 b^3}{9} \ne 0 = f_{X,Y}(0,0).$$

Therefore, X and Y are *not* independent random variables.

Remark. When the joint density function, $f_{X,Y}(x,y)$, is a constant c multiplied by a sum or a difference, like $x+y$, $x^2 - y^2$, and so on, the random variables X and Y cannot be independent. Indeed, it is impossible to write, in particular, that

$$c(x + y) = f(x)g(y),$$

where $f(x)$ depends only on x and $g(y)$ depends only on y.

(c) We calculate

$$f_X(x \mid Y = y) = \frac{f_{X,Y}(x,y)}{f_Y(y)} \overset{(b)}{=} \frac{k(x^2 + y^2)}{k\left(\frac{a^3}{3} + ay^2\right)} = \frac{3(x^2 + y^2)}{a(a^2 + 3y^2)}$$

for $0 \leq x \leq a$ and $0 \leq y \leq b$. Similarly, we find that

$$f_Y(y \mid X = x) = \frac{3(x^2 + y^2)}{b(3x^2 + b^2)}$$

for $0 \leq x \leq a$ and $0 \leq y \leq b$. Finally, we have:

$$f_Y(y \mid X < a/2) = \frac{\int_0^{a/2} k(x^2 + y^2)dx}{\int_0^{a/2} k\left(bx^2 + \frac{b^3}{3}\right)dx} = \frac{k\left\{\frac{x^3}{3} + xy^2 \Big|_0^{a/2}\right\}}{k\left\{b\frac{x^3}{3} + x\frac{b^3}{3} \Big|_0^{a/2}\right\}}$$

$$= \frac{\frac{a^3}{24} + \frac{ay^2}{2}}{\frac{ba^3}{24} + \frac{ab^3}{6}} = \frac{a^2 + 12y^2}{ba^2 + 4b^3} \quad \text{for } 0 \leq y \leq b.$$

(d) By definition,

$$F_{X,Y}(x,y) = \int_0^y \int_0^x k(u^2 + v^2)dudv = \int_0^y k\left(\frac{u^3}{3} + uv^2\right)\Big|_0^x dv$$

$$= k \int_0^y \left(\frac{x^3}{3} + xv^2\right)dv = k\left\{\frac{x^3 y}{3} + \frac{xy^3}{3}\right\} = \frac{xy(x^2 + y^2)}{ab(a^2 + b^2)}$$

for $0 \leq x \leq a$ and $0 \leq y \leq b$. Hence, we deduce that (see Figure 4.1)

$$F_{X,Y}(x,y) = \begin{cases} 0 & \text{if } x < 0 \text{ or } y < 0, \\ \dfrac{xy(x^2 + y^2)}{ab(a^2 + b^2)} & \text{if } 0 \leq x \leq a \text{ and } 0 \leq y \leq b, \\ \dfrac{xb(x^2 + b^2)}{ab(a^2 + b^2)} & \text{if } 0 \leq x \leq a \text{ and } y > b, \\ \dfrac{ay(a^2 + y^2)}{ab(a^2 + b^2)} & \text{if } x > a \text{ and } 0 \leq y \leq b, \\ 1 & \text{if } x > a \text{ and } y > b. \end{cases}$$

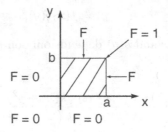

Fig. 4.1. Joint distribution function in Example 4.2.2.

Remark. Corresponding to Formula (3.1) in the one-dimensional case, we have:

$$\frac{\partial^2}{\partial x \partial y} F_{X,Y}(x,y) = f_{X,Y}(x,y)$$

for any point (x,y) at which the function $F_{X,Y}(x,y)$ is differentiable.

Example 4.2.3. Suppose that $X \sim \text{Exp}(\lambda_1)$ and $Y \sim \text{Exp}(\lambda_2)$ are independent random variables. Making use of (4.5), we can write that

$$\begin{aligned}
P[Y > X] &= \int_0^\infty P[Y > X \mid X = x] f_X(x)\, dx \\
&\overset{\text{ind.}}{=} \int_0^\infty P[Y > x] \lambda_1 e^{-\lambda_1 x}\, dx = \int_0^\infty e^{-\lambda_2 x} \lambda_1 e^{-\lambda_1 x}\, dx \\
&= \int_0^\infty \lambda_1 e^{-(\lambda_1 + \lambda_2)x}\, dx = \frac{\lambda_1}{\lambda_1 + \lambda_2}.
\end{aligned} \tag{4.6}$$

Remark. Note that if $\lambda_1 = \lambda_2$, then $P[X < Y] = 1/2$, which actually follows directly by symmetry (and by continuity of the exponential distribution).

4.3 Functions of random vectors

In Chapter 3, we saw that any real-valued function of a random variable is itself a random variable. Similarly, any real-valued function of a random vector is a random variable. More generally, n real-valued functions of a random variable or of a random vector constitute a new random vector of dimension n. The most interesting transformations are the sum, the difference, the product, and the ratio of random variables.

In general, we must be able to calculate the probability function or the density function of the new random variable or vector. In this textbook, we treat the case of a single function g of a two-dimensional random vector (X, Y). We also give important results obtained when the function g is the sum (or a *linear combination*) of n independent random variables.

Sometimes, we only need the mean, for instance, of the new random variable. In that case, it is not necessary to first calculate the probability function or the density function of $g(X, Y)$.

4.3.1 Discrete case

In the particular case when the number of possible values of the pair (X, Y) of random variables is *finite*, we only have to apply the transformation g to each possible value of this pair and to add the probabilities of the points (x, y) that are transformed into the same value of $g(x, y)$.

Example 4.3.1. Consider again the joint probability function in Example 4.1.1:

$y \backslash x$	-1	0	1
0	1/16	1/16	1/16
1	1/16	1/16	2/16
2	2/16	1/16	6/16

Let $Z = XY$. The random variable Z can take on five different values: -2, -1, 0, 1, and 2. The point $(-1, 2)$ corresponds to $z = -2$, $(-1, 1)$ corresponds to $z = -1$, $(1, 1)$ is transformed into $z = 1$, $(1, 2)$ becomes $z = 2$, and all the other points are such that $z = 0$. From the previous table, we obtain that

z	-2	-1	0	1	2	Σ
$p_Z(z)$	2/16	1/16	5/16	2/16	6/16	1

It follows that the mean of Z is given by

$$E[Z] = (-2)\frac{2}{16} + (-1)\frac{1}{16} + 0 + (1)\frac{2}{16} + (2)\frac{6}{16} = \frac{9}{16}.$$

As we mentioned above, if we are only interested in obtaining the expected value of the new random variable Z, then it is not necessary to calculate the function $p_Z(z)$. It suffices to use Formula (4.11) of Section 4.4:

$$E[g(X, Y)] = \sum_{k=1}^{\infty} \sum_{j=1}^{\infty} g(x_k, y_j) p_{X,Y}(x_k, y_j).$$

Here, we obtain that $E[XY] = 9/16$ (see Example 4.4.1), which agrees with the result obtained above for $Z = XY$.

Example 4.3.2. Suppose that we toss two distinct and well-balanced tetrahedrons, whose faces are numbered 1, 2, 3, and 4. Let X_1 (resp., X_2) be the number of the face on which the first (resp., second) tetrahedron lands, and let Y be the *maximum* between X_1 and X_2. What is the probability function of the random variable Y?

Solution. The possible values of Y are 1, 2, 3, and 4. Let A_k (resp., B_k) be the event "the random variable X_1 (resp., X_2) takes on the value k," for $k = 1, \ldots, 4$. By independence of the events A_j and B_k for all j and k, we have:

$$p_Y(1) = P[A_1 \cap B_1] = P[A_1]P[B_1] = \frac{1}{4} \times \frac{1}{4} = \frac{1}{16}.$$

Similarly, by independence and *incompatibility*, we may write that

$$\begin{aligned} p_Y(2) &= P[(A_1 \cap B_2) \cup (A_2 \cap B_1) \cup (A_2 \cap B_2)] \\ &= P[A_1]P[B_2] + P[A_2]P[B_1] + P[A_2]P[B_2] \\ &= \frac{1}{4} \times \frac{1}{4} + \frac{1}{4} \times \frac{1}{4} + \frac{1}{4} \times \frac{1}{4} = \frac{3}{16}. \end{aligned}$$

Next, using the *equiprobability* of the events (and of the intersections), we obtain that

$$\begin{aligned} p_Y(3) &= P[(A_1 \cap B_3) \cup (A_2 \cap B_3) \cup (A_3 \cap B_3) \cup (A_3 \cap B_2) \cup (A_3 \cap B_1)] \\ &= 5 \times \frac{1}{4} \times \frac{1}{4} = \frac{5}{16}. \end{aligned}$$

Finally, because we must have that $\sum_{y=1}^{4} p_Y(y) = 1$, we obtain the following table:

y	1	2	3	4
$p_Y(y)$	1/16	3/16	5/16	7/16

It follows that

y	1	2	3	4
$F_Y(y)$	1/16	1/4	9/16	1

When the number of possible values of the pair (X, Y) is *countably infinite*, it is generally much more difficult to obtain the probability function of the random variable $Z := g(X, Y)$. Indeed, there can be an infinite number of points (x, y) that correspond to the same $z = g(x, y)$, and there can also be an infinite number of different values of Z. However, in the case when the number of possible values of Z is finite, we can sometimes calculate $p_Z(z)$ relatively easily.

Example 4.3.3. Suppose that the joint probability function of the random vector (X, Y) is given by the formula

$$p_{X,Y}(x, y) = \frac{e^{-2}}{x!y!} \quad \text{for } x = 0, 1, \ldots; y = 0, 1, \ldots .$$

Note that X and Y are actually two independent random variables that both follow a Poisson distribution with parameter $\lambda = 1$. Let

$$Z = g(X, Y) := \begin{cases} 1 \text{ if } X = Y, \\ 0 \text{ if } X \neq Y. \end{cases}$$

In this case, Z has a Bernoulli distribution with parameter p, where

$$p := P[X = Y] = \sum_{x=0}^{\infty} \frac{e^{-2}}{(x!)^2}.$$

We find, using a mathematical software package for instance, that the above infinite series converges to $e^{-2} \cdot I_0(2) \simeq 0.3085$, where $I_0(\cdot)$ is a *Bessel function*. We could actually obtain a very good approximation to the exact result by adding the first five terms of the series, because

$$\sum_{x=0}^{4} \frac{e^{-2}}{(x!)^2} \simeq 0.3085$$

as well.

4.3.2 Continuous case

Suppose that we wish to obtain the density function of the transformation $Z := g_1(X, Y)$ of the continuous random vector (X, Y). We consider only the case when it is possible to define an *auxiliary variable* $W = g_2(x, y)$ such that the system

$$z = g_1(x, y),$$
$$w = g_2(x, y)$$

possesses a *unique* solution: $x = h_1(z, w)$ and $y = h_2(z, w)$. The following proposition can then be proved.

Proposition 4.3.1. *Let* (X, Y) *be a continuous random vector and let* $Z = g_1(X, Y)$ *and* $W = g_2(X, Y)$. *Suppose that the functions* $x = h_1(z, w)$ *and* $y = h_2(z, w)$ *have continuous partial derivatives (with respect to z and w) for all (z, w) and that the Jacobian of the transformation:*

$$J(z, w) := \begin{vmatrix} \partial h_1/\partial z & \partial h_1/\partial w \\ \partial h_2/\partial z & \partial h_2/\partial w \end{vmatrix}.$$

is not identical to zero. Then, we can write that

$$f_{Z,W}(z, w) = f_{X,Y}(h_1(z, w), h_2(z, w))|J(z, w)|.$$

It follows that

$$f_Z(z) = \int_{-\infty}^{\infty} f_{X,Y}(h_1(z, w), h_2(z, w))|J(z, w)| \, dw.$$

Remarks. (i) We generally choose a very simple auxiliary variable W, for example, $W = X$.

(ii) In the particular case when $g_1(x, y)$ is a linear transformation of x and y, it suffices to choose another linear transformation of x and y for the partial derivatives of the functions h_1 and h_2 to be continuous. Indeed, these partial derivatives are then constants. Therefore, they are continuous at any point (z, w).

Example 4.3.4. Let $X \sim U(0, 1)$ and $Y \sim U(0, 1)$ be two independent random variables and let $Z = X + Y$. To obtain the density function of Z, we define the auxiliary variable $W = X$. Then, the system

$$z = x + y,$$
$$w = x$$

has the unique solution $x = w$, $y = z - w$. Moreover, the partial derivatives of the functions $h_1(z, w) = w$ and $h_2(z, w) = z - w$ are continuous $\forall (z, w)$ and the Jacobian

$$J(z, w) = \begin{vmatrix} 0 & 1 \\ 1 & -1 \end{vmatrix} = -1$$

is different from zero for all (z, w). Consequently, we can write that

$$f_{Z,W}(z, w) = f_{X,Y}(w, z - w) |-1| \overset{\text{ind.}}{=} f_X(w) f_Y(z - w)$$

$$\implies \quad f_{Z,W}(z, w) = 1 \cdot 1 \quad \text{if } 0 < w < 1 \text{ and } 0 < z - w < 1.$$

Because $0 < z < 2$, the set of possible values of w is the interval $(0, z)$, if $0 < z < 1$, and the interval $(z - 1, 1)$, if $1 \leq z < 2$.

Finally, we have (see Figure 4.2):

$$f_Z(z) = \begin{cases} \displaystyle\int_0^z 1\, dw = z & \text{if } 0 < z < 1, \\[2em] \displaystyle\int_{z-1}^1 1\, dw = 2 - z & \text{if } 1 \leq z < 2. \end{cases}$$

4.3.3 Convolutions

Let X be a discrete random variable whose possible values are $x_1, x_2, \ldots$. The *convolution* of X with itself is obtained by applying the transformation of interest, for instance, the sum, the difference, the product, and so on, to the points $(x_1, x_1), (x_1, x_2), \ldots$, $(x_2, x_1), (x_2, x_2), \ldots$. Therefore, if X can take on n different values, then the transformation must be applied to $n \times n = n^2$ points. We write $X \otimes X$ to denote the *convolution product* of X with itself, $X \oplus X$ for the *convolution sum*, and so on. Observe that obtaining the distribution of $X \otimes X$, for example, is tantamount to finding the distribution of the product $X_1 X_2$, where X_1 and X_2 are two independent random variables having the same distribution as X.

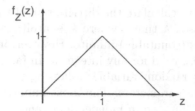

Fig. 4.2. Density function in Example 4.3.4.

Example 4.3.5. Consider the probability function of the random variable X in Example 3.4.1:

x	-1	0	1
$p_X(x)$	$1/4$	$1/4$	$1/2$

Suppose that we want to obtain the distribution of the convolution product of X with itself. We find that the possible results of this convolution are -1, 0, and 1. The points $(-1,1)$ and $(1,-1)$ correspond to -1, the points $(-1,-1)$ and $(1,1)$ to 1, and the five other points to 0. Then, if we define $Y = X \otimes X$, we deduce from the above table that

y	-1	0	1
$p_Y(y)$	$1/4$	$7/16$	$5/16$

because

$$P[Y = -1] = P[X = -1]P[X = 1] + P[X = 1]P[X = -1]$$
$$= 2 \times (1/4)(1/2) = 1/4,$$

and so on.

Remarks. (i) Note that the result obtained is completely different from the probability function of $Z := X^2$ calculated in Example 3.4.1:

z	0	1
$p_Z(z)$	$1/4$	$3/4$

(ii) If we calculate the *convolution difference* of X with itself, we find that the probability function of $D := X \ominus X$ is given by

d	-2	-1	0	1	2
$p_D(d)$	$1/8$	$3/16$	$3/8$	$3/16$	$1/8$

In general, it is difficult to calculate the distribution of the convolution of a discrete random variable X with itself k times, where k is arbitrary, especially if the number of values that X can take is (countably) infinite. However, in a few particular cases we can obtain a general formula, valid for any integer k. In fact, we can prove some results for the sum of independent random variables following the same distribution, but not necessarily having the same parameters. The most important results of this type are the following, where $X_1, \ldots, X_n$ are n independent random variables.

(1) If X_i has a Bernoulli distribution with parameter p, for all i, then we have:

$$\sum_{i=1}^{n} X_i \sim \mathrm{B}(n,p) \quad \text{for } n = 1, 2, \ldots .$$

More generally, if $X_i \sim \mathrm{B}(m_i, p)$, for $i = 1, \ldots, n$, then we find that

$$\sum_{i=1}^{n} X_i \sim \mathrm{B}\left(\sum_{i=1}^{n} m_i, p\right).$$

(2) If $X_i \sim \mathrm{Poi}(\lambda_i)$, for $i = 1, \ldots, n$, then we may write that

$$\sum_{i=1}^{n} X_i \sim \mathrm{Poi}\left(\sum_{i=1}^{n} \lambda_i\right). \tag{4.7}$$

(3) If $X_i \sim \mathrm{Geo}(p)$, for $i = 1, \ldots, n$, then we have:

$$\sum_{i=1}^{n} X_i \sim \mathrm{NB}(n,p).$$

Suppose now that X and Y are two independent *continuous* random variables. Let $Z = X + Y$. We can show that the density function of Z is obtained by computing the convolution product of the density function of X with that of Y. That is, we have:

$$f_Z(z) = f_X(x) * f_Y(y) = \int_{-\infty}^{\infty} f_X(u) f_Y(z - u)\, du. \tag{4.8}$$

We could use this formula to obtain the density function of Z in Example 4.3.4.

As in the discrete case, we can prove some results for the sum (and sometimes for *linear combinations*) of *independent* random variables X_i. We find, in particular, that

(1) if $X_i \sim \mathrm{Exp}(\lambda)$, for $i = 1, \ldots, n$, then

$$\sum_{i=1}^{n} X_i \sim \mathrm{G}(n, \lambda); \tag{4.9}$$

(2) if $X_i \sim \mathrm{G}(\alpha_i, \lambda)$, for $i = 1, \ldots, n$, then

$$\sum_{i=1}^{n} X_i \sim G\left(\sum_{i=1}^{n} \alpha_i, \lambda\right);$$

(3) if $X_i \sim N(\mu_i, \sigma_i^2)$, for $i = 1, \ldots, n$, then

$$\sum_{i=1}^{n} a_i X_i \sim N\left(\sum_{i=1}^{n} a_i \mu_i, \sum_{i=1}^{n} a_i^2 \sigma_i^2\right), \tag{4.10}$$

where a_i is a real constant, for all i.

Remarks. (i) The best way to prove these results is to make use of the *characteristic function*, or of the *moment-generating function*, which are actually particular mathematical expectations. The characteristic function of the random variable X is defined by $E[e^{j\omega X}]$, where $j := \sqrt{-1}$.

(ii) A result of the same type, but for the *product* of independent random variables $X_1, \ldots, X_n$, is the following: if $X_i \sim LN(\mu_i, \sigma_i^2)$, then

$$\prod_{i=1}^{n} X_i^{a_i} \sim LN\left(\sum_{i=1}^{n} a_i \mu_i, \sum_{i=1}^{n} a_i^2 \sigma_i^2\right).$$

4.4 Covariance and correlation coefficient

Definition 4.4.1. *Let (X, Y) be a pair of random variables. We define the* **mathematical expectation** *of the function $g(X, Y)$ by*

$$E[g(X,Y)] = \begin{cases} \displaystyle\sum_{k=1}^{\infty}\sum_{j=1}^{\infty} g(x_k, y_j) p_{X,Y}(x_k, y_j) & \text{(discrete case),} \\[2ex] \displaystyle\int_{-\infty}^{\infty}\int_{-\infty}^{\infty} g(x,y) f_{X,Y}(x,y)\, dx\, dy & \text{(continuous case).} \end{cases} \tag{4.11}$$

Particular cases. (i) If $g(X, Y) = X$, then we have that $E[g(X, Y)] = E[X] \equiv \mu_X$.

(ii) If $g(X, Y) = XY$, we then obtain the formula that enables us to calculate the mathematical expectation of a product, which is used in the calculation of the *covariance* and of the *correlation coefficient*.

(iii) If the function g is a *linear combination* of the random variables X and Y, that is, if we have:

$$g(X, Y) = aX + bY + c,$$

where a, b, and c are real constants, then we easily prove that

$$E[g(X,Y)] = aE[X] + bE[Y] + c$$

(if the expected values exist). This formula may be generalized to the case of linear combinations of n random variables $X_1, \ldots, X_n$. Moreover, if $Y = h(X)$, for example, $Y = X^2$, then the above formula enables us to write that

$$E[aX + bX^2] = aE[X] + bE[X^2],$$

as we did in Chapter 3.

Definition 4.4.2. *The* **covariance** *of X and Y is defined by*

$$\mathrm{COV}[X,Y] \equiv \sigma_{X,Y} = E[(X - \mu_X)(Y - \mu_Y)].$$

Remarks. (i) We can show that

$$\mathrm{COV}[X,Y] = E[XY] - E[X]E[Y].$$

(ii) If X and Y are two *independent* random variables and if we can write that $g(X,Y) = g_1(X)g_2(Y)$, then we have that $E[g(X,Y)] = E[g_1(X)]E[g_2(Y)]$. It follows that if X and Y are independent, then

$$\mathrm{COV}[X,Y] \stackrel{\mathrm{ind.}}{=} E[X]E[Y] - E[X]E[Y] = 0.$$

If the covariance of the random variables X and Y is equal to zero, they are *not necessarily* independent. Nevertheless, we can show that, if X and Y are two random variables having a *normal* distribution, then X and Y are independent *if and only if* $\mathrm{COV}[X,Y] = 0$.

(iii) We have that $\mathrm{COV}[X,X] = E[X^2] - (E[X])^2 = \mathrm{VAR}[X]$. Thus, the variance is a particular case of the covariance. However, contrary to the variance, the covariance may be negative.

(iv) If $g(X,Y)$ is a linear combination of X and Y, then we find that

$$\mathrm{VAR}[g(X,Y)] \equiv \mathrm{VAR}[aX + bY + c] = a^2\mathrm{VAR}[X] + b^2\mathrm{VAR}[Y] + 2ab\mathrm{COV}[X,Y]. \quad (4.12)$$

Note that the constant c does not influence the variance of $g(X,Y)$. Furthermore, if X and Y are independent random variables, then we have:

$$\mathrm{VAR}[aX + bY + c] \stackrel{\mathrm{ind.}}{=} a^2\mathrm{VAR}[X] + b^2\mathrm{VAR}[Y].$$

Finally, Formula (4.12) can be generalized to the case of a linear combination of n (independent or dependent) random variables.

Definition 4.4.3. *The* **correlation coefficient** *of X and Y is given by*

$$\text{CORR}[X,Y] \equiv \rho_{X,Y} = \frac{\text{COV}[X,Y]}{\sqrt{\text{VAR}[X]\text{VAR}[Y]}}.$$

We can show that $-1 \leq \rho_{X,Y} \leq 1$. Moreover, $\rho_{X,Y} = \pm 1$ if and only if we can write that $Y = aX + b$, where $a \neq 0$. More precisely, $\rho_{X,Y} = 1$ (resp., -1) if $a > 0$ (resp., $a < 0$). In fact, $\rho_{X,Y}$ is a measure of the *linear relationship* between X and Y. Finally, if X and Y are independent random variables, then we have that $\rho_{X,Y} = 0$.

In the case when X and Y are random variables having a normal distribution, we have that $\rho_{X,Y} = 0 \Leftrightarrow X$ and Y are independent. This result is very important in practice, because if we *showed* (by means of a *statistical test*) that the two random variables follow (approximately) a normal distribution and if we found that their *sample correlation coefficient* is close to zero, then, in the context of statistical procedures, we can accept that they are independent.

Example 4.4.1. Consider the function $p_{X,Y}$ given by the following table (see Example 4.1.1):

$y \backslash x$	-1	0	1	$p_Y(y)$
0	1/16	1/16	1/16	3/16
1	1/16	1/16	2/16	4/16
2	2/16	1/16	6/16	9/16
$p_X(x)$	4/16	3/16	9/16	1

With the help of this table and the marginal probability functions p_X and p_Y, we calculate

$$E[X] = -1 \times (4/16) + 0 \times (3/16) + 1 \times (9/16) = 5/16,$$
$$E[Y] = 0 \times (3/16) + 1 \times (4/16) + 2 \times (9/16) = 22/16,$$
$$E[X^2] = (-1)^2 \times (4/16) + 0^2 \times (3/16) + 1^2 \times (9/16) = 13/16,$$
$$E[Y^2] = 0^2 \times (3/16) + 1^2 \times (4/16) + 2^2 \times (9/16) = 40/16,$$

$$E[XY] = \sum_{x=-1}^{1} \sum_{y=0}^{2} xy\, p_{X,Y}(x,y)$$

$$= 0 + (-1)(1)(1/16) + (-1)(2)(2/16) + 0 + 0 + 0 + 0$$
$$+ (1)(1)(2/16) + (1)(2)(6/16)$$
$$= -1/16 - 4/16 + 2/16 + 12/16 = 9/16.$$

It follows that

$$\text{VAR}[X] = E[X^2] - (E[X])^2 = \frac{13}{16} - \left(\frac{5}{16}\right)^2 = \frac{183}{(16)^2},$$

$$\text{VAR}[Y] = E[Y^2] - (E[Y])^2 = \frac{40}{16} - \left(\frac{22}{16}\right)^2 = \frac{156}{(16)^2}$$

and

$$\text{COV}[X,Y] = E[XY] - E[X]E[Y] = \frac{9}{16} - \left(\frac{5}{16}\right)\left(\frac{22}{16}\right) = \frac{34}{(16)^2}.$$

Finally, we calculate

$$\rho_{X,Y} \equiv \text{CORR}[X,Y] = \frac{34/(16)^2}{\left(\frac{183}{(16)^2} \cdot \frac{156}{(16)^2}\right)^{1/2}} = \frac{34}{\sqrt{183 \cdot 156}} \simeq 0.2012.$$

Remark. In general, it is advisable to calculate the means $E[X]$, $E[Y]$, and $E[XY]$ before the expected values of the squared variables. Indeed, if $E[XY] - E[X]E[Y] = 0$, then $\text{CORR}[X,Y] = 0$ (if X and Y are not constants, so that their variances are strictly positive).

Example 4.4.2. The joint density function of the continuous random vector (X,Y) is

$$f_{X,Y}(x,y) = \begin{cases} e^{-2y} & \text{if } 0 < x < 2, \ y > 0, \\ 0 & \text{elsewhere.} \end{cases}$$

What is the correlation coefficient of X and Y?

Solution. We can show that, when the joint density function of (X,Y) can be decomposed into a product of a function of x only and a function of y only, the random variables X and Y are independent, provided that there is no relationship between x and y in the set $D_{X,Y}$ of possible values of the pair (X,Y). That is, this set $D_{X,Y}$ is of the form

$$D_{X,Y} = \{(x,y) \in \mathbb{R}^2 : c_1 < x < c_2, k_1 < y < k_2\},$$

where the c_is and the k_is are constants, for $i = 1, 2$.

Here, the possible values of X and Y are not related and we can write that

$$e^{-2y} = g(x)h(y),$$

where $g(x) \equiv 1$ and $h(y) = e^{-2y}$. Therefore, we can conclude that X and Y are independent. It follows that $\text{CORR}[X,Y] = 0$.

We can check that X and Y are indeed independent. We have:

$$f_X(x) = \int_0^\infty e^{-2y}\, dy = -\frac{1}{2}e^{-2y}\Big|_0^\infty = \frac{1}{2} \quad \text{if } 0 < x < 2$$

and

$$f_Y(y) = \int_0^2 e^{-2y}\,dx = 2e^{-2y} \quad \text{if } y > 0.$$

Thus, we have that $f_{X,Y}(x,y) = f_X(x)f_Y(y)$ for every point (x,y), as required. Note that $X \sim U(0,2)$ and $Y \sim \text{Exp}(2)$ in this example.

4.5 Limit theorems

In this section, we present two limit theorems. The first one is useful in statistics, in particular, and the second one is in fact the most important theorem in probability theory.

Theorem 4.5.1. (Law of large numbers) *Suppose that $X_1, X_2, \ldots$ are independent random variables having the same distribution function as the variable X, whose mean μ_X exists. Then, for any constant $\epsilon > 0$, we have:*

$$\lim_{n \to \infty} P\left[\left|\frac{X_1 + \cdots + X_n}{n} - \mu_X\right| > \epsilon\right] = 0.$$

Remarks. (i) This theorem is known, more precisely, as the *weak* law of large numbers. There is also the *strong* law of large numbers, for which the expected value of $|X|$ must exist.

(ii) In practice, the mean μ_X of the random variable X is unknown. To *estimate* it, we gather many (independent) *observations* X_i of X. The above result enables us to assert that the arithmetic mean of these observations converges (*in probability*) to the unknown mean of X.

(iii) We write that the random variables $X_1, X_2, \ldots$ are *i.i.d.* (independent and identically distributed).

Theorem 4.5.2. (Central limit theorem) *Suppose that $X_1, \ldots, X_n$ are independent random variables having the same distribution function as the variable X, whose mean μ_X and variance σ_X^2 exist ($\sigma_X > 0$). Then, the distribution of $S_n := \sum_{i=1}^n X_i$ tends to that of a normal distribution, with mean $n\mu_X$ and variance $n\sigma_X^2$, as n tends to infinity.*

Remarks. (i) Let us define

$$\bar{X} = \sum_{i=1}^n \frac{X_i}{n}.$$

Then, we can assert that the distribution of $\bar{X}$ tends to that of a $N(\mu_X, \sigma_X^2/n)$ distribution.

(ii) In general, if we add up 30 or more independent random variables X_i, then the normal distribution should be a good approximation to the exact (often unknown) distribution of this sum. However, the number of variables that must be added, to obtain a good approximation, actually depends on the degree of *asymmetry* of the distribution of X.

(iii) We can, under certain conditions, generalize the central limit theorem (CLT) to the case when the random variables $X_1, \ldots, X_n$ are not necessarily identically distributed. Indeed, if the mean μ_{X_i} and the variance $\sigma_{X_i}^2$ of X_i exist for all i, then, when n is large enough, we have:

$$\sum_{i=1}^{n} X_i \approx N \left(\sum_{i=1}^{n} \mu_{X_i}, \sum_{i=1}^{n} \sigma_{X_i}^2 \right)$$

and

$$\bar{X} \approx N \left(\frac{1}{n} \sum_{i=1}^{n} \mu_{X_i}, \frac{1}{n^2} \sum_{i=1}^{n} \sigma_{X_i}^2 \right).$$

Example 4.5.1. An American town comprises 10,000 houses and two factories. The demand for drinking water (in gallons) from a given house over an arbitrary day is a random variable D such that $E[D] = 50$ and $\text{VAR}[D] = 400$. In the case of the factories, the demand for drinking water follows (approximately) a $N(10,000, (2000)^2)$ distribution for factory 1 and a $N(25,000, (5000)^2)$ distribution for factory 2. Let D_i, for $i = 1, \ldots, 10,000$, be the demand for drinking water from the ith house and F_i, for $i = 1, 2$, be the demand from factory i. We assume that the random variables D_i and F_i are independent and we set

$$X_d = \sum_{i=1}^{10,000} D_i \qquad \text{(the domestic demand)}$$

and

$$X_t = X_d + F_1 + F_2 \qquad \text{(the total demand).}$$

(a) Find the number a such that $P[X_d \geq a] \simeq 0.01$.

(b) What should the production capacity of the drinking water treatment plant be if we want to be able to satisfy the total demand with probability 0.98?

Solution. (a) By the central limit theorem, we may write that

$$X_d \approx N(10,000\,(50), 10,000\,(20^2)).$$

Remark. We assume that the random variables D_i are independent among themselves.

Then, we have:

$$P[X_d \geq a] = 1 - P[X_d < a] \simeq 1 - P\left[Z < \frac{a - 500,000}{\sqrt{10,000}(20)}\right],$$

where $Z \sim N(0,1)$. It follows that

$$P[X_d \geq a] \simeq 0.01 \quad \Longleftrightarrow \quad P\left[Z < \frac{a - 500,000}{2000}\right] \simeq 0.99.$$

Now, we find in Table B.3, page 279, that $P[Z \leq 2.33] \simeq 0.99$. Thus, we have:

$$a \simeq 500,000 + 2000(2.33) = 504,660.$$

(b) By independence, we may write that $X_t \approx N(\mu, \sigma^2)$, where [see (4.10)]

$$\mu = 500,000 + 10,000 + 25,000 = 535,000$$

and

$$\sigma^2 = (2000)^2 + (2000)^2 + (5000)^2 = 33,000,000.$$

Let c be the capacity of the drinking water treatment plant. We seek the value of c such that $P[X_t \leq c] = 0.98$. Because $P[Z \leq 2.055] \stackrel{\text{Tab. B.3}}{\simeq} 0.98$, proceeding as in part (a) we find that

$$c \simeq 535,000 + \sqrt{33,000,000}(2.055) \simeq 546,805.$$

Remark. We see in this example that it is not necessary to know the exact form of the function p_X or f_X to be able to apply the central limit theorem. It is sufficient to know the mean and the variance of X.

4.6 Exercises for Chapter 4

Solved exercises

Question no. 1
Let

$$p_{X,Y}(x,y) = \frac{1}{6} \quad \text{if } x = 0 \text{ or } 1, \text{ and } y = 0,1 \text{ or } 2.$$

Calculate $p_X(x)$.

Question no. 2
Calculate $f_X(x \mid Y = y)$ if

$$f_{X,Y}(x,y) = x + y \quad \text{for } 0 < x < 1, 0 < y < 1.$$

Question no. 3

Suppose that X and Y are two random variables such that $E[X] = E[Y] = 0$, $E[X^2] = E[Y^2] = 1$, and $\rho_{XY} = 1$. Calculate $\text{COV}[X, Y]$.

Question no. 4

Calculate $P[X + Y > 1]$ if

$$f_{X,Y}(x,y) = 1 \quad \text{for } 0 < x < 1, 0 < y < 1.$$

Question no. 5

Suppose that

$$p_{X,Y}(x,y) = \frac{8}{9}\left(\frac{1}{2}\right)^{x+y} \quad \text{if } x = 0 \text{ or } 1, \text{ and } y = 1 \text{ or } 2.$$

Calculate $E[XY]$.

Question no. 6

Suppose that X and Y are two random variables such that $\text{VAR}[X] = \text{VAR}[Y] = 1$ and $\text{COV}[X, Y] = 1$. Calculate $\text{VAR}[X - 2Y]$.

Question no. 7

Let $X \sim N(0,1)$, $Y \sim N(1,2)$ and $Z \sim N(3,4)$ be independent random variables. What distribution does $W := X - Y + 2Z$ follow? Also give the parameter(s) of this distribution.

Question no. 8

Suppose that $X \sim \text{Poi}(\lambda = 100)$. What other probability distribution can be used to calculate (approximately) $p := P[X \leq 100]$? Also give the parameter(s) of this distribution, as well as the approximate value of p.

Question no. 9

Suppose that X follows a $B(n = 100, p = 0.4)$ distribution. Use a $N(40, 24)$ distribution to calculate approximately $P[X = 40]$.

Question no. 10

We define $Y = \sum_{i=1}^{50} X_i$, where $E[X_i] = 0$, for $i = 1, \ldots, 50$, and the X_is are independent continuous random variables. Calculate approximately $P[Y \geq 0]$.

Question no. 11

The joint probability function, $p_{X,Y}$, of the pair (X, Y) is given by the following table:

$y \backslash x$	-1	0	1
0	1/9	1/9	1/9
2	2/9	2/9	2/9

(a) Are the random variables X and Y independent? Justify.

(b) Evaluate $F_{X,Y}(0, 1/2)$.

(c) Let $Z = X^4$. Calculate $p_Z(z)$.

d) Calculate $E[X^2 Y^2]$.

Question no. 12

Let

$$f_{X,Y}(x, y) = \begin{cases} 2 & \text{if } 0 < x < y, 0 < y < 1, \\ 0 & \text{elsewhere.} \end{cases}$$

Calculate $P[X \geq Y^2]$.

Question no. 13

City buses pass by a certain street corner, between 7:00 a.m. and 7:30 p.m., according to a Poisson process at the (average) rate of four per hour. Let $Y = \sum_{k=1}^{50} X_k$, where X_k is the total number of buses that pass during the kth 15-minute time period, from 7:00 a.m.

(a) What is the *exact* distribution of Y and its parameter(s)?

(b) What other probability distribution can approximate the distribution of Y? Justify and give the parameter(s) of this distribution as well.

Question no. 14

We consider the discrete random variable X whose probability function is given by

x	0	1	2
$p_X(x)$	1/2	1/4	1/4

Suppose that X_1 and X_2 are two independent random variables having the same distribution as X. Calculate $P[X_1 = X_2]$.

Question no. 15

The table below gives the function $p_{X,Y}(x, y)$ of the pair (X, Y) of discrete random variables:

$y \backslash x$	0	1	3	4
1	0.1	0.1	0	0.2
2	0.3	0	0.2	0.1

Calculate $P[\{X < 5\} \cap \{Y < 2\}]$.

Question no. 16

Calculate the covariance of X_1 and X_2 if

$$f_{X_1, X_2}(x_1, x_2) = \begin{cases} 2 - x_1 - x_2 & \text{for } 0 < x_1 < 1, 0 < x_2 < 1, \\ 0 & \text{elsewhere.} \end{cases}$$

Question no. 17

Suppose that $Y = 1/X$, where X is a discrete random variable such that

x	1	2
$p_X(x)$	1/3	2/3

We define $W = Y_1 - Y_2$, where Y_1 and Y_2 are two independent random variables identically distributed as Y. Calculate $p_W(w)$.

Question no. 18

Let $X_1, \ldots, X_n$ be independent random variables, where X_i has an exponential distribution with parameter $\lambda = 2$, for $i = 1, \ldots, n$. Use the central limit theorem to find the value of n for which

$$P\left[\sum_{i=1}^{n} X_i > \frac{n}{2} + 1\right] \simeq 0.4602.$$

Question no. 19

A bus passes by a certain street corner every morning around 9:00 a.m. Let X be the difference (in minutes) between the time instant at which the bus passes and 9:00 a.m. We suppose that X has approximately a $N(\mu = 0, \sigma^2 = 25)$ distribution. We consider two independent days. Let X_k be the value of the random variable X on the kth day, for $k = 1, 2$.

(a) Calculate the probability $P[X_1 - X_2 > 15]$.

(b) Find the joint density function $f_{X_1, X_2}(x_1, x_2)$.

(c) Calculate (i) $P[X_1 = 2 \mid X_1 > 1]$ and (ii) $P[X_1 < 2 \mid X_1 = 1]$.

Question no. 20

An assembly comprises 100 sections. The length of each section (in centimeters) is a random variable with mean 10 and variance 0.9. Furthermore, the sections are independent. The technical specification for the total length of the assembly is 1000 cm $\pm$ 30 cm. What is approximately the probability that the assembly fails to meet the specification in question?

Question no. 21

Let

$$f_{X,Y}(x, y) = \begin{cases} 3x^2 e^{-x} y\,(1 - y) & \text{if } x > 0, 0 < y < 1, \\ 0 & \text{elsewhere.} \end{cases}$$

(a) Calculate the functions $f_X(x)$ and $f_Y(y)$. What is the distribution of X and that of Y?

(b) Are X and Y independent random variables? Justify.

(c) Calculate the kurtosis of X.

(d) Calculate the skewness of Y.

Question no. 22

The following table gives the joint probability function $p_{X,Y}(x, y)$ of the pair (X, Y):

$y\backslash x$	0	1	2
-1	1/9	0	1/9
0	2/9	0	2/9
1	0	1/3	0

(a) Find $p_X(x)$ and $p_Y(y)$.

(b) Are X and Y independent random variables? Justify.

(c) Calculate (i) $p_Y(y \mid X = 1)$ and (ii) $p_Y(y \mid X \leq 1)$.

(d) Calculate the correlation coefficient of X and Y.

(e) Let $W = \max\{X, Y\}$. Find $p_W(w)$.

Question no. 23

We consider the pair (X, Y) of discrete random variables whose joint probability function $p_{X,Y}(x, y)$ is given by

$y\backslash x$	1	2	3
2	1/12	1/6	1/12
3	1/6	0	1/6
4	0	1/3	0

Calculate $P[X + Y \leq 4 \mid X \leq 2]$.

Question no. 24

Use a normal distribution to calculate approximately the probability that, among 10,000 (independent) random digits, the digit "7" appears more than 968 times.

Question no. 25

A number X is taken at random in the interval $(0, 1)$, and next a number Y is taken at random in the interval $(0, X]$, so that

$$f_{X,Y}(x, y) = \begin{cases} 1/x & \text{if } 0 < x < 1, 0 < y \leq x, \\ 0 & \text{elsewhere.} \end{cases}$$

(a) Show that

$$E[X^r Y^s] = \frac{1}{(s+1)(r+s+1)}$$

for $r, s = 0, 1, 2, \ldots$.

(b) Check the formula in part (a) for $r = 2$ and $s = 0$ by directly calculating $E[X^2]$.

(c) Use part (a) to calculate the correlation coefficient of X and Y.

Question no. 26

The following table gives part of the function $p_{X,Y}(x,y)$ of the pair (X,Y) of discrete random variables:

$y \backslash x$	0	1	2	$p_Y(y)$
-1	1/16		1/16	1/4
0				1/2
1		0		1/4
$p_X(x)$	1/4			1

We also have:

y	-1	0	1
$p_Y(y \mid X = 2)$	1/8	3/8	1/2

(a) Find $P[X = 2]$.

(b) Complete the table of the function $p_{X,Y}(x,y)$.

(c) We set $W = Y + 1$. The distribution of W is then a particular case of one of the discrete distributions seen in Chapter 3. Find this distribution and give its parameter(s).

Question no. 27

The joint density function of the pair (X,Y) of continuous random variables is given by

$$f_{X,Y}(x,y) = \begin{cases} \frac{1}{2}xy & \text{if } 0 < y < x < 2, \\ 0 & \text{elsewhere.} \end{cases}$$

(a) Calculate $E[1/XY]$.

(b) Calculate $E[X^2]$.

(c) What is the median, x_m, of the random variable X?

Question no. 28

A device is constituted of two independent components connected in parallel. The lifetime X (in years) of component no. 1 follows an exponential distribution with parameter $\lambda = 1/2$, whereas the lifetime Y (in years) of component no. 2 has a Weibull distribution with parameters $\lambda = 2$ and $\beta = 2$. That is,

$$f_Y(y) = 4ye^{-2y^2} \quad \text{for } y > 0.$$

Calculate the probability that the device lasts less than one year.

Question no. 29

We take 100 numbers at random in the interval $[0,1]$. Let S be the sum of these 100 numbers. Use the central limit theorem to calculate approximately the probability $P[45 \le S < 55]$.

Question no. 30

The number of floods that occur in a certain region over a given year is a random variable having a Poisson distribution with parameter $\alpha = 2$, independently from one year to the other. Moreover, the time period (in days) during which the ground is flooded, at the time of an arbitrary flood, is an exponential random variable with parameter $\lambda = 1/5$. We assume that the durations of the floods are independent. Use the central limit theorem to calculate (approximately) the probability that

(a) over the course of the next 50 years, there will be at least 80 floods in this region (without making a continuity correction);

(b) the total time during which the ground will be flooded over the course of the next 50 floods will be smaller than 200 days.

Exercises

Question no. 1

Telephone calls arrive at an exchange according to a Poisson process with rate λ per minute. We know, from past experience, that the probability of receiving exactly one call during a one-minute period is three times that of receiving no calls during the same time period. We consider 100 consecutive one-minute time periods and we designate by U the number of periods during which no calls were received.

(a) Use a normal approximation to calculate $P[U = 5]$.

(b) Use the central limit theorem to calculate approximately

$$P\left[\frac{1}{100}\sum_{i=1}^{100} X_i \geq 3.1\right],$$

where X_i is the number of calls received during the ith one-minute period, for $i = 1,\ldots,100$.

Question no. 2

Let

$$f_{X,Y}(x,y) = \begin{cases} kx & \text{if } 0 < x < 1, 0 < y < x, \\ 0 & \text{elsewhere} \end{cases}$$

be the joint density function of the random vector (X,Y).

(a) Find the constant k.

(b) Obtain the marginal density functions of X and Y.

(c) Calculate VAR$[X]$ and VAR$[Y]$.

(d) Calculate the correlation coefficient of X and Y.

Question no. 3

In a bank, an automatic teller machine (ATM) enables the customers to withdraw $50 or $100 banknotes. It may also happen that a given customer cannot withdraw any money if her account is without funds or if the customer in question made an error when using the ATM. The number X of customers using the ATM in a five-minute interval is a random variable whose probability function $p_X(x)$ is

x	0	1	2
$p_X(x)$	0.3	0.5	0.2

Furthermore, we observed that the total amount Y of money withdrawn in a five-minute interval is a random variable whose conditional probability function $p_Y(y \mid X = x)$ is given by

y	0	50	100	150	200
$p_Y(y \mid X = 0)$	1	0	0	0	0
$p_Y(y \mid X = 1)$	0.1	0.7	0.2	0	0
$p_Y(y \mid X = 2)$	0.01	0.14	0.53	0.28	0.04

(a) Are the random variables X and Y independent? Justify.

(b) Calculate the probability $P[X = 1, Y = 100]$.

(c) Calculate the probability $P[Y = 0]$.

(d) Find the average number of customers using the ATM in a one-hour period.

Question no. 4

A private club decides to organize a *charity casino night*. The organizers decide to

- ask their members to cover the overhead costs;
- to admit only 1000 players, each of them with the same initial stake θ (in thousands of dollars);
- to choose games such that the gross winnings X_i (in thousands of dollars) of the ith player are uniformly distributed on the interval $(0, 3\theta/2)$.

Indication. We have that the mean of a $U(0, 3\theta/2)$ distribution is $3\theta/4$ and its variance is equal to $3\theta^2/16$.

(a) Let Y be the total gross winnings of the 1000 players. Give the approximate distribution of Y, as well as its parameters.

(b) Determine the amount θ that each player must pay in order that the net profit (in thousands of dollars) of the casino be greater than 50 with probability 0.95.

Question no. 5

A certain freeway has three access roads: A, B, and C (see Figure 4.3). The number of cars accessing the freeway over a one-hour period, via the three access roads, is defined by random variables denoted by X_A, X_B, and X_C and having the following characteristics:

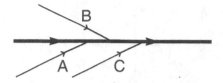

Fig. 4.3. Figure for Exercise no. 5.

	X_A	X_B	X_C
Mean	800	1000	600
Standard deviation	40	50	30

Let us designate by X the total number of cars accessing the freeway over a one-hour period.

(a) Calculate

(i) the mean of X and

(ii) the standard deviation of X, assuming that the random variables X_A, X_B, and X_C are pairwise independent;

(iii) the probability that the random variable X takes on a value between 2300 and 2500 if we suppose that the variables X_A, X_B, and X_C are independent and (approximately) normally distributed;

(iv) the probability that X is greater than 2500, under the same assumptions as above.

(b) Let Y be the number of times that X is greater than or equal to 2500 (under the same assumptions as above) over 100 (independent) one-hour periods.

(i) Give the distribution of Y and its parameters.

(ii) Calculate, using an approximation based on a normal distribution, the probability that the random variable Y is greater than or equal to 10.

(c) Calculate

(i) the mean of X and

(ii) the standard deviation of X if we suppose that the random variables X_A, X_B, and X_C are normally distributed and that the correlation coefficients of the three pairs of random variables are $\mathrm{CORR}[X_A, X_B] = 1/2$, $\mathrm{CORR}[X_A, X_C] = 4/5$, and $\mathrm{CORR}[X_B, X_C] = -1/2$.

Question no. 6

The joint density function of the pair (X, Y) of random variables is defined by (see Figure 4.4):

$$f_{X,Y}(x, y) = \begin{cases} 3/4 & \text{if } -1 \leq x \leq 1,\ x^2 \leq y < 1, \\ 0 & \text{elsewhere.} \end{cases}$$

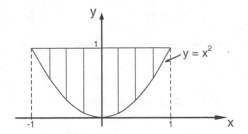

Fig. 4.4. Figure for Exercise no. 6.

(a) Calculate
   (i) the marginal density functions of X and Y;
   (ii) the correlation coefficient of X and Y.

(b) Are the random variables X and Y independent? Justify.

Question no. 7
   Suppose that

$$f_{X,Y}(x,y) = \begin{cases} 4/\pi \text{ if } 0 < x \leq 1, 0 < y \leq \sqrt{2x - x^2}, \\ 0 \quad \text{elsewhere.} \end{cases}$$

Find the conditional density function of Y, given that $X = x$.

Question no. 8
   Calculate the mathematical expectation of $(X + Y)^2$ if

$$f_{X,Y}(x,y) = \begin{cases} \frac{1}{8}(x + y) \text{ for } 0 \leq x \leq 2, 0 \leq y \leq 2, \\ 0 \quad \text{elsewhere.} \end{cases}$$

Question no. 9
   Suppose that X and Y are two random variables such that $\text{VAR}[X] = \text{VAR}[Y] = 1$. We set $Z = X - \frac{3}{4}Y$. Calculate the correlation coefficient of X and Z if $\text{COV}[X, Z] = 1/2$.

Question no. 10
   A fair coin is tossed until "heads" is obtained, then until "tails" is obtained. If we assume that the successive tosses are independent, what is the probability that the coin has to be tossed exactly nine times?

Question no. 11
   Suppose that $X_1 \sim N(2, 4)$, $X_2 \sim N(4, 2)$, and $X_3 \sim N(4, 4)$ are independent random variables. Calculate the 75th percentile of the random variable $Y := X_1 - 2X_2 + 4X_3$.

Question no. 12

The lifetime of a certain type of tire follows (approximately) a normal distribution with mean 25,000 km and standard deviation 5000 km. Two (independent) tires are taken at random. What is the probability that one of the two tires lasts at least 10,000 km more than the other?

Question no. 13

A factory produces articles whose average weight is equal to 1.62 kg, with a standard deviation of 0.05 kg. What is (approximately) the probability that the total weight of a batch of 100 articles is between 161.5 kg and 162.5 kg?

Question no. 14

The joint density function of the pair (X, Y) of random variables is

$$f_{X,Y}(x, y) = \begin{cases} x + y & \text{if } 0 \leq x \leq 1, 0 \leq y \leq 1, \\ 0 & \text{elsewhere.} \end{cases}$$

We find that $E[X] = 7/12$ and $\text{VAR}[X] = 11/144$. Calculate the correlation coefficient of X and Y.

Question no. 15

Let $X \sim N(2, 1)$ and $Y \sim N(4, 4)$ be two independent random variables. Calculate $P[|2X - Y| < \sqrt{8}]$.

Question no. 16

Let X_i, for $i = 1, 2, \ldots, 100$, be independent random variables having a gamma distribution with parameters $\alpha = 9$ and $\lambda = 1/3$. Calculate approximately $P[\bar{X} \geq 26]$, where $\bar{X} := \frac{1}{100} \sum_{i=1}^{100} X_i$.

Question no. 17

Let X be the number of "do" loops in a FORTRAN program and let Y be the number of attempts needed by a beginner to get a working program. Suppose that the joint probability function of (X, Y) is given by the following table:

$x \backslash y$	1	2	3
0	0.05	0.15	0.10
1	0.10	0.20	0.10
2	0.15	0.10	0.05

(a) Calculate $E[XY]$.

(b) Evaluate the probability $P[Y \geq 2 \mid X = 1]$.

(c) Are the random variables X and Y independent? Justify.

Question no. 18
    Let
$$f_{X,Y}(x,y) = \begin{cases} 6x \text{ if } 0 < x < 1, x < y < 1, \\ 0 \text{ elsewhere} \end{cases}$$

be the joint density function of (X,Y).

(a) Calculate the marginal density functions of X and Y.

(b) Evaluate the probability $P[XY < 1/4]$.

Question no. 19
    A device is made up of two independent components. One of the components is placed on *standby* and begins to operate when the other one fails. The lifetime (in hours) of each component follows an exponential distribution with parameter $\lambda = 1/2000$. Let X be the lifetime of the device.

(a) Give the distribution of X and its parameters.

(b) What is the mean of X?

Question no. 20
    The weight (in kilograms) of manufactured items follows approximately a normal distribution with parameters $\mu = 1$ and $\sigma^2 = 0.02$. We take 100 items at random. Let X_j be the weight of the jth item, for $j = 1, 2, \ldots, 100$. We suppose that the X_js are independent random variables.

(a) Calculate $P[X_1 - X_2 < 0.05]$.

(b) Find the number b such that $P[X_1 + X_2 < b] = 0.025$.

(c) Calculate approximately, using a normal distribution, the probability that exactly 70 of the 100 items considered have a weight smaller than 1.072 kg.

Question no. 21
    Let
$$f_{X,Y}(x,y) = \begin{cases} e^{-2x} \text{ if } x > 0, 0 < y < 2, \\ 0 \quad \text{elsewhere} \end{cases}$$

be the joint density function of the random vector (X,Y).

(a) Find $f_X(x)$ and $f_Y(y)$.

(b) What is the correlation coefficient of X and Y? Justify.

(c) What is the 50th percentile of Y?

(d) Calculate $P[Y < e^X]$.

Question no. 22
    The time T (in years) elapsed between two major power failures in a particular region has an exponential distribution with mean 1.5. The duration X (in hours) of these major power failures follows approximately a normal distribution with mean 4 and standard deviation 2. We assume that the failures occur independently of one another.

(a) Given that there were no major power failures during the last year, what is the probability that there will be no major power failures over the next nine months?

(b) How long, at most, do 95% of the major power failures last?

(c) Calculate the probability that the duration of the next major power failure and that of the following one differ by at most 30 minutes.

(d) Calculate the probability that the longest major power failure, among the next three, lasts less than five hours.

(e) Use the central limit theorem to calculate (approximately) the probability that the 30th major power failure from now occurs within the next 50 years.

Question no. 23

Suppose that

$$f_{X,Y}(x,y) = \begin{cases} 2 & \text{if } 0 \le x \le 1, 0 \le y \le x, \\ 0 & \text{elsewhere} \end{cases}$$

is the joint density function of the random vector (X, Y).

(a) Find the marginal density functions of X and Y.

(b) Calculate $P[X - Y < 1/2]$.

(c) Are X and Y independent? Justify.

(d) Calculate $E[XY]$.

Question no. 24

Let X be the number of customers of a car salesman over a one-day period. Suppose that X has a Poisson distribution with parameter $\lambda = 3$. Furthermore, suppose that one customer in five, on average, buys a car (on a given visit), independently of the other customers. Let Y be the number of cars sold by the salesman in one day.

(a) Given that the salesman had five customers during a given day, what is the probability that he sold exactly two cars?

(b) What is the average number of cars sold by the salesman in a one-day period? Justify.

Indication. We have that $E[Y] = \sum_{x=0}^{\infty} E[Y \mid X = x]P[X = x]$. Moreover, knowing that $X = x$, Y is a binomial random variable.

(c) What is the probability that the salesman sells no cars during a given day?

Indication. We have that $\sum_{x=0}^{\infty} \frac{k^x}{x!} = e^k$.

(d) Knowing that the salesman sold no cars during a given day, what is the probability that he had no customers?

Question no. 25

Let $X_1, X_2, \ldots, X_{50}$ be independent random variables having an exponential distribution with parameter $\lambda = 2$.

(a) Calculate $P[X_1^2 > 4 \mid X_1 > 1]$.

(b) Let $S = \sum_{i=1}^{50} X_i$.

(i) Give the exact probability distribution of S, as well as its parameter(s).

(ii) Calculate approximately, using the central limit theorem, $P[S < 24]$.

Question no. 26

Let $X_1 \sim N(0,1)$ and $X_2 \sim N(1,3)$ be two independent random variables.

(a) Calculate $P[|X_1 - X_2| > 1]$.

(b) What is the 90th percentile of $Y := X_1 + X_2$?

Question no. 27

The joint density function of the random vector (X, Y, Z) is given by

$$f_{X,Y,Z}(x,y,z) = \begin{cases} k[(x/y) + z] & \text{if } 0 < x < 1,\, 1 < y < e,\, -1 < z < 1, \\ 0 & \text{elsewhere.} \end{cases}$$

(a) Find the constant k.

(b) Show that

$$f_{X,Y}(x,y) = \begin{cases} 2x/y & \text{if } 0 < x < 1,\, 1 < y < e, \\ 0 & \text{elsewhere.} \end{cases}$$

(c) Are X and Y independent random variables? Justify.

(d) Calculate the mathematical expectation of Y/X.

Question no. 28

Let X be a random variable following a gamma distribution with parameters $\alpha = 25$ and $\lambda = 1/2$.

(a) Calculate the probability $P[X < 40]$ by making use of a Poisson distribution.

(b) Use the central limit theorem to calculate (approximately) $P[40 \le X \le 50]$. Justify the use of the central limit theorem.

Question no. 29

Let X be a random variable having a binomial distribution with parameters $n = 100$ and $p = 1/2$, and let Y be a random variable following a normal distribution with parameters $\mu = 50$ and $\sigma^2 = 25$.

(a) Calculate approximately $P[X < 40]$.

(b) Calculate the probability $P[X = Y]$.

(c) What is the 33rd percentile of Y?

Question no. 30

The table below presents the joint probability distribution of the random vector (X, Y):

$x \backslash y$	1	2	3
0	1/9	2/9	1/9
1	1/18	1/9	1/18
2	1/6	1/18	1/9

Calculate $E[2XY]$.

Question no. 31

Let

$$f_{X,Y}(x,y) = \begin{cases} 4xy & \text{if } 0 \leq x \leq 1, 0 \leq y \leq 1, \\ 0 & \text{elsewhere} \end{cases}$$

be the joint density function of the random vector (X, Y). Calculate the probability $P[X^2 + Y^2 < 1/4]$.

Question no. 32

Let X_1, X_2, and X_3 be random variables such that (i) X_1 and X_2 are independent, (ii) $\text{VAR}[X_i] = 2$, for $i = 1, 2, 3$, (iii) $\text{COV}[X_1, X_3] = 1/2$, and (iv) $\text{COV}[X_2, X_3] = 1$. Calculate $\text{VAR}[X_1 + X_2 + 2X_3]$.

Question no. 33

The lifetime (in years) of a certain machine follows approximately a $N(5, 4)$ distribution. Use the central limit theorem to calculate (approximately) the probability that at most 10, among 30 (independent) machines of this type, last at least six years.

Question no. 34

The joint density function of the random vector (X, Y) is given by

$$f_{X,Y}(x,y) = \begin{cases} 1/\pi & \text{if } x^2 + y^2 \leq 1, \\ 0 & \text{elsewhere.} \end{cases}$$

(a) Check that $f_{X,Y}(x, y)$ is a valid joint density function.

(b) Calculate the marginal density functions $f_X(x)$ and $f_Y(y)$.

(c) Are X and Y independent? Justify.

(d) Calculate $P[X^2 + Y^2 \geq 1/4]$.

Question no. 35

A fair die is rolled 30 times, independently. Let X be the number of 6s obtained and Y be the sum of all the numbers obtained.

(a) Use a Poisson distribution to calculate (approximately) $P[X > 5]$ (even if the probability of success is relatively large).

(b) Use the central limit theorem to calculate (approximately) $P[100 \leq Y < 111]$.

Indication. If W is the number obtained on an arbitrary roll of a fair die, then we have that $E[W] = 7/2$ and $\text{VAR}[W] = 35/12$.

Question no. 36
    Let
$$f_X(x) = \begin{cases} 1/10 \text{ if } 0 \le x \le 10, \\ 0 \quad \text{elsewhere} \end{cases}$$
and
$$p_Y(y) = \begin{cases} 1/10 \text{ if } y = 1, 2, \ldots, 10, \\ 0 \quad \text{otherwise.} \end{cases}$$

Suppose that X and Y are independent random variables. Calculate (a) the probability $P[X > Y]$ and (b) VAR$[XY]$.

Question no. 37
    Suppose that X and Y are two random variables such that VAR$[X] =$ VAR$[Y] = 1$ and COV$[X, Y] = 2/3$. For what value of k is the correlation coefficient of X and $Z := X + kY$ equal to 2/3?

Question no. 38
    An electronic device is made up of ten components whose lifetime (in months) follows an exponential distribution with mean 50. Suppose that the components operate independently of one another. Let T be the lifetime of the device. Obtain the density function of T if

(a) the components are connected in series;

(b) the components are connected in parallel;

(c) the components are placed in standby redundancy. That is, only one component operates at a time and, when it fails, it is immediately replaced by another component (if there remains at least one working component).

Question no. 39
    Electric light bulbs bought to illuminate an outside rink have an average lifetime of 3000 hours, with a standard deviation of 339 hours, independently from one light bulb to the other. Suppose that the lifetime of the light bulbs follows approximately a normal distribution.

(a) If it is more economical to replace all the light bulbs when 20% among them are burnt out, rather than to change the light bulbs when needed, after how many hours should we replace them?

(b) Suppose that only the burnt-out light bulbs have been replaced after t_1 hours, where t_1 is the time when 20% of the light bulbs should be burnt out. Find the percentage of light bulbs that will be burnt out after $\frac{1}{2}t_1$ additional hours.

Question no. 40
    The continuous random vector (X, Y) has the following joint density function:
$$f_{X,Y}(x, y) = \begin{cases} 6(1 - x - y) \text{ if } x > 0, \, 0 < y < 1 - x, \\ 0 \quad\quad\quad \text{elsewhere.} \end{cases}$$

(a) Calculate the marginal density function of Y.

(b) Find the 40th percentile of Y.

(c) Let $Z = Y^3$. Obtain the density function of Z.

(d) Calculate the probability $P[X < Y]$.

Question no. 41

A fair die is tossed twice, independently. Let X be the number of 5s and Y be the number of 6s obtained. Calculate

(a) the joint probability function $p_{X,Y}(x,y)$, for $0 \le x+y \le 2$;

(b) the function $F_{X,Y}(1/2, 3/2)$;

(c) the standard deviation of 2^X;

(d) the correlation coefficient of X and Y.

Question no. 42

Let

x_1	−2	−1	1	2
$p_{X_1}(x_1)$	1/3	1/6	1/3	1/6

and

x_2	0	1
$p_{X_2}(x_2 \mid X_1 = -2)$	1/2	1/2
$p_{X_2}(x_2 \mid X_1 = -1)$	1/2	1/2
$p_{X_2}(x_2 \mid X_1 = 1)$	1	0
$p_{X_2}(x_2 \mid X_1 = 2)$	0	1

(a) Calculate
   (i) the marginal probability function of X_2;
   (ii) the probability $p_{X_2}(x_2 \mid \{X_1 = -2\} \cup \{X_1 = 2\})$, for $x_2 = 0$ and 1.

(b) Let $Y = 2X_1 + X_1^2$. Find the distribution function of Y.

Question no. 43

The duration X (in hours) of the major breakdowns of a given subway system follows approximately a normal distribution with mean $\mu = 2$ and standard deviation $\sigma = 0.75$. We assume that the durations of the various breakdowns are independent random variables.

(a) Calculate (exactly) the probability that the duration of each of more than 40 of the next 50 major breakdowns is smaller than three hours.

Remark. This question requires the use of a pocket calculator or a software package.

(b) Use a normal distribution to calculate approximately the probability in part (a).

Question no. 44

Suppose that $X_1, \ldots, X_9$ are independent random variables having an exponential distribution with parameter $\lambda = 1/2$.

(a) Calculate the probability

$$P\left[\frac{X_1 + X_2 + X_3}{X_4 + \cdots + X_9} < 1.5\right].$$

(b) Let $Y = X_1 + X_2$.
 (i) Calculate $P[Y \leq y \mid X_1 = x_1]$, where $x_1 > 0$ and $y > 0$.
 (ii) Obtain the conditional density function $f_Y(y \mid X_1 = x_1)$.

Question no. 45
We consider a discrete random variable X having a hypergeometric distribution with parameters $N = 10$, $n = 5$, and $d = 2$.

(a) We define $Y = X^2$. Calculate the correlation coefficient of X and Y.

(b) Let $X_1, X_2, \ldots, X_8$ be independent random variables following the same distribution as X. We define $Z = X_1 + \cdots + X_8$. Calculate the probability $P[8 \leq Z \leq 12]$.

Remark. This question requires the use of a software package.

Question no. 46
Suppose that the joint probability function of the random vector (X, Y) is given by

$$p_{X,Y}(x, y) = \begin{cases} \dbinom{x}{y} \dfrac{e^{-1}(1/2)^x}{x!} & \text{if } x = 0, 1, 2, \ldots; \ y = 0, 1, \ldots, x, \\ 0 & \text{otherwise.} \end{cases}$$

(a) Obtain the functions $p_X(x)$, $p_Y(y)$, and $p_Y(y \mid X = 35)$.

(b) Calculate the probability $P[12 < Y \leq 18 \mid X = 35]$
 (i) exactly (with the help of a software package, if possible);
 (ii) using an approximation based on a normal distribution.

(c) Calculate the probability $P[X \leq 2 \mid X \geq 2]$.

Question no. 47
A system is made up of three components, C_1, C_2, and C_3, connected in parallel. The lifetime T_1 (in years) of component C_1 follows (approximately) a normal distribution with parameters $\mu = 4$ and $\sigma^2 = 2.25$. In the case of component C_2, its lifetime T_2 has an exponential distribution with parameter $\lambda = 1/4$. Finally, the lifetime T_3 of component C_3 has a gamma distribution with parameters $\alpha = 2$ and $\lambda = 1/2$. Furthermore, we assume that the random variables T_1, T_2, and T_3 are independent.

(a) Calculate the probability that the system operates for more than one year.

(b) We consider 500 systems similar to the one described above. Calculate, assuming that these 500 systems are independent, the probability that 2, 3, 4, or 5 among them are down after one year
 (i) exactly (with the help of a software package or a pocket calculator);
 (ii) using an approximation based on a Poisson distribution.

(c) Suppose that at first only components C_1 and C_2 are active. Component C_3 is on standby and begins to operate as soon as C_1 and C_2 are both down, or after one year if C_1 or C_2 still operates at that time. Suppose also that if C_1 or C_2 still operates after one year, then $T_3 \sim G(2, 1/2)$; otherwise T_3 has an exponential distribution with parameter $\lambda = 1$. Calculate the probability that component C_3 operates for at least two years.

Question no. 48

Suppose that

$$f_{X,Y}(x,y) = \begin{cases} 1/8 & \text{if } x \geq 0, \ y \geq 0, \ 0 \leq x + y \leq 4, \\ 0 & \text{elsewhere} \end{cases}$$

is the joint density function of the random vector (X, Y).

(a) Obtain the marginal density function $f_X(x)$.

(b) Calculate (i) the expected value of X, (ii) its variance, (iii) its skewness β_1, and (iv) its kurtosis β_2.

(c) Calculate the correlation coefficient of X and Y. Are the random variables X and Y independent? Justify.

Question no. 49

In a particular region, the daily temperature X (in degrees Celsius) during the month of September has a normal distribution with parameters $\mu = 15$ and $\sigma^2 = 25$. Calculate, using an approximation based on a normal distribution, the probability that the temperature exceeds 17 degrees Celsius on exactly 10 days over the course of September.

Question no. 50

Consider the joint density function

$$f_{X,Y}(x,y) = \begin{cases} 90x^2 y(1-y) & \text{if } 0 < y < 1, 0 < x < y, \\ 0 & \text{elsewhere.} \end{cases}$$

(a) Calculate the marginal density functions $f_X(x)$ and $f_Y(y)$.

(b) Are X and Y independent random variables? Justify.

(c) Calculate the covariance of X and Y.

Question no. 51

Let X_1 and X_2 be two discrete random variables whose joint probability function is given by

$$p_{X_1,X_2}(x_1, x_2) = \frac{2}{x_1! x_2! (2 - x_1 - x_2)!} \left(\frac{1}{4}\right)^{x_1} \left(\frac{1}{3}\right)^{x_2} \left(\frac{5}{12}\right)^{2-x_1-x_2}$$

if $x_1 \in \{0, 1, 2\}$, $x_2 \in \{0, 1, 2\}$, and $x_1 + x_2 \leq 2$ [and $p_{X_1,X_2}(x_1, x_2) = 0$, otherwise].

(a) Let $Y_1 = X_1 + X_2$ and $Y_2 = X_1 - X_2$. Find the probability functions of Y_1 and Y_2.

(b) Calculate the function $p_{Y_2}(y_2 \mid Y_1 = 2)$.

Question no. 52

A number X is taken at random in the interval $[-1, 1]$, and then a number Y is taken at random in the interval $[-1, X]$.

(a) Find $f_{X,Y}(x, y)$ and $f_Y(y)$.

(b) Calculate (i) $E[(X + 1)Y]$ and (ii) $E[Y]$.

(c) Use part (b) to calculate $\mathrm{COV}[X, Y]$.

Question no. 53

The storage tank of a gas station is usually filled every Monday. The capacity of the storage tank is equal to 20,000 liters. The gas station owner is told, on a given Monday, that there will be no gasoline delivery the next Monday. What is the probability that the gas station will not be able to satisfy the demand for a two-week period (with the 20,000 liters) if the weekly demand (in thousands of liters) follows

(a) an exponential distribution with parameter $\lambda = 1/10$?

(b) a gamma distribution with parameters $\alpha = 5$ and $\lambda = 1/2$?

Question no. 54

A random variable X has the following probability function:

x	-1	0	1
$p_X(x)$	$1/8$	$3/4$	$1/8$

Let X_1 and X_2 be two independent random variables distributed as X. We set $Y = X_2 - X_1$.

(a) Obtain the joint probability function of the pair (X_1, X_2).

(b) Calculate the correlation coefficient of X_1 and Y.

(c) Are the random variables X_1 and Y independent? Justify.

Question no. 55

Let $X_1, \ldots, X_{10}$ be independent random variables having an exponential distribution with parameter $\lambda = 1$. We define $Y = \sum_{i=1}^{10} X_i$.

(a) Evaluate, *without* making use of the central limit theorem, the probability $P[Y < 5]$.

(b) Use the central limit theorem to evaluate $P[Y \geq 10]$.

Multiple choice questions

Question no. 1

Let $X \sim N(0, 1)$ and $Y \sim N(1, 4)$ be two random variables such that $\mathrm{COV}[X, Y] = 1$. Calculate $P[X + Y < 12]$.

(a) 0   (b) 0.6915   (c) 0.8413   (d) 0.9773   (e) 1

Question no. 2
    Calculate $P[3 \le X + Y < 6]$ if $X \sim \text{Poi}(1)$ and $Y \sim \text{Poi}(2)$ are independent random variables.

(a) 0.269   (b) 0.493   (c) 0.543   (d) 0.726   (e) 0.916

Question no. 3
    Use the approximation of the binomial distribution by a normal distribution to calculate $P[X \le 12]$, where $X \sim \text{B}(n = 25, p = 1/2)$.

(a) 0.4207   (b) 0.4681   (c) 0.5   (d) 0.5319   (e) 0.5793

Question no. 4
    Let

$$f_{X,Y}(x, y) = \begin{cases} \frac{1}{4\pi} & \text{if } x^2 + y^2 \le 4, \\ 0 & \text{elsewhere.} \end{cases}$$

Find $f_X(x)$.

(a) $\frac{1}{2\pi}\sqrt{4 - x^2}$ if $-2 \le x \le 2$ (b) $\frac{1}{2\pi}\sqrt{4 - x^2}$ if $0 \le x \le 2$
(c) $\frac{1}{4\pi}\sqrt{4 - x^2}$ if $-2 \le x \le 2$ (d) $\frac{1}{4\pi}\sqrt{4 - x^2}$ if $0 \le x \le 2$
(e) $\frac{1}{4\pi}\sqrt{4 - x^2}$ if $-2 \le x \le y$

Question no. 5
    Suppose that $p_X(x \mid Y = y) = 1/3$, for $x = 0, 1, 2$ and $y = 1, 2$, and that $p_Y(y) = 1/2$, for $y = 1, 2$. Calculate $p_{X,Y}(1, 2)$.

(a) 1/6   (b) 1/3   (c) 1/2   (d) 2/3   (e) 1

Question no. 6
    Let X be a random variable such that $E[X^n] = 1/2$, for $n = 1, 2, \ldots$. We set $Y = X^2$. Calculate $\rho_{X,Y}$.

(a) 0   (b) 1/4   (c) 1/2   (d) 3/4   (e) 1

Question no. 7
    Suppose that the random variable X is such that $E[X] = \text{VAR}[X] = 1$. Calculate (approximately) the probability $P\left[\sum_{i=1}^{49} X_i < 56\right]$, where $X_1, X_2, \ldots, X_{49}$ are independent random variables distributed as X.

(a) 0.5   (b) 0.6554   (c) 0.8413   (d) 0.8643   (e) 1

Question no. 8
    We define $W = 3X + 2Y - Z$, where X, Y, and Z are independent random variables such that $\sigma_X^2 = 1$, $\sigma_Y^2 = 4$, and $\sigma_Z^2 = 9$. Calculate σ_W.

(a) $\sqrt{2}$ (b) 4   (c) $\sqrt{20}$ (d) $\sqrt{34}$ (e) 10

Question no. 9

We consider the joint density function

$$f_{X,Y}(x,y) = \begin{cases} 1/4 \text{ if } 0 < x < 2, 0 < y < 2, \\ 0 \text{ elsewhere.} \end{cases}$$

Calculate $P[X > 2Y]$.

(a) 1/8   (b) 1/4   (c) 1/2   (d) 3/4   (e) 7/8

Question no. 10

Calculate $P[2X < Y]$ if

$$f_{X,Y}(x,y) = \begin{cases} 2 \text{ for } 0 \leq x \leq y \leq 1, \\ 0 \text{ elsewhere.} \end{cases}$$

(a) 0   (b) 1/4   (c) 1/2   (d) 3/4   (e) 1

Question no. 11

We define $X = \max\{X_1, X_2\}$, where X_1 and X_2 are the numbers obtained by simultaneously rolling two fair dice. That is, X is the greater of the two numbers observed. Calculate $E[X]$.

(a) 91/36   (b) 3.5   (c) 4   (d) 161/36   (e) 4.5

Question no. 12

Suppose that

$$f_X(x \mid Y = y) = \begin{cases} \frac{1}{2y} \text{ if } 0 < x < 2y, \\ 0 \text{ elsewhere} \end{cases}$$

and

$$f_Y(y) = \begin{cases} \frac{1}{2} \text{ if } 0 < y < 2, \\ 0 \text{ elsewhere.} \end{cases}$$

Find $f_{X,Y}(x,y)$.

(a) $\frac{1}{8}$ if $0 < x < 4$ and $0 < y < 2$ (b) $\frac{1}{4y}$ if $0 < x < 4$ and $0 < y < 2$

(c) $\frac{1}{4y}$ if $0 < x < 2y$ and $0 < y < 2$ (d) $\frac{1}{2y}$ if $0 < x < 2y$ and $0 < y < 2$

(e) $\frac{1}{y}$ if $0 < x < 4$ and $0 < y < 2$

Question no. 13

Let

x	-2	0	2
$p_X(x)$	1/8	3/4	1/8

be the probability function of the random variable X. We define $Y = -X^2$. Calculate the correlation coefficient of X and Y.

(a) -1 (b) $-1/2$ (c) 0   (d) 1/2   (e) 1

Question no. 14

Suppose that X and Y are two independent random variables. We define two other random variables by $R = aX + b$ and $S = cY + d$. For what values of a, b, c, and d are the variables R and S uncorrelated (i.e., $\rho_{R,S} = 0$)?

(a) none   (b) $a = b = 1$ (c) $b = d = 0$ (d) $a = c = 1$, $b = d = 0$ (e) all

Question no. 15

Suppose that X_1 and X_2 are two independent random variables uniformly distributed on the interval $[0, 1]$. Let X be the smaller of the two random variables. Calculate $P[X > 1/4]$.

(a) 1/16   (b) 1/8   (c) 1/4   (d) 9/16   (e) 3/4

Question no. 16

Calculate $P[X_1 + X_2 < 2]$ if X_1 and X_2 are two independent random variables having an exponential distribution with parameter $\lambda = 1$.

(a) 0.324   (b) 0.405   (c) 0.594   (d) 0.676   (e) 0.865

Question no. 17

Let $X_1, \ldots, X_{36}$ be independent random variables, where X_i follows a gamma distribution with parameters $\alpha = 2$ and $\lambda = 3$, for all i. Calculate (approximately) $P[2/3 < \bar{X} < 3/4]$, where $\bar{X} := \frac{1}{36} \sum_{i=1}^{36} X_i$.

(a) 0.218   (b) 0.355   (c) 0.360   (d) 0.497   (e) 0.855

Question no. 18

Suppose that $X_1, \ldots, X_n$ are independent $N(0, 1)$ random variables. What is the smallest value of n for which $P[-0.1n < \sum_{i=1}^{n} X_i < 0.1n] \geq 0.95$?

(a) 19   (b) 20   (c) 271   (d) 384   (e) 385

Question no. 19

Let $X_1 \sim N(0, 1)$, $X_2 \sim N(-1, 1)$, and $X_3 \sim N(1, 1)$ be independent random variables. Calculate $P[|X_1 + 2X_2 - 3X_3| > 5]$.

(a) 0.004   (b) 0.496   (c) 0.5   (d) 0.504   (e) 0.996

Question no. 20

Calculate approximately, by means of a normal distribution, $P\left[\sum_{i=1}^{100} X_i \leq 251\right]$, where $X_1, \ldots, X_{100}$ are independent random variables such that X_i has a binomial distribution with parameters $n = 10$ and $p = 1/4$, for $i = 1, \ldots, 100$.

(a) 0.50   (b) 0.51   (c) 0.53   (d) 0.56   (e) 0.59

5

Reliability

In many applied fields, particularly in most engineering disciplines, it is important to be able to calculate the probability that a certain device or system will be active at a given time instant, or over a fixed period of time. We already considered many exercises on *reliability* theory in Chapters 2 to 4. In Chapter 2, it was understood that we were calculating the reliability of a system at a given time instant t_0, knowing the reliability of each of its components at t_0. In order to calculate the probability that a machine will operate without failure for a given amount of time, we need to know the distribution of its lifetime or of the lifetime of its components. It becomes a problem on random variables or vectors. In this chapter, we present in detail the main concepts of reliability theory.

5.1 Basic notions

There are many possible interpretations of the word *reliability*. In this textbook, it always corresponds to the probability of functioning correctly at a given time instant or over a given period of time. Moreover, in the current chapter, we are mainly interested in the reliability over a certain time interval $[0, t]$.

Definition 5.1.1. *Let X be a nonnegative random variable representing the lifetime (or time to failure) of a system or a device. The probability*

$$R(x) = P[X > x] \quad [= 1 - F_X(x)] \quad \text{for } x \geq 0$$

*is called the **reliability** function or **survival** function of the system.*

Remarks. (i) The function $R(x)$ can also be denoted by $S(x)$. The notation $\bar{F}_X(x) = 1 - F_X(x)$ is used as well.

M. Lefebvre, *Basic Probability Theory with Applications*, Springer Undergraduate Texts in Mathematics and Technology, DOI: 10.1007/978-0-387-74995-2_5,
© Springer Science + Business Media, LLC 2009

(ii) Most often, it is assumed that the random variable X is continuous. However, in some applications, the lifetime is measured in number of cycles. Therefore, X is then a discrete (integer-valued, to be precise) random variable. Furthermore, if we accept the possibility that a device may be defective, then X could take on the value 0 and be a *mixed type* random variable.

(iii) All discrete distributions considered in Section 3.2 could serve as reliability models. In the case of the continuous distributions, we must limit ourselves to the ones that are always nonnegative. Therefore, the normal distribution cannot be an *exact* reliability model. Nonetheless, depending on the values of the parameters μ and σ, it can be a good *approximate* model for the survival time of a machine. Furthermore, we can consider the *truncated* normal distribution, defined for $x \geq 0$.

A useful measure of the dependability of a system is its *mean lifetime* $E[X]$. In the context of reliability theory, $E[X]$ is called the *mean time to failure* of the system.

Definition 5.1.2. *The symbol MTTF (which stands for* **Mean Time To Failure***) denotes the expected value of the lifetime X of a system. If the system can be repaired, we also define the symbols MTBF (***Mean Time Between Failures***) and MTTR (***Mean Time To Repair***). We have that $MTBF = MTTF + MTTR$.*

Remarks. (i) Suppose that we are interested in the lifetime X of a car. It is obvious that, except in case of a very major failure, the car will be repaired when it breaks down. When we calculate the quantity $MTBF$, we assume that, after having been "repaired," a system is as good as new. Of course, in the case of a car, this is not exactly true, because cars *age* and wear.

(ii) To distinguish between critical and noncritical failures, we can use the more precise term *Mean Time Between Critical Failures* ($MTBCF$). Then, $MTBF$ could be interpreted as the mean time between failures of any type, that is, critical or noncritical. In the context of a computer or data transmission system, we also have the *Mean Time Between System Aborts* ($MTBSA$).

To calculate the mean lifetime of a system, we can, of course, use the definition of the expected value of a random variable. However, it is sometimes simpler to proceed as in the following proposition.

Proposition 5.1.1. *Let X be a nonnegative random variable. Then, we have:*

$$E[X] = \begin{cases} \sum_{k=0}^{\infty} P[X > k] = \sum_{k=0}^{\infty} R(k) \text{ if } X \in \{0, 1, \ldots\}, \\ \int_0^{\infty} P[X > x]\,dx = \int_0^{\infty} R(x)\,dx \text{ if } X \in [0, \infty). \end{cases}$$

Proof. Consider first the case when X is an integer-valued random variable. We have:

$$E[X] := \sum_{j=0}^{\infty} j\,P[X=j] = \sum_{j=1}^{\infty} j\,P[X=j] = \sum_{j=1}^{\infty} \sum_{k=1}^{j} P[X=j]$$

$$= \sum_{k=1}^{\infty} \sum_{j=k}^{\infty} P[X=j] = \sum_{k=1}^{\infty} P[X \geq k] = \sum_{k=0}^{\infty} P[X > k].$$

Similarly, if X (is continuous and) belongs to the interval $[0, \infty)$, we can write that

$$E[X] := \int_{0}^{\infty} t f_X(t)\,dt = \int_{0}^{\infty} \int_{0}^{t} f_X(t)\,dx dt$$

$$= \int_{0}^{\infty} \int_{x}^{\infty} f_X(t)\,dt dx = \int_{0}^{\infty} P[X > x]\,dx.$$

∎

Remarks. (i) It is not necessary that the discrete random variable X can take on all nonnegative integers. It is sufficient that the set of possible values of X be included in $\{0, 1, \ldots\}$. Likewise, in the continuous case, X must take its values in the interval $[0, \infty)$.

(ii) Often, the formulas in the proposition do not simplify the calculation of the expected value of X. For example, it is more complicated to calculate the mean of a Poisson random variable from the first formula than from the definition.

Example 5.1.1. Let X be a geometric random variable with parameter p in the interval $(0, 1)$. Its possible values are the integers $1, 2 \ldots$ and its mean is equal to $1/p$. We saw (on p. 64) that

$$P[X > k] = (1 - p)^k \quad \text{for } k = 0, 1, \ldots .$$

It follows that (indeed)

$$E[X] = \sum_{k=0}^{\infty} (1-p)^k = \frac{1}{1 - (1 - p)} = \frac{1}{p}.$$

Example 5.1.2. If $X \sim \text{Exp}(\lambda)$, we find (see Example 3.5.2) that

$$P[X > x] = \int_{x}^{\infty} \lambda e^{-\lambda t}\,dt = e^{-\lambda x} \quad \text{for } x \geq 0.$$

Hence,

$$E[X] = \int_{0}^{\infty} e^{-\lambda x}\,dx = -\frac{e^{-\lambda x}}{\lambda}\Big|_{0}^{\infty} = \frac{1}{\lambda}.$$

In Chapter 4, we defined various conditional functions, for example, the function $f_X(x \mid Y = y)$, where (X, Y) is a continuous random vector. We can also define functions of the type $f_X(x \mid A_X)$, where A_X is an event that involves only the random variable X. One such particular *conditional density function* is important in reliability theory.

Definition 5.1.3. *Suppose that the lifetime T of a system is a continuous nonnegative random variable. The* **failure rate function** *(or* **hazard rate function***) $r(t)$ of the system is defined by*

$$r(t) = f_T(t \mid T > t) := \lim_{s \downarrow t} \frac{d}{ds} F_T(s \mid T > t) \quad \text{for } t \geq 0.$$

Remarks. (i) The function $r(t)$, multiplied by dt, can be interpreted as the probability that a machine, which is t time units old and still operating, will break down in the interval $(t, t + dt]$. Indeed, we have:

$$f_T(t \mid T > t) = \lim_{dt \downarrow 0} \frac{P[t < T \leq t + dt \mid T > t]}{dt}.$$

(ii) We assume that the *conditional distribution function* $F_T(s \mid T > t)$ is differentiable at $s \in (t, t + dt]$.

(iii) We must take the limit as s decreases to t in the definition, because we have that $F_T(t \mid T > t) \equiv P[T \leq t \mid T > t] = 0$. However, we don't have to take any limit to calculate $f_T(s \mid T > t)$ from $F_T(s \mid T > t)$, for $s > t$.

Proposition 5.1.2. *We have:*

$$r(t) = \frac{f_T(t)}{1 - F_T(t)} = -\frac{R'(t)}{R(t)} \quad \text{for } t \geq 0.$$

Proof. By definition,

$$F_T(s \mid T > t) = P[T \leq s \mid T > t] = \frac{P[\{T \leq s\} \cap \{T > t\}]}{P[T > t]}$$

$$= \begin{cases} 0 & \text{if } s \leq t, \\ \dfrac{F_T(s) - F_T(t)}{1 - F_T(t)} & \text{if } s > t. \end{cases}$$

Hence,

$$f_T(s \mid T > t) := \frac{d}{ds} F_T(s \mid T > t) = \frac{f_T(s)}{1 - F_T(t)} \quad \text{if } s > t.$$

Taking the limit as s decreases to t, we obtain that

$$r(t) := f_T(t \mid T > t) = \frac{f_T(t)}{1 - F_T(t)}.$$

Finally, because

$$R'(t) = \frac{d}{dt}[1 - F_T(t)] = -f_T(t),$$

we also have:

$$r(t) = -\frac{R'(t)}{R(t)} \quad \text{for } t \geq 0.$$

■

Remark. In the discrete case, the failure rate function is given by

$$r(k) = \frac{p_X(k)}{\sum_{j=k}^{\infty} p_X(j)} \quad \text{for } k = 0, 1, \dots .$$

Note that $0 \leq r(k) \leq 1$ for any k, whereas $r(t) \geq 0$ in the continuous case.

Example 5.1.3. One of the most commonly used models in reliability theory is the exponential distribution, mainly because of its memoryless property (see p. 76). This property implies that for a system whose lifetime is exponentially distributed, the failure rate function is *constant*. Indeed, if $X \sim \text{Exp}(\lambda)$, we have (see the previous example):

$$r(t) = \frac{\lambda e^{-\lambda t}}{e^{-\lambda t}} = \lambda \quad \text{for } t \geq 0.$$

In practice, this is generally not realistic. There are some applications though for which this is acceptable. For example, it seems that the lifetime of an electric fuse that cannot melt only partially is approximately exponentially distributed. The time between the failures of a system made up of a very large number of independent components connected in series can also follow approximately an exponential distribution, if we assume, in particular, that every time a component fails it is immediately replaced by a new one. However, in most cases, the exponential distribution should only be used for t in a finite interval $[t_1, t_2]$.

Example 5.1.4. The geometric distribution is the equivalent of the exponential distribution in discrete time. It also possesses the memoryless property. Because $P[X \geq k] = P[X > k - 1]$, we calculate (see Example 5.1.1)

$$r(k) = \frac{(1-p)^{k-1}p}{(1-p)^{k-1}} = p \quad \text{for } k = 1, 2, \dots .$$

Therefore, the failure rate function $r(k)$ is a constant in this case too, as expected.

The failure rate function of a given distribution is a good indicator of the value of this distribution as a model in reliability theory. In most applications, $r(t)$ should be a strictly increasing function of t, at least when t is large enough.

Definition 5.1.4. *If the random variable X is such that its failure rate function $r_X(t)$ or $r_X(k)$ is* increasing *(resp., decreasing) in t or k, then X is said to have an* **increasing failure rate** *(resp.,* **decreasing failure rate***) distribution.*

Notation. We use the acronym *IFR* (resp., *DFR*) for Increasing Failure Rate (resp., Decreasing Failure Rate).

Now, making use of Proposition 5.1.2, we obtain that

$$\int_0^t r(s)ds = -\int_0^t \frac{R'(s)}{R(s)} ds = -\ln R(t) + \ln R(0).$$

Moreover, the random variable T [with failure rate function $r(t)$] being continuous and nonnegative, we may write that $R(0) := P[T > 0] = 1$. Hence, we may state the following proposition.

Proposition 5.1.3. *There is a one-to-one relationship between the functions $R(t)$ and $r(t)$:*

$$R(t) = \exp\left\{ -\int_0^t r(s)ds \right\}.$$

Remark. The proposition implies that the exponential distribution is the *only* continuous distribution having a constant failure rate function.

Example 5.1.5. We can show that the failure rate function $r(t)$ of a lognormal distribution starts at zero [because $\lim_{t \downarrow 0} f_T(t) = 0$], next it increases to a maximum, and then

$$\lim_{t \to \infty} r(t) = 0.$$

So, we must conclude that the lognormal distribution is *not* a good model for the lifetime of a device that is subject to wear, at least not for t large. Indeed, the failure rate should generally increase with t, as mentioned above.

Example 5.1.6. The normal distribution $N(\mu, \sigma^2)$ should not be used to model the lifetime of a system, unless μ and σ are such that the probability that the random variable takes on a negative value is negligible. For any values of μ and σ, we can define the *truncated* normal distribution as follows:

$$f_X(x) = \frac{1}{\sqrt{2\pi}\sigma c} \exp\left\{ -\frac{(x-\mu)^2}{2\sigma^2} \right\} \quad \text{for } x \geq 0,$$

where c is a constant such that $\int_0^\infty f_X(x)dx = 1$. That is,

$$c = \left[\int_0^\infty f_Y(y)dy \right]^{-1} = \frac{1}{1 - \Phi(-\mu/\sigma)},$$

where $Y \sim N(\mu, \sigma^2)$. We can write that $X \equiv Y \mid \{Y \geq 0\}$. Note that if $\mu = 0$, then $c = 2$.

We find that the failure rate function of a truncated normal distribution is strictly increasing, which makes it an interesting model for many applications.

Example 5.1.7. The Weibull distribution (see p. 77) is a really important model in reliability theory and *fatigue* analysis. We have:

$$R(t) = \int_t^\infty \lambda \beta x^{\beta-1} \exp\left(-\lambda x^\beta\right) dx = \exp\left(-\lambda t^\beta\right).$$

It follows that

$$r(t) = \frac{\lambda \beta t^{\beta-1} \exp\left(-\lambda t^\beta\right)}{\exp\left(-\lambda t^\beta\right)} = \lambda \beta t^{\beta-1} \quad \text{for } t \geq 0.$$

Therefore, the Weibull distribution is DFR if $\beta < 1$ and IFR if $\beta > 1$. When $\beta = 1$, we retrieve the exponential distribution.

Although it is true that the failure rate function $r(t)$ should increase as t increases, for large enough values of t, in many situations it is first a decreasing function of t. For example, the *mortality rate* of children does indeed decrease at first. There is a greater risk that a baby will die at birth or shortly thereafter than when it is six months old, in particular. When the child grows older, the death rate is more or less constant for some time, whereas it increases for adults. Therefore, the function $r(t)$ looks like a *bathtub* (see Figure 5.1). As mentioned above, the exponential distribution should only be used

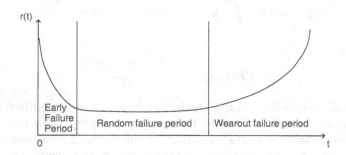

Fig. 5.1. Failure rate function having the shape of a bathtub.

for t such that $t_1 \leq t \leq t_2 < \infty$. It is valid for the flat portion of the bathtub.

Suppose that the lifetime X is defined as follows:

$$X = c_1 X_1 + c_2 X_2 + c_3 X_3,$$

where X_i has a Weibull distribution and $c_i > 0$, for $i = 1, 2, 3$. We assume that $c_1 + c_2 + c_3 = 1$. Then, the linear combination of Weibull random variables is called a *mixed Weibull distribution*. The bathtub shape can be obtained by choosing the parameter β of X_1 smaller than 1, that of X_2 equal to 1, and that of X_3 greater than 1. Note that X_2 is actually an exponential random variable.

Next, sometimes we are interested in the probability that a system will fail during a particular time interval.

Definition 5.1.5. *The* **interval failure rate** *of a system in the interval* $(t_1, t_2]$ *is denoted by* $FR(t_1, t_2)$ *and is defined by*

$$FR(t_1, t_2) = \frac{P[t_1 < T \le t_2 \mid T > t_1]}{t_2 - t_1} = \frac{R(t_1) - R(t_2)}{R(t_1)} \frac{1}{(t_2 - t_1)}$$

for $0 \le t_1 < t_2 < \infty$, *where* T *is a continuous random variable.*

Remark. We have:

$$r(t_1) = \lim_{t_2 \downarrow t_1} FR(t_1, t_2).$$

Example 5.1.8. Suppose that $T \sim \text{Exp}(\lambda)$. First, we calculate the conditional density function $f_T(t \mid T > t_1)$. We find that

$$f_T(t \mid T > t_1) = \frac{f_T(t)}{P[T > t_1]} = \frac{\lambda e^{-\lambda t}}{e^{-\lambda t_1}} = \lambda e^{-\lambda(t - t_1)} \quad \text{for } t > t_1$$

[and $f_T(t \mid T > t_1) = 0$ for $t \le t_1$]. It follows that

$$P[t_1 < T \le t_2 \mid T > t_1] = \int_{t_1}^{t_2} \lambda e^{-\lambda(t - t_1)} \, dt = -e^{-\lambda(t - t_1)} \Big|_{t_1}^{t_2} = 1 - e^{-\lambda(t_2 - t_1)}.$$

Hence, we have:

$$FR(t_1, t_2) = \frac{1 - e^{-\lambda(t_2 - t_1)}}{t_2 - t_1}.$$

Actually, we could have obtained this result at once from the reliability function $R(t) = e^{-\lambda t}$. However, we wanted to give the formula for the density function of the *shifted exponential distribution*. A shifted distribution can be used as a model in reliability theory in the following situation: suppose that a man buys a device for which there is a guarantee period of length $t_1 > 0$. Then, the buyer is sure that the device in question will last at least t_1 time units (it may be repaired or replaced if it fails before the end of the guarantee period).

Notice that the function $FR(t_1, t_2)$ actually depends only on the difference $t_2 - t_1$. Therefore, the probability that the system will fail in a given interval depends only on the length of this interval, which is a consequence of the memoryless property of the exponential distribution. Finally, making use of l'Hospital's rule, we obtain that

$$\lim_{t_2 \downarrow t_1} FR(t_1, t_2) = \lim_{t_2 \downarrow t_1} \frac{1 - e^{-\lambda(t_2 - t_1)}}{t_2 - t_1} = \lim_{\epsilon \downarrow 0} \frac{1 - e^{-\lambda \epsilon}}{\epsilon} = \lim_{\epsilon \downarrow 0} \frac{e^{-\lambda \epsilon} \lambda}{1} = \lambda,$$

as should be.

Example 5.1.9. Let

$$f_T(t) = te^{-t} \quad \text{for } t > 0.$$

The random variable T has a gamma distribution with parameters $\alpha = 2$ and $\lambda = 1$. We calculate

$$R(t) = \int_t^\infty se^{-s}\,ds = -se^{-s}\Big|_t^\infty + \int_t^\infty e^{-s}\,ds = (t+1)e^{-t} \quad \text{for } t > 0,$$

where we used l'Hospital's rule to evaluate the limit $\lim_{s\to\infty} se^{-s}$. We have:

$$r(t) = \frac{te^{-t}}{(t+1)e^{-t}} = \frac{t}{t+1} = 1 - \frac{1}{t+1},$$

which is an increasing function for all values of t. Actually, the gamma distribution is IFR for any $\alpha > 1$ (and DFR if $0 < \alpha < 1$).

From the function $R(t)$, we deduce that

$$FR(t_1, t_2) = \left[1 - \frac{(t_2+1)}{(t_1+1)}e^{-(t_2-t_1)}\right]\frac{1}{t_2 - t_1}.$$

Finally, we define another quantity of interest in reliability theory.

Definition 5.1.6. *The* **average failure rate** *of a system over an interval* $[t_1, t_2]$ *is given by*

$$AFR(t_1, t_2) = \frac{\int_{t_1}^{t_2} r(t)\,dt}{t_2 - t_1} = \frac{\ln[R(t_1)] - \ln[R(t_2)]}{t_2 - t_1}.$$

Remark. This quantity is an example of a *temporal average*.

Example 5.1.10. If T has a Weibull distribution, we have (see Example 5.1.7):

$$\int_{t_1}^{t_2} r(t)\,dt = \int_{t_1}^{t_2} \lambda\beta t^{\beta-1}\,dt = \lambda(t_2^\beta - t_1^\beta),$$

so that

$$AFR(t_1, t_2) = \frac{\lambda(t_2^\beta - t_1^\beta)}{t_2 - t_1}.$$

If $\beta = 1$, we have that $AFR(t_1, t_2) \equiv \lambda$, whereas $\beta = 2$ implies that $AFR(t_1, t_2)$ is equal to $\lambda(t_1 + t_2)$. Note that, when $\beta = 2$, the average failure rate is three times as large in the interval $(t, 2t)$ than from 0 to t.

5.2 Reliability of systems

In this section, we consider systems constituted of at least two subsystems or components that may be connected *in series* or *in parallel*. When components are connected in parallel, we must distinguish between *active redundancy* and *standby* (or *passive*) *redundancy*. We assume that the components making up the systems considered cannot be repaired. When a system fails, it will remain down indefinitely. In fact, it would resume operating if the failed components were replaced by new ones. However, here we are only interested in the time elapsed until the *first* failure of the system.

5.2.1 Systems in series

Consider n subsystems operating independently of one another. Let $R_k(t)$ be the reliability function of subsystem k, for $k = 1, \ldots, n$. If the subsystems are connected in series, then each subsystem must be active for the system to operate. Therefore, the lifetime T of the system is such that

$$T > t \quad \Longleftrightarrow \quad T_k > t \quad \text{for all } k,$$

where T_k is the lifetime of subsystem k. It follows that the reliability function of the system is given by (see Proposition 5.1.3)

$$R(t) = \prod_{k=1}^{n} R_k(t) = \exp\left\{ -\int_0^t [r_1(s) + \cdots + r_n(s)] \, ds \right\}.$$

Remarks. (i) In Chapter 2, we could have asked for the probability that a system made up of n independent components connected in series operates at a given time instant t_0, given the reliability of each component at t_0. Define the events $S = $ "the system operates at time t_0" and $B_k = $ "component k operates at time t_0." We have:

$$P[S] = P\left[\bigcap_{k=1}^{n} B_k \right] \overset{\text{ind.}}{=} \prod_{k=1}^{n} P[B_k].$$

We assumed that the components cannot be repaired, thus this is actually a particular case of the previous formula. Indeed, we may write that $P[S] = R(t_0)$ and $P[B_k] = R_k(t_0)$. We must also assume that no components were replaced in the interval $(0, t_0)$.

(ii) When the components making up a system are connected in series, we must assume that they operate independently of one another, because the system fails as soon as one of its components fails. For example, suppose that there are only two components, denoted by A and B. We cannot imagine that there is a certain probability $p_1(t)$ [resp., $p_2(t)$] that component B will be active at time t if A is active (resp., down) at time t. If component A is down at time t, then the system stopped operating when A failed and remained down henceforth. Furthermore, in continuous time, the probability that components A and B will fail exactly at the same time instant is equal to zero. Therefore, assuming that a component cannot fail while the system is down, if A is down, then B must be active. Unless the lifetime of component B follows an exponential distribution (or a geometric distribution, in discrete time), when component A is replaced by a new one, component B should be replaced as well, if we want the system to be as good as new. Similarly, we cannot suppose that the lifetime of component A has a distribution that depends on the lifetime of B.

(iii) We can write that

$$T = \min\{T_1, T_2, \ldots, T_n\}.$$

The *minimum* of a sequence of random variables is a special case of what is known as *order statistics*.

Proposition 5.2.1. *Suppose that T_1 and T_2 are independent exponential random variables with parameters λ_1 and λ_2, respectively. We have:*

$$T := \min\{T_1, T_2\} \sim Exp(\lambda_1 + \lambda_2).$$

Proof. We calculate

$$P[T > t] \overset{\text{ind.}}{=} P[T_1 > t]P[T_2 > t] = e^{-\lambda_1 t}e^{-\lambda_2 t} = e^{-(\lambda_1 + \lambda_2)t} \quad \text{for } t \geq 0.$$

It follows that

$$f_T(t) = \frac{d}{dt}\{1 - P[T > t]\} = \frac{d}{dt}\{1 - e^{-(\lambda_1 + \lambda_2)t}\} = (\lambda_1 + \lambda_2)e^{-(\lambda_1 + \lambda_2)t}$$

for $t \geq 0$. ∎

Remark. The proposition can be generalized as follows: if $T_k \sim Exp(\lambda_k)$, for $k = 1, \ldots, n$, and if the T_ks are independent random variables, then $T := \min\{T_1, T_2, \ldots, T_n\} \sim Exp(\lambda_1 + \lambda_2 + \cdots + \lambda_n)$. Therefore, if n independent components having exponentially distributed lifetimes are placed in series, it is equivalent to having a single component whose lifetime follows an exponential distribution with parameter λ equal to the sum of the λ_ks. Note that because $r_k(t) \equiv \lambda_k$, for $k = 1, \ldots, n$, we have:

$$R(t) = \exp\left\{-\int_0^t (\lambda_1 + \cdots + \lambda_n)\, ds\right\} = \exp[-(\lambda_1 + \cdots + \lambda_n)t] = e^{-\lambda t},$$

where $\lambda := \lambda_1 + \cdots + \lambda_n$.

Example 5.2.1. If T_k is uniformly distributed on the interval $[0,1]$, for $k = 1, \ldots, n$, then

$$R_k(t) = \int_t^1 1 \, ds = 1 - t \quad \text{for } 0 \le t \le 1.$$

It follows that

$$R(t) = \prod_{k=1}^n R_k(t) = \prod_{k=1}^n (1 - t) = (1 - t)^n \quad \text{for } 0 \le t \le 1.$$

Because $T_k \le 1$ for all k, we can write that $R(t) = 0$ if $t > 1$.

5.2.2 Systems in parallel

Active redundancy

We now consider systems constituted of at least two subsystems connected *in parallel*. Assume first that all subsystems, which may contain one or many components, operate from the initial time $t = 0$. This is called *active redundancy*. Then, the whole system will operate as long as there is at least one active subsystem remaining. We can write that the lifetime T of the system is the *maximum* of the random variables $T_1, \ldots, T_n$, where n is the number of subsystems placed in parallel. It follows that

$$T \le t \quad \Longleftrightarrow \quad T_k \le t \quad \text{for } k = 1, \ldots, n.$$

Hence, if the subsystems operate independently of one another, we have:

$$R(t) = 1 - \prod_{k=1}^n [1 - R_k(t)].$$

Remark. When the subsystems are connected in parallel, we may consider the case where they do not operate independently. This case was already considered in Chapter 2, for instance, in Example 2.4.1.

Example 5.2.2. A device comprises two components connected in parallel and operating independently of each other. The lifetime T_k of component k has an exponential distribution with parameter λ_k, for $k = 1, 2$. It follows that

$$P[T \le t] = P[\{T_1 \le t\} \cap \{T_2 \le t\}] \stackrel{\text{ind.}}{=} P[T_1 \le t] P[T_2 \le t]$$
$$= (1 - e^{-\lambda_1 t})(1 - e^{-\lambda_2 t}) \quad \text{for } t \ge 0,$$

where T is the total lifetime of the system. Hence, the reliability function of the system is

$$R(t) = e^{-\lambda_1 t} + e^{-\lambda_2 t} - e^{-(\lambda_1+\lambda_2)t}.$$

Note that the maximum of independent exponential random variables does not follow an exponential distribution (not even if $\lambda_1 = \lambda_2$), because

$$f_T(t) = \frac{d}{dt}P[T \le t] = \lambda_1 e^{-\lambda_1 t} + \lambda_2 e^{-\lambda_2 t} - (\lambda_1 + \lambda_2)e^{-(\lambda_1+\lambda_2)t} \quad \text{for } t \ge 0.$$

We have:

$$E[T] = \int_0^\infty R(t)\,dt = \frac{1}{\lambda_1} + \frac{1}{\lambda_2} - \frac{1}{\lambda_1 + \lambda_2}.$$

In the special case when $\lambda_1 = \lambda_2 = \lambda$, we obtain that $E[T] = 1.5/\lambda$. That is, the fact of installing two identical components in parallel, in this example, increases the mean time to failure of the device by 50%.

Example 5.2.3. Suppose, in the preceding example, that the probability that component no. 2 is active at a fixed time instant $t_0 > 0$ is equal to $e^{-\lambda_{21} t_0}$ if component no. 1 too is active at time t_0, and to $e^{-\lambda_{22} t_0}$ if component no. 1 is down at time t_0. Then, we can write (because the exponential distribution is continuous, so that $P[T_1 = t_0] = 0$) that

$$P[T \le t_0] = P[\{T_1 \le t_0\} \cap \{T_2 \le t_0\}] = P[T_2 \le t_0 \mid T_1 \le t_0]\,P[T_1 \le t_0]$$
$$= (1 - e^{-\lambda_{22} t_0})(1 - e^{-\lambda_1 t_0}).$$

Moreover, we have:

$$P[T_2 \le t_0] = P[T_2 \le t_0 \mid T_1 \le t_0]\,P[T_1 \le t_0] + P[T_2 \le t_0 \mid T_1 > t_0]\,P[T_1 > t_0]$$
$$= (1 - e^{-\lambda_{22} t_0})(1 - e^{-\lambda_1 t_0}) + (1 - e^{-\lambda_{21} t_0})e^{-\lambda_1 t_0}.$$

In general, the constant λ_{22} should actually be a function of the exact time at which component no. 1 failed. In Example 2.4.1, we provided the numerical probabilities that component B operates at an unspecified time instant, given that component A does or does not operate at that time instant. Note that the sum $e^{-\lambda_{21} t_0} + e^{-\lambda_{22} t_0}$ can take on any value in the interval $[0, 2]$.

Remark. By *conditioning* on the failure time of component no. 1, we can show that

$$P[T_2 > t] = \int_0^\infty P[T_2 > t \mid T_1 = \tau]\,f_{T_1}(\tau)\,d\tau$$
$$= \int_0^t P[T_2 > t \mid T_1 = \tau]\,f_{T_1}(\tau)\,d\tau + \int_t^\infty P[T_2 > t \mid T_1 = \tau]\,f_{T_1}(\tau)\,d\tau.$$

Suppose that T_2 has an exponential distribution with parameter λ_2 as long as component no. 1 operates, and an exponential distribution with parameter λ_3 (which should be greater than λ_2) from the moment when component no. 1 fails. Then, by the memoryless property of the exponential distribution, we can write that

$$P[T_2 > t \mid T_1 = \tau] = \begin{cases} e^{-\lambda_2 \tau} e^{-\lambda_3 (t-\tau)} & \text{if } 0 < \tau \le t, \\ e^{-\lambda_2 t} & \text{if } \tau > t. \end{cases}$$

If the lifetime of component no. 2 is actually independent of that of component no. 1, so that $\lambda_3 = \lambda_2$, we obtain:

$$P[T_2 > t \mid T_1 = \tau] = \begin{cases} e^{-\lambda_2 \tau} e^{-\lambda_2 (t-\tau)} = e^{-\lambda_2 t} & \text{if } 0 < \tau \le t, \\ e^{-\lambda_2 t} & \text{if } \tau > t. \end{cases}$$

That is,

$$P[T_2 > t \mid T_1 = \tau] = e^{-\lambda_2 t} \quad \text{for } t \ge 0 \text{ and any } \tau > 0.$$

We have:

$$P[T_2 > t] = \int_0^\infty e^{-\lambda_2 t} f_{T_1}(\tau) d\tau = e^{-\lambda_2 t} \int_0^\infty f_{T_1}(\tau) d\tau = e^{-\lambda_2 t},$$

as should be.

Passive redundancy

Suppose now that a system comprises n subsystems (numbered from 1 to n) connected in parallel, but that only one subsystem operates at a time. At first, only subsystem no. 1 is active. When it fails, subsystem no. 2 relieves it, and so forth. This type of redundancy is called *passive* (or *standby*) redundancy.

Remarks. (i) It is understood that there is a device that sends signals to the system instructing it to activate subsystem no. 2 when the first one fails, and so on. In practice, this device itself can fail. However, we assume in this book that the signaling device remains 100% reliable over an indefinite time period. We also assume that the subsystems placed in standby mode cannot fail before they are activated, although we could actually have two failure time distributions: a *dormant* failure distribution and an *active* failure distribution.

(ii) Because the subsystems operate one after the other, it is natural to assume (unless otherwise stated) that their lifetimes are independent random variables.

The total lifetime T of the system is obviously given by

$$T = T_1 + T_2 + \cdots + T_n,$$

so that its mean time to failure is

$$E[T] = \sum_{k=1}^n E[T_k].$$

Moreover, because the subsystems operate independently of one another, we have:

$$\text{VAR}[T] \overset{\text{ind.}}{=} \sum_{k=1}^{n} \text{VAR}[T_k].$$

In general, it is not easy to find an explicit expression for the reliability function of the system, because the density function of T is the convolution of the density functions of the random variables $T_1, \ldots, T_n$. In the particular case when $T_k \sim \text{Exp}(\lambda)$, for all k, we know (see Subsection 4.3.3) that T has a gamma distribution with parameters n and λ. Furthermore, making use of Formula (3.3.2), we can write that

$$R(t) := P[T > t] = P[\text{Poi}(\lambda t) \leq n - 1] = \sum_{k=0}^{n-1} e^{-\lambda t} \frac{(\lambda t)^k}{k!} \quad \text{for } t \geq 0.$$

Example 5.2.4. A system is made up of two identical (and independent) components arranged in standby redundancy. If the lifetime T_k of each component follows a uniform distribution on the interval $(0, 1)$, then (see Example 4.3.4) we can write that

$$F_T(t) = \begin{cases} \displaystyle\int_0^t s\, ds = \frac{t^2}{2} & \text{if } 0 < t < 1, \\[2mm] \displaystyle\frac{1}{2} + \int_1^t (2 - s)\, ds = 2t - \frac{t^2}{2} - 1 & \text{if } 1 \leq t < 2. \end{cases}$$

It follows that

$$R(t) = 1 - F_T(t) = \begin{cases} 1 - \dfrac{t^2}{2} & \text{if } 0 < t < 1, \\[2mm] 2 + \dfrac{t^2}{2} - 2t & \text{if } 1 \leq t < 2 \end{cases}$$

[and $R(t) = 0$ if $t \geq 2$].

Example 5.2.5. Suppose that, in the previous example, $T_1 \sim \text{Exp}(\lambda_1)$ and $T_2 \sim \text{Exp}(\lambda_2)$, where $\lambda_1 \neq \lambda_2$. Then, using (4.8), we can write that

$$\begin{aligned} f_T(t) &= f_{T_1}(t_1) * f_{T_2}(t_2) = \int_{-\infty}^{\infty} f_{T_1}(u) f_{T_2}(t - u)\, du \\[2mm] &= \int_0^t \lambda_1 e^{-\lambda_1 u} \lambda_2 e^{-\lambda_2(t-u)}\, du = e^{-\lambda_2 t} \int_0^t \lambda_1 \lambda_2 e^{-(\lambda_1 - \lambda_2)u}\, du \\[2mm] &\overset{\lambda_1 \neq \lambda_2}{=} \frac{\lambda_1 \lambda_2}{\lambda_1 - \lambda_2} \left(e^{-\lambda_2 t} - e^{-\lambda_1 t} \right) \quad \text{for } t \geq 0. \end{aligned}$$

It follows that

$$R(t) = \int_t^\infty f_T(s)\,ds = \frac{\lambda_1 \lambda_2}{\lambda_1 - \lambda_2}\left(\frac{e^{-\lambda_2 t}}{\lambda_2} - \frac{e^{-\lambda_1 t}}{\lambda_1}\right)$$

$$= \frac{\lambda_1 e^{-\lambda_2 t} - \lambda_2 e^{-\lambda_1 t}}{\lambda_1 - \lambda_2} \quad \text{for } t \geq 0.$$

Note that, making use of l'Hospital's rule, we obtain:

$$\lim_{\lambda_2 \to \lambda_1} f_T(t) = \lambda_1^2 \lim_{\lambda_2 \to \lambda_1} \frac{e^{-\lambda_2 t}(-t) - 0}{0 - 1} = \lambda_1^2 t e^{-\lambda_1 t} \quad \text{for } t \geq 0.$$

That is, $T \sim G(\alpha = 2, \lambda_1)$, as should be. We also have:

$$\lim_{\lambda_2 \to \lambda_1} R(t) \overset{\text{L'Hos.}}{=} \lim_{\lambda_2 \to \lambda_1} \frac{\lambda_1 e^{-\lambda_2 t}(-t) - e^{-\lambda_1 t}}{0 - 1} = e^{-\lambda_1 t}(\lambda_1 t + 1) \quad \text{for } t \geq 0.$$

5.2.3 Other cases

Suppose that a system is made up of n subsystems and that at least k working subsystems are needed for the system to operate, where $0 < k \leq n$. This is called a *k-out-of-n system*. Note that a series system is the particular case when $k = n$, whereas a parallel system (with active redundancy) corresponds to the case when $k = 1$.

In general, we cannot give a simple formula for the reliability function $R(t)$ of the system. However, if all the subsystems are independent and have the same reliability function $R_1(t)$, then the function $R(t)$ is given by

$$R(t) = P[N \geq k], \quad \text{where } N \sim B(n, p = R_1(t)).$$

That is,

$$R(t) = \sum_{i=k}^n \binom{n}{i} [R_1(t)]^i [1 - R_1(t)]^{n-i} = 1 - \sum_{i=0}^{k-1} \binom{n}{i} [R_1(t)]^i [1 - R_1(t)]^{n-i}.$$

Remark. We assume that the subsystems operate independently of one another. In practice, the lifetimes of the working components often depend on the total number of active components. For example, suppose that an airplane has four engines, but that it can fly and land with only two of them. If two engines fail while the airplane is flying, more load will be put on the two remaining engines, so that their lifetimes are likely to be shorter.

Example 5.2.6. Consider a 2-out-of-3 system for which the lifetime of subsystem i follows an exponential distribution with parameter $\lambda_i = i$, for $i = 1, 2, 3$. To obtain the reliability function of the system, we use the following formula: if A_1, A_2, and A_3 are independent events, then

$$P[(A_1 \cap A_2) \cup (A_1 \cap A_3) \cup (A_2 \cap A_3)]$$

$$= P[A_1 \cap A_2] + P[A_1 \cap A_3] + P[A_2 \cap A_3] - 3P[A_1 \cap A_2 \cap A_3]$$
$$+ P[A_1 \cap A_2 \cap A_3]$$
$$\overset{\text{ind.}}{=} P[A_1]P[A_2] + P[A_1]P[A_3] + P[A_2]P[A_3] - 2P[A_1]P[A_2]P[A_3].$$

Let $A_i = \{T_i > t\}$, so that

$$P[A_i] = P[T_i > t] = e^{-it} \quad \text{for } i = 1, 2, 3.$$

Then, we may write that

$$R(t) = e^{-3t} + e^{-4t} + e^{-5t} - 2e^{-6t} \quad \text{for } t \geq 0.$$

Remark. We can also write that

$$R(t) = P[A_1 \cap A_2 \cap A_3'] + P[A_1 \cap A_2' \cap A_3] + P[A_1' \cap A_2 \cap A_3]$$
$$+ P[A_1 \cap A_2 \cap A_3]$$
$$\overset{\text{ind.}}{=} P[A_1]P[A_2](1 - P[A_3]) + P[A_1](1 - P[A_2])P[A_3]$$
$$+ (1 - P[A_1])P[A_2]P[A_3] + P[A_1]P[A_2]P[A_3]$$
$$= P[A_1]P[A_2] + P[A_1]P[A_3] + P[A_2]P[A_3] - 2P[A_1]P[A_2]P[A_3],$$

as above. That is, we decompose the event $\{T > t\}$ into four incompatible cases: *exactly two subsystems operate at time t, or* the three subsystems operate.

Next, the system shown in Figure 5.2 is called a *bridge system*. It operates at time t if and only if at least one of the following events occurs:

- $A_1 =$ "components nos. 1 and 4 are active at time t;"
- $A_2 =$ "components nos. 2 and 5 are active at time t;"
- $A_3 =$ "components nos. 1, 3, and 5 are active at time t;"
- $A_4 =$ "components nos. 2, 3, and 4 are active at time t."

Because the events $A_1, \ldots, A_4$ are neither independent nor incompatible, we need the formula for the probability of the union of four arbitrary events:

$$P[A_1 \cup A_2 \cup A_3 \cup A_4] = P[A_1] + P[A_2] + P[A_3] + P[A_4] - P[A_1 \cap A_2]$$
$$- P[A_1 \cap A_3] - P[A_1 \cap A_4] - P[A_2 \cap A_3] - P[A_2 \cap A_4] - P[A_3 \cap A_4]$$
$$+ P[A_1 \cap A_2 \cap A_3] + P[A_1 \cap A_2 \cap A_4] + P[A_1 \cap A_3 \cap A_4]$$
$$+ P[A_2 \cap A_3 \cap A_4] - P[A_1 \cap A_2 \cap A_3 \cap A_4]. \tag{5.1}$$

In the special case when the five components in the bridge system operate independently and all have the same reliability function $R_1(t)$, we can easily calculate the reliability function of the system.

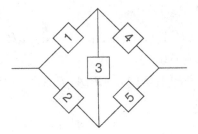

Fig. 5.2. A bridge system.

Finally, as we did in Chapter 2, we can consider systems made up of a number of subsystems connected in series and others connected in parallel.

Example 5.2.7. A system is constituted of two subsystems placed in series. The first subsystem comprises two components connected in parallel, and the second subsystem contains a single component. Suppose that the three components operate independently. Let $R_i(t)$ be the reliability function of component i, for $i = 1, 2, 3$. Then, the reliability function of the system is given by

$$R(t) = \{1 - [1 - R_1(t)][1 - R_2(t)]\} R_3(t) = [R_1(t) + R_2(t) - R_1(t)R_2(t)] R_3(t).$$

That is, we make use of the formulas for both series and parallel systems at the same time.

5.3 Paths and cuts

When a system consists of a large (enough) number of components, a first task in reliability theory is to find the various sets of active components that will enable the system to operate. These sets are called *paths*. Conversely, we can try to determine the sets of components, called *cuts*, which, when the components they comprise all are down, entail the failure of the whole system.

Let X_i be the Bernoulli random variable that represents the state of component i at a fixed time instant $t_0 \geq 0$. More precisely, $X_i = 1$ (resp., $X_i = 0$) if component i is active (resp., down) at time t_0, for $i = 1, \ldots, n$. The random variable X_i is actually the *indicator variable* of the event "component i is active at time t_0." To calculate the reliability of the system at time t_0, we need the value of the probability $p_i :=$ $P[X_i]$, for $i = 1, \ldots, n$. However, to determine the paths and cuts of a system, we only have to consider the particular values x_i taken by the corresponding random variables. Furthermore, we assume that the fact that the system operates or not at time t_0 depends only on the state of its components at that time instant.

Definition 5.3.1. *The function*

$$H(x_1, \ldots, x_n) = \begin{cases} 1 \ \textit{if the system operates,} \\ 0 \ \textit{if the system is down} \end{cases}$$

is called the **structure function** *of the system. The vector* $\mathbf{x} := (x_1, \ldots, x_n)$ *is the* **state vector** *of the system.*

Remarks. (i) The function $H(\mathbf{X}) = H(X_1, \ldots, X_n)$ is itself a Bernoulli random variable.
(ii) Let $\mathbf{0} = (0, \ldots, 0)$ and $\mathbf{1} = (1, \ldots, 1)$. We assume that $H(\mathbf{0}) = 0$ and $H(\mathbf{1}) = 1$. That is, if all the components of the system have failed, then the system is down and, conversely, if all the components are active, then the system is operating.

Notation. Consider two n-dimensional vectors: $\mathbf{x} = (x_1, \ldots, x_n)$ and $\mathbf{y} = (y_1, \ldots, y_n)$. We write:

$$\mathbf{x} \geq \mathbf{y} \quad \text{if } x_i \geq y_i \text{ for } i = 1, \ldots, n$$

and

$$\mathbf{x} > \mathbf{y} \quad \text{if } x_i \geq y_i \text{ for } i = 1, \ldots, n \text{ and } x_i > y_i \text{ for at least one } i.$$

Definition 5.3.2. *A structure function* $H(\mathbf{x})$ *such that* $H(\mathbf{0}) = 0$, $H(\mathbf{1}) = 1$, *and*

$$H(\mathbf{x}) \geq H(\mathbf{y}) \quad \textit{if } \mathbf{x} \geq \mathbf{y} \tag{5.2}$$

is said to be **monotonic.**

In this book, we assume that the structure functions $H(\mathbf{x})$ of the systems considered are monotonic.

Example 5.3.1. In the case of a series system (made up of n components), we have:

$$H(\mathbf{x}) = 1 \quad \Longleftrightarrow \quad x_i = 1 \ \forall i \quad \left(\Longleftrightarrow \quad \sum_{i=1}^{n} x_i = n \right),$$

whereas if the n components are connected in parallel, then

$$H(\mathbf{x}) = 1 \quad \Longleftrightarrow \quad \sum_{i=1}^{n} x_i \geq 1.$$

In general, for a k-out-of-n system, we can write that

$$H(x_1, \ldots, x_n) = \begin{cases} 1 \text{ if } \sum_{i=1}^{n} x_i \geq k, \\ 0 \text{ if } \sum_{i=1}^{n} x_i < k. \end{cases}$$

Remarks. (i) We can also express the structure function $H(\mathbf{x})$ as follows:

$$H(x_1, \ldots, x_n) = \begin{cases} \min\{x_1, \ldots, x_n\} \text{ for a series system,} \\ \max\{x_1, \ldots, x_n\} \text{ for a parallel system.} \end{cases}$$

(ii) In the present section, when we write that the components are connected in parallel, we assume that they all operate from the initial time. That is, they are in *active* redundancy.

To calculate the value of the structure function of an arbitrary system, the following formulas are useful:

$$\min\{x_1, \ldots, x_n\} = \prod_{i=1}^{n} x_i$$

and

$$\max\{x_1, \ldots, x_n\} = 1 - \prod_{i=1}^{n}(1 - x_i),$$

which are valid when $x_i = 0$ or 1, for $i = 1, \ldots, n$.

Definition 5.3.3. *A* **path vector** *is any vector* $\mathbf{x}$ *for which* $H(\mathbf{x}) = 1$. *If, besides,* $H(\mathbf{y}) = 0$ *for all vectors* $\mathbf{y}$ *such that* $\mathbf{y} < \mathbf{x}$, *then* $\mathbf{x}$ *is called a* **minimal path vector**. *Moreover, with every minimal path vector* $\mathbf{x} = (x_1, \ldots, x_n)$ *we associate a set* $MP := \{k \in \{1, \ldots, n\} : x_k = 1\}$ *called a* **minimal path set**.

Definition 5.3.4. *If* $H(\mathbf{x}) = 0$, *the state vector* $\mathbf{x}$ *is said to be a* **cut vector**. *If, in addition,* $H(\mathbf{y}) = 1$ *when* $\mathbf{y} > \mathbf{x}$, *then* $\mathbf{x}$ *is a* **minimal cut vector**. *Furthermore, the set* $MC := \{k \in \{1, \ldots, n\} : x_k = 0\}$, *where* $\mathbf{x} = (x_1, \ldots, x_n)$ *is a minimal cut vector, is called a* **minimal cut set**.

Remarks. (i) In some books, the definition of a *path* (resp., *cut*) (set) corresponds to that of a *minimal path set* (resp., *minimal cut set*) here.

(ii) A minimal path set is a group of components such that when they are all active the system operates, but if at least one of this group of components fails, then the system too fails. Conversely, if all the components in a minimal cut set are down, the system is down as well, but if at least one component of the minimal cut set is replaced by an active component, then the system will operate.

Example 5.3.2. A series system made up of n components has a single minimal path set, namely the set $MP = \{1, 2, \ldots, n\}$ (because all components must operate for the system to function). It has n minimal cut sets, which are all the sets containing exactly one component: $\{1\}$, $\ldots$, $\{n\}$. Note that when we write that $MC = \{1\}$, it implies that components $2, \ldots, n$ are active. Moreover, the state vector $(0, 0, 1, 1, \ldots, 1)$ is a cut vector, but not a *minimal* cut vector, because if we replace only component no. 1 by an active component, the system will remain down.

Conversely, in the case of a parallel system comprising n components, the minimal path sets are $\{1\}, \ldots, \{n\}$, whereas there is only one minimal cut set: $\{1, 2, \ldots, n\}$.

Example 5.3.3. We can generalize the results of the previous example as follows: in a k-out-of-n system, there are $\binom{n}{k}$ minimal path sets. That is, we can choose any set of k components among the n. The number of minimal cut sets is given by $\binom{n}{n-k+1}$. Indeed, if exactly $n - k + 1$ components are down, then the system will resume operating if one of them is replaced by an active component.

Example 5.3.4. The bridge system in Figure 5.2 has four minimal path sets, as indirectly mentioned above: $\{1, 4\}$, $\{2, 5\}$, $\{1, 3, 5\}$, and $\{2, 3, 4\}$. It also has four minimal cut sets: $\{1, 2\}$, $\{4, 5\}$, $\{1, 3, 5\}$, and $\{2, 3, 4\}$. Notice that $\{1, 3, 5\}$ and $\{2, 3, 4\}$ are both minimal path and minimal cut sets.

Now, suppose that an arbitrary system has r minimal path sets. Let

$$\pi_j(x_1, \ldots, x_n) = \prod_{i \in MP_j} x_i \quad \text{for } j = 1, \ldots, r.$$

That is, $\pi_j(x_1, \ldots, x_n) = 1$ if all the components in the minimal path set MP_j function, and $\pi_j(x_1, \ldots, x_n) = 0$ otherwise. Because a system operates if and only if all the components in at least one of its minimal path sets are active, we can represent the structure function of the system in question as follows:

$$H(\mathbf{x}) = 1 - \prod_{j=1}^{r} [1 - \pi_j(\mathbf{x})].$$

This formula implies that a given system can be considered as being equivalent to the one obtained by connecting its minimal path sets in parallel.

Likewise, if an arbitrary system has s minimal cut sets, we can write that

$$H(x_1, \ldots, x_n) = 1 - \prod_{m=1}^{s} \gamma_m(x_1, \ldots, x_n),$$

where

$$\gamma_m(x_1, \ldots, x_n) := 1 - \prod_{i \in MC_m} (1 - x_i) \quad \text{for } m = 1, \ldots, s.$$

We have that $\gamma_m(x_1, \ldots, x_n)$ is equal to 0 if all the components in the minimal cut set MC_m are down, and to 1 otherwise. This time, we can say that a given system and the one made up of its minimal cut sets connected in series are equivalent.

Example 5.3.5. From the preceding example, we deduce that the bridge system in Figure 5.2 is equivalent to either the system depicted in Figure 5.3 or that in Figure 5.4.

Because $H(\mathbf{X})$, where $\mathbf{X} := (X_1, \ldots, X_n)$, is a Bernoulli random variable, the reliability of the system, at the fixed time instant t_0, is given by

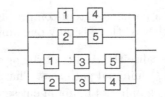

Fig. 5.3. A bridge system represented as a parallel system made up of its minimal path sets.

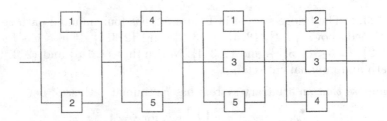

Fig. 5.4. A bridge system represented as a series system made up of its minimal cut sets.

$$R(t_0) = P[H(\mathbf{X}) = 1] = E[H(\mathbf{X})].$$

If we let

$$p_i = P[X_i = 1] \quad \text{(at time } t_0) \quad \text{for } i = 1, \ldots, n$$

and if we assume that the components operate independently of one another, then we can write that

$$R(t_0) = \begin{cases} \prod_{i=1}^{n} p_i & \text{for a series system,} \\ 1 - \prod_{i=1}^{n}(1 - p_i) & \text{for a parallel system.} \end{cases}$$

Moreover, if $p_i = p$ for all i, then

$$R(t_0) = \sum_{i=k}^{n} \binom{n}{i} p^i (1 - p)^{n-i}$$

in the case of a k-out-of-n system. These formulas are simply particular cases of the corresponding ones in Section 5.2.

When a system comprises many components, we can at least try to obtain bounds for its reliability $R(t_0)$ at time t_0. It can be shown that

$$\prod_{m=1}^{s} P[\gamma_m(\mathbf{X}) = 1] \leq R(t_0) \leq 1 - \prod_{j=1}^{r} \{1 - P[\pi_j(\mathbf{X}) = 1]\},$$

where

$$P[\gamma_m(\mathbf{X}) = 1] = 1 - \prod_{i \in MC_m} (1 - p_i)$$

and

$$P[\pi_j(\mathbf{X}) = 1] = \prod_{i \in MP_j} p_i.$$

Example 5.3.6. Suppose that $p_i \equiv 0.9$ for the bridge system in Figure 5.2. Making use of Equation (5.1), we find (assuming that the components are independent) that

$$R(t_0) = 2(0.9)^2 + 2(0.9)^3 - 5(0.9)^4 + 2(0.9)^5 = 0.97848.$$

Indeed, we have that $P[A_i] = (0.9)^2$, for $i = 1, 2$, and $P[A_i] = (0.9)^3$, for $i = 3, 4$. Moreover, the probability of any intersection in (5.1) is equal to $(0.9)^k$, where k is the number of distinct components involved in the intersection in question. For example, $A_1 \cap A_2$ occurs if and only if components 1, 2, 4, and 5 are active, so that $P[A_1 \cap A_2] = (0.9)^4$.

Now, we deduce from Example 5.3.4 that the lower bound for the probability $R(t_0)$ is given by

$$\left[1 - (0.1)^2\right]^2 \left[1 - (0.1)^3\right]^2 \simeq 0.97814$$

and the upper bound is

$$1 - \left\{\left[1 - (0.9)^2\right]^2 \left[1 - (0.9)^3\right]^2\right\} \simeq 0.99735.$$

Notice that in this example the lower bound, in particular, is very precise.

5.4 Exercises for Chapter 5

Solved exercises

Question no. 1

Suppose that the lifetime X of a certain system can be expressed as $X = Y^2$, where Y is a random variable uniformly distributed on the interval $(0, 1)$. Find the reliability function of the system.

Question no. 2

The time to failure (in years) for a given device is a random variable X having an exponential distribution with parameter $\lambda = 1/2$. When the device fails, it is repaired. The repair time Y (in days) is a random variable such that

$$P[Y > y] = \begin{cases} e^{-y} & \text{if } X < 2 \text{ and } y \geq 0, \\ e^{-y/2} & \text{if } X \geq 2 \text{ and } y \geq 0. \end{cases}$$

Calculate the quantity $MTBF$ for this device.

Question no. 3

Let T be a continuous random variable having the following probability density function:

$$f_T(t) = \frac{1}{\lambda} \exp \left\{ -\frac{e^t - 1}{\lambda} + t \right\} \quad \text{if } t \geq 0$$

[and $f_T(t) = 0$ if $t < 0$], where λ is a positive parameter. We say that T has a (particular) *extreme value distribution*. Calculate the failure rate function of a device whose lifetime is distributed as T.

Question no. 4

Suppose that the lifetime (in cycles) of a system is a Poisson random variable with parameter $\lambda > 0$. Is the failure rate function $r(k)$ increasing or decreasing at $k = 0$?

Question no. 5

If a system has a lifetime X that is uniformly distributed on the interval $[0, 1]$, what is its average failure rate over the interval $[0, 1/2]$?

Question no. 6

A system comprises two (independent) components connected in series. The lifetime X_k of component k has an exponential distribution with parameter λ_k, for $k = 1, 2$. Use the formula [see (4.5)]

$$P[X_2 < X_1] = \int_{-\infty}^{\infty} P[X_2 < x_1 \mid X_1 = x_1] f_{X_1}(x_1) dx_1,$$

which is valid for arbitrary continuous random variables X_1 and X_2, to calculate the probability that the first breakdown of the system will be caused by a failure of component no. 2.

Question no. 7

We consider a system made up of two components connected in parallel and operating independently of each other. Let T be the total lifetime of the system, and let T_k be the lifetime of component k, for $k = 1, 2$. Suppose that $T_k \sim \text{Exp}(\lambda_k)$. What is the probability that both components are still active at time $t_0 > 0$, given that the system operates at that time instant?

Question no. 8

We have two identical brand A components and two identical brand B components at our disposal. To build a certain device, we must connect a brand A and a brand B component in series. Suppose that the reliability of each component is equal to 0.9 (at the initial time) and that they all operate independently of one another. Is it better to build two distinct devices and hope that at least one of them will work, or to build a device made up of two subsystems connected in series, with the first (resp., second) subsystem comprising the two brand A (resp., B) components placed in parallel?

Question no. 9

Calculate the structure function of the system represented in Figure 2.14, p. 46, in terms of the indicator variables x_k, for $k = 1, \ldots, 4$.

Question no. 10

Find the minimal path sets of the system represented in Figure 2.14, p. 46, and express the structure function $H(x_1, x_2, x_3, x_4)$ in terms of the functions $\pi_j(x_1, x_2, x_3, x_4)$.

Exercises

Question no. 1

We want to build a system made up of two components connected in parallel, followed by a component connected in series. Suppose that we have three components at our disposal and that any arrangement of the components inside the system is admissible. If the reliability of component no. k, at a fixed time instant $t_0 > 0$, is equal to p_k, for $k = 1, 2, 3$, and if $0 < p_1 < p_2 < p_3 < 1$, what arrangement of the three components gives the largest probability that the system is active at time t_0?

Question no. 2

The lifetime X of a certain machine has the following probability density function:

$$f_X(x) = \begin{cases} 1/x & \text{if } 1 \leq x \leq e, \\ 0 & \text{elsewhere.} \end{cases}$$

Calculate the failure rate function $r(x)$, for $1 \leq x \leq e$. Is the distribution of X an IFR or DFR distribution? Justify.

Question no. 3

Suppose that the failure rate function $r(t)$ of a given system is $r(t) = 1 \; \forall t \geq 0$. Find the probability that the system will fail in the interval $[2, 3]$, given that it still operates at time $t = 1$.

Question no. 4

A device has a lifetime T that follows an exponential distribution with parameter Λ, where Λ is a random variable uniformly distributed over the interval $[1, 3]$. Calculate the reliability function of the device.

Question no. 5

Let

$$f_T(t) = \begin{cases} \frac{1}{2}t^2 e^{-t} & \text{if } t > 0, \\ 0 & \text{elsewhere,} \end{cases}$$

where T denotes the lifetime of a certain system. Calculate the interval failure rate of this system in the interval $(0, 1]$.

Question no. 6

Suppose that the lifetime T of a system has a $G(4,1)$ distribution. That is,

$$f_T(t) = \frac{1}{6}t^3 e^{-t} \quad \text{for } t \geq 0.$$

Calculate the reliability function of the system in question at time $t = 4$, given that it is still active at time $t = 2$.

Question no. 7

Calculate the failure rate function at time $t = 1$ of a system whose lifetime T is distributed as $|Z|$, where $Z \sim N(0,1)$.

Question no. 8

Let

$$p_X(k) = \frac{1}{N} \quad \text{for } k = 1, 2, \ldots, N$$

be the probability mass function of the lifetime X (in cycles) of a particular system. Calculate $r_X(k)$, for $k = 1, \ldots, N$. Does X have an IFR or a DFR distribution? Justify.

Question no. 9

Consider the interval failure rate of a given system in the interval $(n, n+1]$, for $n = 0, 1, \ldots, 9$. Calculate the average interval failure rate of this system if $T \sim U[0, 10]$.

Question no. 10

Assume that T has a Weibull distribution with parameters $\lambda > 0$ and $\beta > 0$. What is the average failure rate in the interval $[0, \tau]$ if τ is a random variable distributed as the square root of a $U(0,1)$ distribution and (a) $\beta = 2$? (b) $\beta = 3$?

Question no. 11

Three independent components are connected in parallel. Suppose that the lifetime of component no. k has an exponential distribution with parameter λ_k, for $k = 1, 2, 3$. What is the probability that component no. 3 will not be the first one to fail?

Question no. 12

A system comprises two components that operate independently of each other, both from the initial time. Suppose that the lifetime X_k (in cycles) of component no. k has a geometric distribution with parameter $p_k = 1/2$, for $k = 1, 2$. Find the probability that both components will fail during the same cycle.

Question no. 13

Three independent components are connected in series. Suppose that the lifetime T_k of the kth component has a uniform distribution on the interval $(0, k+1)$, for $k = 1, 2, 3$. What is the value of the reliability function of the system (made up of these three components) at time $t = 1$, given that at least one of the three components is not down at time $t = 1$?

Question no. 14

Consider the system represented in Figure 2.13, p. 45. Suppose that the components A have a lifetime that follows an exponential distribution with parameter λ_A, and the lifetime of component B (resp., C) is exponentially distributed with parameter λ_B (resp., λ_C). Calculate the reliability function of the system, assuming that the components operate independently of one another.

Question no. 15

A system consists of two components operating independently of each other and connected in parallel. Assume that the lifetime T_k of the kth component is exponentially distributed with parameter k, for $k = 1, 2$. What is the probability that the system is still operating at time $t = 2$, given that exactly one of its components is down at time $t = 1$?

Question no. 16

We have four independent components having an exponentially distributed lifetime at our disposal. The expected lifetime of the kth component is equal to $1/k$, for $k = 1, 2, 3, 4$. The components are used to build a system made up of two subsystems connected in series. Each subsystem comprises two components connected in parallel. What is the expected lifetime of the system if the first subsystem comprises the first and the second component (see Figure 5.5)?

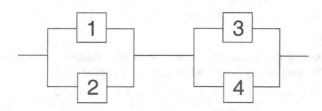

Fig. 5.5. Figure for Exercise no. 16.

Question no. 17

Let the components of the system in Figure 2.13, p. 45, be numbered as follows: component $C = 1$, components $A = 2$, 3, and 4, and component $B = 5$. Find the minimal path sets and minimal cut sets of this system.

Question no. 18

Suppose that the reliability of the (independent) components in the preceding question is as in (unsolved) Exercise no. 8 of Chapter 2 (at a fixed time instant $t_0 > 0$). Use the minimal path sets and minimal cut sets found in the previous question to calculate the lower and upper bounds for the reliability of the system. Compare with the exact answer.

Question no. 19

A certain system made up of four independent components operates if and only if component no. 1 is active and at least two of the other three components operate. What are the minimal path sets and the minimal cut sets of this system?

Question no. 20

In the preceding question, (a) what is the probability that the system operates at time t_0 if the reliability of each component is equal to p at t_0? (b) What is the reliability function of the system at time t if the lifetime of each component is an $\text{Exp}(\theta)$ random variable?

Multiple choice questions

Question no. 1

Suppose that $X \sim \text{Exp}(1)$ and $Y \sim U(0,1)$ are independent random variables. We define $Z = \min\{X,Y\}$. Find the failure rate function $r_Z(t)$ of Z for $0 < t < 1$.

(a) 1   (b) 2   (c) $\dfrac{1}{1-t}$ (d) $\dfrac{2-t}{1-t}$ (e) $\dfrac{2}{1-t}$

Question no. 2

The lifetime T of a device has a lognormal distribution with parameters $\mu = 10$ and $\sigma^2 = 4$. That is, $\ln T \sim N(10,4)$. Find the reliability of the device at time $t = 200$.

(a) 0.01   (b) 0.05   (c) 0.5   (d) 0.95   (e) 0.99

Question no. 3

For what values of the parameter α is a beta distribution with parameters α (> 0) and $\beta = 1$ an IFR distribution everywhere in the interval $(0,1)$?

(a) any $\alpha > 0$ (b) none   (c) $\alpha \geq 1$ (d) $\alpha < 1$ (e) $\alpha \geq 2$

Question no. 4

Suppose that the lifetime T of a system is uniformly distributed on the interval $(0, B)$, where B is a random variable having an exponential distribution with parameter $\lambda > 0$. Find the reliability function of the system.

(a) $\dfrac{1}{\lambda} - t$ (b) $\lambda - t$ (c) $1 - \dfrac{t}{\lambda}$ (d) $1 - \lambda t$ (e) $\lambda(1-t)$

Question no. 5

Calculate the average failure rate over the interval $[0,2]$ of a system whose lifetime T has the following probability density function:

$$f_T(t) = \begin{cases} \lambda e^{-\lambda(t-1)} & \text{if } t \geq 1, \\ 0 & \text{if } t < 1, \end{cases}$$

where λ is a positive parameter.

(a) $\dfrac{\lambda - 1}{2}$ (b) $\dfrac{\lambda}{2}$ (c) $\lambda - 1$ (d) λ (e) 2λ

Question no. 6

Two independent components are connected in series. The lifetime T_k of component no. k has an exponential distribution with parameter λ_k, for $k = 1, 2$. When a component fails, it is repaired. Suppose that the lifetime T_k^* of a repaired component is twice shorter, on average, than that of a new component, so that $T_k^* \sim \mathrm{Exp}(2\lambda_k)$, for $k = 1, 2$. Find the probability that component no. 1 will cause the first two failures of the series system.

(a) $\dfrac{\lambda_1^2}{(\lambda_1 + \lambda_2)^2}$ (b) $\dfrac{\lambda_1^2}{(\lambda_1 + 2\lambda_2)^2}$ (c) $\dfrac{\lambda_1^2}{(\lambda_1 + 2\lambda_2)(\lambda_1 + \lambda_2)}$

(d) $\dfrac{\lambda_1^2}{(2\lambda_1 + \lambda_2)(\lambda_1 + \lambda_2)}$ (e) $\dfrac{\lambda_1^2}{(2\lambda_1 + \lambda_2)^2}$

Question no. 7

We consider a system made up of two independent components placed in standby redundancy. The lifetime T_1 of component no. 1 has a uniform distribution on the interval $(0, 2)$, whereas the lifetime T_2 of component no. 2 is exponentially distributed with parameter $\lambda = 2$. Moreover, suppose that component no. 2 relieves the first one as soon as it fails if $T_1 < 1$, or at time $t = 1$ if component no. 1 is still active at that time instant. What is the average lifetime of the system?

(a) 9/4   (b) 5/2   (c) 11/4   (d) 3   (e) 13/4

Question no. 8

A certain 15-out-of-20 system is such that all the (independent) components have a probability equal to 3/4 of being active at time $t_0 > 0$.
(i) Calculate the probability p that the system operates at $t = t_0$.
(ii) Use a Poisson approximation to calculate the probability p in (i).

(a) (i) 0.4148; (ii) 0.4405   (b) (i) 0.6172; (ii) 0.5343
(c) (i) 0.6172; (ii) 0.6160   (d) (i) 0.7858; (ii) 0.6380
(e) (i) 0.7858; (ii) 0.7622

Question no. 9

Consider the system in Figure 5.6. How many (i) minimal path sets and (ii) minimal cut sets are there?

(a) (i) 4; (ii) 4   (b) (i) 5; (ii) 4   (c) (i) 5; (ii) 5   (d) (i) 6; (ii) 4
(e) (i) 6; (ii) 5

Question no. 10

Suppose that, in the preceding question, the components operate independently of one another and all have a 3/4 probability of being active at time $t_0 > 0$. Give a lower bound for the reliability of the system at time t_0.

(a) 0.1279   (b) 0.3164   (c) 0.6836   (d) 0.8721   (e) 0.8789

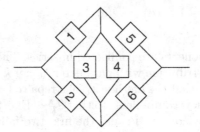

Fig. 5.6. Figure for multiple choice question no. 9.

6

Queueing

An important application of probability theory is the field known as *queueing theory*.
This field studies the behavior of waiting lines or queues. Telecommunication engineers
and computer scientists are particularly interested in queueing theory to solve problems
concerned with the efficient allocation and use of resources in wireless and computer
networks, for instance. In general, the models considered in this chapter are such that
the *arrivals* in the *queueing system* and the *departures* from this system both constitute
Poisson processes, which were defined in Chapter 3 (p. 69). Poisson processes are actually
particular *continuous-time Markov chains*, which are the subject of the first section of
the present chapter. Next, the case when there is a single *server* in the *queueing system*
and that when the number of servers is greater than or equal to two is studied separately.

6.1 Continuous-time Markov chains

Definition 6.1.1. *A **stochastic process** (or **random process**) is a set $\{X(t), t \in T\}$
of random variables $X(t)$, where T is a subset of $\mathbb{R}$.*

Remarks. (i) The *deterministic* variable t is often interpreted as *time* in the problems
considered. In this chapter, we are interested in *continuous-time* stochastic processes,
so that the set T is generally the interval $[0, \infty)$.

(ii) The set of all possible values of the random variables $X(t)$ is called the *state space*
of the stochastic process.

A very important class of stochastic processes for the applications is the class of
those known as *Markov processes*.

M. Lefebvre, *Basic Probability Theory with Applications*, Springer Undergraduate Texts in Mathematics
and Technology, DOI: 10.1007/978-0-387-74995-2_6,
© Springer Science + Business Media, LLC 2009

Definition 6.1.2. *If we can write that*

$$P[X(t_n) \leq x_n \mid X(t), \forall t \leq t_{n-1}] = P[X(t_n) \leq x_n \mid X(t_{n-1})], \qquad (6.1)$$

*where $t_{n-1} < t_n$, we say that the stochastic process $\{X(t), t \in T\}$ is a **Markov process** (or **Markovian process**).*

Remark. The preceding equation, known as the *Markov property*, means that the *future* of the process depends only on its *present state*. That is, assuming that the set T is the interval $[0, \infty)$, the *history* of the process in the interval $[0, t_{n-1})$ is not needed to calculate the distribution of the random variable $X(t_n)$, where $t_n > t_{n-1}$, *if* the value of $X(t_{n-1})$ is known.

Definition 6.1.3. *If the possible values taken by the various random variables $X(t)$ are assumed to be at most countably infinite, so that $X(t)$ is a discrete random variable for any fixed value of the variable t, then we say that $\{X(t), t \in T\}$ is a **discrete-state** stochastic process.*

Now, let τ_i be the time that the continuous-time and discrete-state Markovian process $\{X(t), t \geq 0\}$ spends in a given state i before making a transition to any other state. We deduce from the Markov property that

$$P[\tau_i > s + t \mid \tau_i > s] = P[\tau_i > t] \quad \forall s, t \geq 0 \qquad (6.2)$$

(otherwise the *future* would depend on the *past*). This equation implies that the continuous random variable τ_i is exponentially distributed. Indeed, only the exponential distribution possesses this *memoryless property* (see p. 76) in continuous time.

Remarks. (i) We denote the parameter of the random variable τ_i by ν_i, for any i. In the general case, ν_i depends on the corresponding state i. However, in the case of the Poisson process with rate λ, we have that $\nu_i = \lambda$ for all i.

(ii) We also deduce from the Markov property that the state that will be visited when the process leaves its current state i must be *independent* of the total time τ_i that the process spent in i before making a transition.

Definition 6.1.4. *The continuous-time and discrete-state stochastic process $\{X(t), t \geq 0\}$ is called a **continuous-time Markov chain** if*

$$P[X(t) = j \mid X(s) = i, X(r) = x_r, 0 \leq r < s] = P[X(t) = j \mid X(s) = i]$$

for all $t \geq s$ and for all states i, j, x_r.

Remarks. (i) We assume that the Markov chains considered have *time-homogeneous* transition probabilities. That is, if $t \geq s \geq 0$ and $\tau \geq 0$, we may write that

$$P[X(t) = j \mid X(s) = i] := p_{i,j}(t - s) \Leftrightarrow P[X(\tau + t) = j \mid X(\tau) = i] = p_{i,j}(t).$$

That is, the probability that the process moves from state i to state j in a given time interval depends only on the length of this time interval. This assumption is made in most textbooks and is realistic in many applications. The function $p_{i,j}(t)$ is known as the *transition function* of the continuous-time Markov chain.

(ii) If $p_{i,j}(t) > 0$ for some $t \geq 0$ and $p_{j,i}(t^*) > 0$ for some $t^* \geq 0$, we say that states i and j *communicate*. If all states communicate, the chain is said to be *irreducible*.

(iii) In the context of queueing theory, the $X(t)$s are nonnegative integer-valued random variables. That is, the state space of the stochastic process $\{X(t), t \geq 0\}$ is the set $\{0, 1, \ldots\}$. Under this assumption, we can write that

$$\sum_{j=0}^{\infty} p_{i,j}(t) = 1 \quad \forall i \in \{0, 1, \ldots\}.$$

Indeed, whatever the state of the process at a fixed time $\tau \geq 0$ is, it must be in some state at time $\tau + t$, where $t \geq 0$. Note that we have:

$$p_{i,j}(0) = \delta_{i,j} := \begin{cases} 1 \text{ if } i = j, \\ 0 \text{ if } i \neq j \end{cases}$$

for all states $i, j \in \{0, 1, \ldots\}$.

Notation. We denote by $\rho_{i,j}$ the probability that the continuous-time Markov chain $\{X(t), t \geq 0\}$, when it leaves its current state i, goes to state j, for $i, j \in \{0, 1, \ldots\}$.

We have, by definition, $\rho_{i,i} = 0$ for all states i and

$$\sum_{j=0}^{\infty} \rho_{i,j} = 1 \quad \forall i \in \{0, 1, \ldots\}.$$

Definition 6.1.5. *Let* $\{X(t), t \geq 0\}$ *be a continuous-time Markov chain with state space* $\{0, 1, 2, \ldots\}$. *If*

$$\rho_{i,j} = 0 \quad \text{when } |j - i| > 1 \tag{6.3}$$

the process is called a **birth and death process**. *Moreover, if*

$$\rho_{i,i+1} = 1 \quad \text{for all } i,$$

then $\{X(t), t \geq 0\}$ *is a* **pure birth process**, *whereas in the case when*

$$\rho_{i,i-1} = 1 \quad \text{for all } i \in \{1, 2, \ldots\},$$

we say that $\{X(t), t \geq 0\}$ *is a* **pure death process**.

We deduce from the definition that a birth and death process is such that

$$\rho_{0,1} = 1 \quad \text{and} \quad \rho_{i,i+1} + \rho_{i,i-1} = 1 \quad \text{for } i \in \{1, 2, \ldots\}.$$

That is, when the process is in state $i \geq 1$, the next state visited will necessarily be $i+1$ or $i-1$.

Remark. The state space of the birth and death process can be a finite set $\{0, 1, 2, \ldots, c\}$. Then, we have that $\rho_{c,c-1} = 1$.

In queueing theory, the state of the process at a fixed time instant will generally be the number of individuals in the queueing system at that time. When $\{X(t), t \geq 0\}$ goes from state i to $i + 1$, we say that an *arrival* occurred, and if it moves from i to $i - 1$, then a *departure* took place. We assume that, when the chain is in state i, the time A_i needed for a new arrival to occur is a random variable having an $\text{Exp}(\lambda_i)$ distribution, for $i \in \{0, 1, \ldots\}$. Furthermore, A_i is assumed to be *independent* of the random time $D_i \sim \text{Exp}(\mu_i)$ until the next departure, for $i \in \{1, 2, \ldots\}$.

Proposition 6.1.1. *The total time τ_i that the birth and death process $\{X(t), t \geq 0\}$ spends in state i, on a given visit to that state, is an exponentially distributed random variable with parameter*

$$\nu_i = \begin{cases} \lambda_0 & \text{if } i = 0, \\ \lambda_i + \mu_i & \text{if } i = 1, 2, \ldots. \end{cases}$$

Proof. When $i = 0$, we simply have that $\tau_0 \equiv A_0$, so that $\tau_0 \sim \text{Exp}(\lambda_0)$. For $i = 1, 2, \ldots$, we can write that $\tau_i = \min\{A_i, D_i\}$. The result then follows from Proposition 5.2.1. ∎

Remark. We also have [see (4.6)] that

$$\rho_{i,i+1} = P[A_i < D_i] = \frac{\lambda_i}{\lambda_i + \mu_i} \quad \text{if } i > 0$$

and

$$\rho_{i,i-1} = P[D_i < A_i] = \frac{\mu_i}{\lambda_i + \mu_i} \quad \text{if } i > 0.$$

Definition 6.1.6. *The parameters λ_i, for $i = 0, 1, \ldots$, are called the* **birth** *(or* **arrival***) rates of the birth and death process $\{X(t), t \geq 0\}$, whereas the parameters μ_i, for $i = 1, 2, \ldots$, are the* **death** *(or* **departure***) rates of the process.*

Example 6.1.1. In addition to being a continuous-time Markov chain, the Poisson process $\{N(t), t \geq 0\}$ is a particular *counting process*. That is, $N(t)$ denotes the total number of events in the interval $[0, t]$. Because only the number of events is recorded, and not whether these events were arrivals or departures, $\{N(t), t \geq 0\}$ is an example of a pure birth process. It follows that $p_{i,j}(t) = 0$ if $j < i$. Furthermore, using the fact that the increments of the Poisson process are stationary, we can write that

$$p_{i,j}(t) \equiv P[N(\tau + t) = j \mid N(\tau) = i] = P[N(t) = j - i]$$

$$= P[\text{Poi}(\lambda t) = j - i] = e^{-\lambda t}\frac{(\lambda t)^{j-i}}{(j-i)!} \quad \text{for } j \geq i \geq 0.$$

The time τ_i that $\{N(t), t \geq 0\}$ spends in any state $i \in \{0, 1, \ldots\}$ follows an exponential distribution with parameter λ. Indeed, we have:

$$P[\tau_0 > t] = P[N(t) = 0] = e^{-\lambda t} \quad \text{for } t \geq 0,$$

which implies that $\tau_0 \sim \text{Exp}(\lambda)$. Next, because the Poisson process has independent and stationary increments, we can then assert that $\tau_i \sim \text{Exp}(\lambda)$, for $i = 1, 2, \ldots$, as well.

In general, it is very difficult to calculate explicitly the transition function $p_{i,j}(t)$. Therefore, we have to express the quantities of interest, such as the average number of customers in a given queueing system, in terms of the *limiting probabilities* of the stochastic process $\{X(t), t \geq 0\}$.

Definition 6.1.7. *Let $\{X(t), t \geq 0\}$ be an irreducible continuous-time Markov chain. The quantity*

$$\pi_j := \lim_{t \to \infty} p_{i,j}(t) \quad \text{for all } j \in \{0, 1, \ldots\}$$

*is called the **limiting probability** that the process will be in state j when it is in* **equilibrium**.

Remarks. (i) We assume that the limiting probabilities π_j exist and are independent of the initial state i.

(ii) The π_js also represent the proportion of time that the continuous-time Markov chain spends in state j, over a long period of time.

It can be shown that the limiting probabilities π_j satisfy the following system of linear equations:

$$\pi_j \nu_j = \sum_{i \neq j} \pi_i \nu_i \rho_{i,j} \quad \forall j \in \{0, 1, \ldots\}. \tag{6.4}$$

To obtain the π_js, we can solve the preceding system, under the condition

$$\sum_{j=0}^{\infty} \pi_j = 1. \tag{6.5}$$

Remarks. (i) The various equations in (6.4) are known as the *balance equations* of the stochastic process $\{X(t), t \geq 0\}$, because we can interpret them as follows: the *departure rate* from state j must be equal to the *arrival rate* to j, for all j.

(ii) If $\{X(t), t \geq 0\}$ is a birth and death process with state space $\{0, 1, \ldots\}$, the balance equations are:

$$\text{state } j \; \underline{\text{departure rate from } j} = \underline{\text{arrival rate to } j}$$

$$
\begin{array}{cl}
0 & \lambda_0 \pi_0 = \mu_1 \pi_1 \\
1 & (\lambda_1 + \mu_1)\pi_1 = \mu_2 \pi_2 + \lambda_0 \pi_0 \\
\vdots & \qquad\qquad \vdots \; \vdots \; \vdots \\
k \, (\geq 1) & (\lambda_k + \mu_k)\pi_k = \mu_{k+1}\pi_{k+1} + \lambda_{k-1}\pi_{k-1}
\end{array}
$$

The basic models in queueing theory are particular birth and death processes. For this class of processes, we can give the general solution of the balance equations.

Theorem 6.1.1. *If $\{X(t), t \geq 0\}$ is an irreducible birth and death process with state space $\{0, 1, \ldots\}$, then the limiting probabilities are given by*

$$
\pi_j =
\begin{cases}
\dfrac{1}{1 + \sum_{k=1}^{\infty} \Pi_k} & \text{for } j = 0, \\[2mm]
\Pi_j \pi_0 & \text{for } j = 1, 2, \ldots,
\end{cases}
\tag{6.6}
$$

where

$$
\Pi_k := \frac{\lambda_0 \lambda_1 \cdots \lambda_{k-1}}{\mu_1 \mu_2 \cdots \mu_k} \quad \text{for } k \geq 1.
$$

Remark. The limiting probabilities exist if and only if the sum $\sum_{k=1}^{\infty} \Pi_k$ converges. In the case when the state space of $\{X(t), t \geq 0\}$ is *finite*, the sum in question always converges, so that the existence of the limiting probabilities is guaranteed.

Example 6.1.2. Suppose that the birth and death rates of the birth and death process $\{X(t), t \geq 0\}$ with state space $\{0, 1, 2\}$ are given by

$$
\lambda_0 = \lambda_1 = \lambda \quad \text{and} \quad \mu_1 = \mu, \quad \mu_2 = 2\mu.
$$

Write the balance equations of the system and solve them to obtain the limiting probabilities.

Solution. We have:

$$\text{state } j \; \underline{\text{departure rate from } j} = \underline{\text{arrival rate to } j}$$

$$
\begin{array}{cl}
0 & \lambda \pi_0 = \mu \pi_1 \\
1 & (\lambda + \mu)\pi_1 = 2\mu \pi_2 + \lambda \pi_0 \\
2 & 2\mu \pi_2 = \lambda \pi_1
\end{array}
$$

Because this system of equations is simple, we can solve it easily. We deduce from the equation for state 0 that

$$\pi_1 = \frac{\lambda}{\mu}\pi_0.$$

Similarly, the equation for state 2 implies that

$$\pi_2 = \frac{\lambda}{2\mu}\pi_1 = \left(\frac{\lambda}{2\mu}\right)\left(\frac{\lambda}{\mu}\right)\pi_0.$$

It follows that

$$\pi_0 + \frac{\lambda}{\mu}\pi_0 + \left(\frac{\lambda}{2\mu}\right)\left(\frac{\lambda}{\mu}\right)\pi_0 = 1.$$

That is,

$$\pi_0 = \left[1 + \frac{\lambda}{\mu} + \left(\frac{\lambda}{2\mu}\right)\left(\frac{\lambda}{\mu}\right)\right]^{-1},$$

so that

$$\pi_1 = \left(\frac{\lambda}{\mu}\right)\left[1 + \frac{\lambda}{\mu} + \left(\frac{\lambda}{2\mu}\right)\left(\frac{\lambda}{\mu}\right)\right]^{-1}$$

and

$$\pi_2 = \left(\frac{\lambda}{2\mu}\right)\left(\frac{\lambda}{\mu}\right)\left[1 + \frac{\lambda}{\mu} + \left(\frac{\lambda}{2\mu}\right)\left(\frac{\lambda}{\mu}\right)\right]^{-1}.$$

Remarks. (i) We can check that the equation for state 1, which we did not need to solve the system of linear equations, is also satisfied by the solution obtained above.

(ii) Because $\{X(t), t \geq 0\}$ is a particular birth and death process, we can also appeal to Theorem 6.1.1 to find the limiting probabilities. We have:

$$\Pi_1 := \frac{\lambda}{\mu} \quad \text{and} \quad \Pi_2 := \frac{\lambda \times \lambda}{\mu \times 2\mu},$$

from which we retrieve the formulas for π_0, π_1, and π_2.

6.2 Queueing systems with a single server

Let $X(t)$ designate the number of *customers* in a *queueing system* at time t. If we assume that the times A_n between the arrivals of successive customers and the service times S_n of customers are independent exponential random variables, then the process $\{X(t), t \geq 0\}$ is a continuous-time Markov chain. Moreover, in most cases, we also assume that the customers arrive one at a time and are served one at a time. It follows that $\{X(t), t \geq 0\}$ is a birth and death process. The arrivals of customers in the system constitute a Poisson process. It can be shown that the departures from the system *in equilibrium* constitute a Poisson process as well. Such a queueing system is denoted by $M/M/s$, where s is the number of *servers* in the system. In the present section, s is equal to 1.

Remarks. (i) We used the word *customers* above. However, customers in a queueing system may actually be machines in a repair shop, jobs in a computer system, or airplanes arriving or departing from an airport, among others.

(ii) To be precise, we should specify that the random variables S_n are independent of the A_ns. Furthermore, the S_ns are identically distributed random variables, and so are the A_ns.

(iii) The notation M for the arrival process (and the departure process) is used because the Poisson process is *Markovian*.

We are interested in the *average number* of customers and the *average time* that an arbitrary customer spends in the queueing system, when it is in *equilibrium* or in *stationary regime*.

Notations. We denote, respectively, by $\bar{N}$, $\bar{N}_Q$, and $\bar{N}_S$ the (total) average number of customers in the system in equilibrium, the average number of customers who are waiting in line, and the average number of customers being served. Moreover, $\bar{T}$ is the (total) average time that an arbitrary customer spends in the system, $\bar{Q}$ is the average waiting time of an arbitrary customer, and $\bar{S}$ is the average service time of an arbitrary customer.

We have that $\bar{N} = \bar{N}_Q + \bar{N}_S$ and $\bar{T} = \bar{Q} + \bar{S}$. As mentioned in the previous section, we express the various quantities of interest in terms of the limiting probabilities π_n of the stochastic process $\{X(t), t \geq 0\}$.

Definition 6.2.1. *Let $N(t)$, for $t \geq 0$, be the number of arrivals in the system in the interval $[0, t]$. The quantity*

$$\lambda_a := \lim_{t \to \infty} \frac{N(t)}{t} \tag{6.7}$$

is called the **average arrival rate** *of customers in the system.*

Remarks. (i) It can be shown that

$$\lim_{t \to \infty} \frac{N(t)}{t} = \frac{1}{E[A_n]}.$$

In our case, we assume that $A_n \sim \text{Exp}(\lambda)$, for $n = 0, 1, \ldots$, so that the stochastic process $\{N(t), t \geq 0\}$ is a Poisson process with rate $\lambda > 0$. It follows that $\lambda_a = \lambda$.

(ii) When the system capacity is infinite, all the arriving customers can enter the system. However, in practice, the capacity of any system is finite. Therefore, we also consider the *average entering rate* of customers into the system, which is denoted by λ_e. In the case when the system capacity is equal to a constant c $(< \infty)$, we have that $\lambda_e = \lambda(1 - \pi_c)$, because $(1 - \pi_c)$ is the (limiting) probability that an arriving customer will be allowed to enter the system. Note that, in fact, even if the system capacity is assumed to be infinite, some arriving customers may decide not to enter the system if they find that the queue length is too long, for instance. So, in general, λ_e is smaller than or equal to λ.

(iii) Let $D(t)$ denote the number of departures from the queueing system in the interval $[0, t]$. We assume that

$$\lambda_d := \lim_{t \to \infty} \frac{D(t)}{t} = \lambda_e.$$

To analyze a given queueing system, we often start by computing its limiting probabilities π_n. Next, we try to obtain the quantities $\bar{N}$, $\bar{N}_Q$, and so on, in terms of the π_ns. Furthermore, we can use a *cost equation* to establish a relation between $\bar{N}$ and $\bar{T}$. Indeed, if we assume that an arbitrary customer pays \$1 per time unit that she spends in the system (either waiting to be served or being served), then it can be shown that

$$\bar{N} = \lambda_e \cdot \bar{T}. \tag{6.8}$$

This equation is known as *Little's formula* (or Little's law). It is valid if we assume that both λ_e and $\bar{T}$ exist and are finite. Moreover, we have:

$$\bar{N} = \lim_{t \to \infty} \frac{1}{t} \int_0^t X(s)\,ds$$

and, if T_k denotes the time spent in the system by the kth customer,

$$\bar{T} = \lim_{t \to \infty} \frac{\sum_{k=1}^{N(t)} T_k}{N(t)}.$$

Remarks. (i) Little's formula holds for very general systems, in particular, for the $M/M/s$ systems, with finite or infinite capacity, that are studied in this book.

(ii) When t is large enough for the process to be in stationary regime, we may write that

$$\bar{N} = E[X(t)].$$

Similarly, we have:

$$\bar{N}_S = \lambda_e \cdot \bar{S}. \tag{6.9}$$

It follows, using the fact that $\bar{N} = \bar{N}_Q + \bar{N}_S$ and $\bar{T} = \bar{Q} + \bar{S}$, that

$$\bar{N}_Q = \lambda_e \cdot \bar{Q}.$$

In the case of the $M/M/s$ model, $\lambda_e = \lambda$ and the service times S_n are assumed to be i.i.d. exponentially distributed random variables with parameter μ. Hence, we deduce that $\bar{S} = E[S_n] = 1/\mu$. Equation (6.9) then implies that $\bar{N}_S = \lambda/\mu$.

6.2.1 The $M/M/1$ model

The most basic queueing system is the $M/M/1$ model. In this model, we assume that the successive arrivals of customers constitute a Poisson process with rate λ and that the service times S_n are independent $\text{Exp}(\mu)$ random variables. Furthermore, the S_ns are independent of the interarrival times of customers. Finally, the system capacity is infinite and we take for granted that all arriving customers decide to enter the system, whatever the state of the system upon their arrival is.

The stochastic process $\{X(t), t \geq 0\}$, where $X(t)$ denotes the number of customers in the system at time $t \geq 0$, is an irreducible birth and death process. Indeed, because the birth rates $\lambda_n \equiv \lambda$ and the death rates $\mu_n \equiv \mu$ are positive for any value of n, all states communicate. We find that the balance equations for the $M/M/1$ queue are (see p. 196):

state j	departure rate from j = arrival rate to j
0	$\lambda \pi_0 = \mu \pi_1$
$n \, (\geq 1)$	$(\lambda + \mu)\pi_n = \lambda \pi_{n-1} + \mu \pi_{n+1}$

We can solve the previous system of linear equations, under the condition $\sum_{n=0}^{\infty} \pi_n = 1$, to obtain the limiting probabilities. However, Theorem 6.1.1 gives us the solution almost at once. We calculate

$$\Pi_k = \underbrace{\frac{\lambda \lambda \cdots \lambda}{\mu \mu \cdots \mu}}_{k \text{ times}} = \left(\frac{\lambda}{\mu}\right)^k \quad \text{for } k = 1, 2, \ldots . \tag{6.10}$$

It follows that

$$S^* := \sum_{k=1}^{\infty} \Pi_k = \frac{\lambda/\mu}{1 - (\lambda/\mu)} < \infty \quad \text{if and only if} \quad \rho := \frac{\lambda}{\mu} < 1.$$

Remarks. (i) The quantity ρ is called the *traffic intensity* or the *utilization rate* of the system. Because $1/\mu$ is the average service time of an arbitrary customer and λ is the average arrival rate of customers, the condition $\rho < 1$ means that the customers must not arrive more rapidly than the rate at which they are served or, equivalently, more rapidly than the average time it takes to serve one customer, if we want the system to reach a *stationary* (or *steady-state*) *regime*. When $\rho \geq 1$, we can assert that the queue length will increase indefinitely.

(ii) In Chapter 2, we used Venn diagrams to represent sample spaces and events. In queueing theory, we draw a *state transition diagram* to describe a given system. The possible states of the system are depicted by circles. To indicate that a transition from state i to state j is possible, we draw an arrow from the circle corresponding to state i to the one representing j. We also write above (or under) each arrow the rate of the transition in question (see Figure 6.1). Once the appropriate state transition diagram has been drawn, it is a simple matter to write the balance equations of the system.

Next, we deduce from Theorem 6.1.1 that, if $\rho < 1$,

$$\pi_0 = \frac{1}{1 + S^*} = \left(\frac{1}{1 - (\lambda/\mu)}\right)^{-1} = 1 - \frac{\lambda}{\mu} = 1 - \rho$$

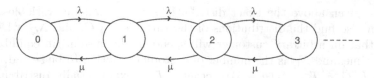

Fig. 6.1. State transition diagram for the $M/M/1$ model.

and

$$\pi_j = \Pi_j \pi_0 = \left(\frac{\lambda}{\mu}\right)^j \left(1 - \frac{\lambda}{\mu}\right) \quad \text{for } j = 1, 2, \dots .$$

That is,

$$\pi_k = \rho^k (1 - \rho) \quad \forall k \geq 0. \tag{6.11}$$

Making use of the limiting probabilities and Formula (1.7) with $a = 1 - \rho$ and $r = \rho$, we can write that

$$\bar{N} := \sum_{k=0}^{\infty} k \pi_k = \sum_{k=0}^{\infty} k \rho^k (1 - \rho) = \frac{(1 - \rho)\rho}{(1 - \rho)^2} = \frac{\rho}{1 - \rho}. \tag{6.12}$$

Remarks. (i) Note that $\lim_{\rho \uparrow 1} \bar{N} = \infty$, which reflects the fact that the queue length grows indefinitely if $\rho = 1$ (and, *a fortiori*, if $\rho > 1$).

(ii) If we let N denote the (random) number of customers in the system in equilibrium, so that $\bar{N} = E[N]$, we can write that $N_1 := N + 1$ has a geometric distribution with parameter $p := 1 - \rho$.

Now, we deduce from Little's formula (6.8) that

$$\bar{T} = \frac{\bar{N}}{\lambda} = \frac{\rho}{\lambda(1 - \rho)} = \frac{1}{\mu - \lambda}. \tag{6.13}$$

Because $\bar{S} = 1/\mu$ and $\bar{N}_S = \lambda/\mu = \rho$, as already mentioned above, it follows that

$$\bar{Q} = \bar{T} - \bar{S} = \frac{1}{\mu - \lambda} - \frac{1}{\mu} = \frac{\lambda}{\mu(\mu - \lambda)} \tag{6.14}$$

and

$$\bar{N}_Q = \bar{N} - \bar{N}_S = \frac{\rho}{1 - \rho} - \rho = \frac{\rho^2}{1 - \rho}. \tag{6.15}$$

Remark. Let N_S be the random variable that denotes the number of customers being served, when the system is in stationary regime. Because there is only one server, N_S has a Bernoulli distribution with parameter $p_0 := 1 - \pi_0$, which yields

$$\bar{N}_S = E[N_S] = p_0 = 1 - \pi_0 = 1 - (1 - \rho) = \rho,$$

as stated above.

We have given above the exact distributions of the random variables S, N, and N_S. We can also find the distributions of the variables T, Q, and N_Q, where T is the total time that an arbitrary customer will spend in the system (in equilibrium), Q is his waiting time, and N_Q is the number of customers waiting to be served. We already found that $E[T] = \bar{T} = 1/(\mu - \lambda)$. Actually, T is exponentially distributed and the quantity $\mu - \lambda$ is its parameter.

Proposition 6.2.1. *The total time T spent by an arbitrary customer in an M/M/1 queue in equilibrium is an exponentially distributed random variable with parameter $\mu - \lambda$.*

Proof. To prove the result, we condition on the number K of customers already in the system upon the arrival of the customer of interest. We can write that

$$P[T \le t] = \sum_{k=0}^{\infty} P[T \le t \mid K = k]P[K = k].$$

Now, by the *memoryless property* of the exponential distribution, if $K = k$, then the random variable T is the sum of $k + 1$ independent random variables, all having an $\mathrm{Exp}(\mu)$ distribution. Indeed, the service time of the customer being served (if $k > 0$), from the moment when the customer of interest enters the system, also has an exponential distribution with parameter μ. Using (4.9), we can write that

$$T \mid \{K = k\} \sim \mathrm{G}(k + 1, \mu).$$

Next, we can show that when the arrival process is a Poisson process, the probability $P[K = k]$ that an arbitrary customer finds k customers in the system in equilibrium upon his arrival is equal to the limiting probability π_k that there are k customers in the system (in equilibrium). It follows that

$$P[K = k] = \pi_k = \rho^k (1 - \rho) \quad \text{for } k = 0, 1, \dots.$$

Hence, we have:

$$
\begin{aligned}
P[T \le t] &= \sum_{k=0}^{\infty} \left[\int_0^t \mu e^{-\mu\tau} \frac{(\mu\tau)^k}{k!} d\tau \right] \rho^k (1 - \rho) \\
&\overset{\rho = \lambda/\mu}{=} (\mu - \lambda) \sum_{k=0}^{\infty} \int_0^t e^{-\mu\tau} \frac{(\lambda\tau)^k}{k!} d\tau = (\mu - \lambda) \int_0^t e^{-\mu\tau} \sum_{k=0}^{\infty} \frac{(\lambda\tau)^k}{k!} d\tau \\
&= (\mu - \lambda) \int_0^t e^{-\mu\tau} e^{\lambda\tau} d\tau = 1 - e^{-(\mu-\lambda)t},
\end{aligned}
$$

which implies that

$$f_T(t) = \frac{d}{dt} P[T \le t] = (\mu - \lambda) e^{-(\mu-\lambda)t} \quad \text{for } t \ge 0. \quad \blacksquare$$

The waiting time, Q, of an arbitrary customer entering the system is a random variable of *mixed* type. Using the fact that $P[K = k] = \pi_k$, where K is defined above, we have:

$$P[Q = 0] = P[K = 0] = \pi_0 = 1 - \rho.$$

In the case when $K = k > 0$, we can write that $Q \sim G(k, \mu)$. Then, conditioning on all possible values of the random variable K, we obtain that

$$P[Q \leq t] = 1 - \left(\frac{\lambda}{\mu}\right) e^{-(\mu - \lambda)t} \quad \text{for } t \geq 0.$$

Remark. We find that $Q \mid \{Q > 0\} \sim \text{Exp}(\mu - \lambda)$. That is, $R := Q \mid \{Q > 0\}$ and the total time T spent by an arbitrary customer in the system are identically distributed random variables. Note that an exponential random variable being continuous, we may define it in the interval $[0, \infty)$ or $(0, \infty)$ indifferently.

Finally, the number N_Q of customers waiting in line when the system is in stationary regime can be expressed as follows:

$$N_Q = \begin{cases} 0 & \text{if } N = 0, \\ N - 1 & \text{if } N = 1, 2, \dots . \end{cases}$$

Hence, we may write that

$$P[N_Q = 0] = P[N = 0] + P[N = 1] = \pi_0 + \pi_1 = (1 + \rho)(1 - \rho)$$

and

$$P[N_Q = k] = P[N = k + 1] = \pi_{k+1} = \rho^{k+1}(1 - \rho) \quad \text{if } k = 1, 2, \dots .$$

Remarks. (i) Because we obtained the distributions of all the random variables of interest, we could calculate their respective variances.

(ii) The random variables Q and S are independent. However, N_Q and N_S are *not* independent. Indeed, we may write that

$$N_S = 0 \quad \Longrightarrow \quad N_Q = 0$$

(because if nobody is being served, then nobody is waiting in line either).

Example 6.2.1. Suppose that at a fixed time instant $t_0 > 0$, the number $X(t_0)$ of customers in an $M/M/1$ queue in stationary regime is smaller than or equal to 3. Calculate the expected value of the random variable $X(t_0)$, as well as its variance, if $\lambda = \mu/2$.

Solution. Because $\rho = 1/2$, the limiting probabilities of the system are:

$$\pi_k = \rho^k(1 - \rho) = (1/2)^{k+1} \quad \text{for } k = 0, 1, 2, \dots .$$

It follows that

$$P[X(t_0) \le 3] = \sum_{k=0}^{3}(1/2)^{k+1} = \frac{1}{2} + \frac{1}{4} + \frac{1}{8} + \frac{1}{16} = \frac{15}{16}.$$

Therefore, under the condition $X(t_0) \le 3$, we have:

$$\pi_0 = \frac{1/2}{15/16} = \frac{8}{15}, \quad \pi_1 = \frac{4}{15}, \quad \pi_2 = \frac{2}{15} \quad \text{and} \quad \pi_3 = \frac{1}{15}.$$

It is then easy to obtain the mean and the variance of $X(t_0)$. We calculate

$$E[X(t_0)] = 0 + 1 \times \frac{4}{15} + 2 \times \frac{2}{15} + 3 \times \frac{1}{15} = \frac{11}{15}$$

and

$$E[X^2(t_0)] = 0 + 1^2 \times \frac{4}{15} + 2^2 \times \frac{2}{15} + 3^2 \times \frac{1}{15} = \frac{21}{15},$$

so that

$$\text{VAR}[X(t_0)] = \frac{21}{15} - \left(\frac{11}{15}\right)^2 = \frac{194}{225}.$$

Remark. The fact that $X(t_0) \le 3$ does not mean that the system capacity is $c = 3$. When $c = 3$, the random variable $X(t)$ must necessarily be smaller than or equal to 3 for all values of t, whereas in the previous example the number of customers in the system *at a fixed time instant* was smaller than 4. The case when the system capacity is finite is the subject of the next subsection.

Example 6.2.2. Often a particular queueing model is a more or less simple transformation of the basic $M/M/1$ queue. For example, suppose that the server in an otherwise $M/M/1$ queue always waits until there are (at least) two customers in the system before beginning to serve them, exactly two at a time, at an exponential rate μ. Then, the stochastic process $\{X(t), t \ge 0\}$, where $X(t)$ denotes the number of customers in the system at time $t \ge 0$, is still a continuous-time Markov chain. However, it is no longer a birth and death process. To obtain the limiting probabilities of the process, we must solve the appropriate balance equations, under the condition $\sum_{k=0}^{\infty} \pi_k = 1$. These balance equations are (see Figure 6.2):

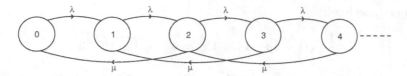

Fig. 6.2. State transition diagram for the queueing model in Example 6.2.2.

state j	departure rate from j = arrival rate to j
0	$\lambda \pi_0 \overset{(0)}{=} \mu \pi_2$
1	$\lambda \pi_1 \overset{(1)}{=} \lambda \pi_0 + \mu \pi_3$
2	$(\lambda + \mu) \pi_2 \overset{(2)}{=} \lambda \pi_1 + \mu \pi_4$
$k \ (\geq 3)$	$(\lambda + \mu) \pi_k = \lambda \pi_{k-1} + \mu \pi_{k+2}$

A solution of the equation for $k \geq 3$ can be obtained by assuming that $\pi_k = a^{k-2}\pi_2$, for $k = (2), 3, 4, \ldots$, where a is a constant such that $0 < a < 1$. The equation in question becomes:

$$(\lambda + \mu)a^{k-2}\pi_2 = \lambda a^{k-3}\pi_2 + \mu a^k \pi_2.$$

Given that π_2 cannot be equal to zero, we can write that

$$(\lambda + \mu)a = \lambda + \mu a^3.$$

That is, we must solve a third-degree polynomial equation. We find that $a = 1$ is an obvious root. It follows that

$$(\lambda + \mu)a = \lambda + \mu a^3 \quad \Longleftrightarrow \quad (a - 1)(\mu a^2 + \mu a - \lambda) = 0.$$

Hence, the other two roots are:

$$a = -\frac{1}{2} \pm \frac{\sqrt{\mu^2 + 4\mu\lambda}}{2\mu}.$$

Because $a > 0$, we deduce that

$$a = -\frac{1}{2} + \frac{\sqrt{\mu^2 + 4\mu\lambda}}{2\mu} = \frac{1}{2}\left(\sqrt{1 + 4\rho} - 1\right).$$

Remarks. (i) The solution $a = 1$ must be discarded, because it would imply that $\pi_k = \pi_2$, for $k = 2, 3, \ldots$, so that the condition $\sum_{k=0}^{\infty} \pi_k = 1$ could not be fulfilled.

(ii) We must also have:

$$\frac{1}{2}\left(\sqrt{1+4\rho}-1\right) < 1 \quad \Longleftrightarrow \quad \sqrt{1+4\rho} < 3.$$

Hence, we deduce that the limiting probabilities exist (if and) only if $\rho < 2$ or, equivalently, if $\lambda < 2\mu$. That is, the arrival rate of the customers must not exceed their service rate. This is the same condition for the existence of the limiting probabilities as the one in the case of the $M/M/2$ model, as shown in the next section.

To complete this example, we use the equations for states 0 and 1 above to express π_0 and π_1 in terms of π_2. The equation (0) yields at once that $\pi_0 = (\mu/\lambda)\pi_2$, whereas (0) and (1) together imply that

$$\lambda\pi_1 = \mu\pi_2 + \mu\pi_3 = \mu\pi_2 + \mu a\pi_2 \quad \Longrightarrow \quad \pi_1 = \frac{\mu}{\lambda}(a+1)\pi_2.$$

Then, the condition $\sum_{k=0}^{\infty}\pi_k = 1$ enables us to obtain an explicit expression for π_2 (from which we deduce the value of π_k, for $k = 0, 1, 3, 4, \ldots$). We have:

$$1 = \sum_{k=0}^{\infty}\pi_k = \frac{\mu}{\lambda}\pi_2 + \frac{\mu}{\lambda}(a+1)\pi_2 + \sum_{k=2}^{\infty}a^{k-2}\pi_2$$

$$= \pi_2\left[\frac{\mu}{\lambda}(a+2) + \sum_{k=0}^{\infty}a^k\right] = \pi_2\left[\frac{\mu}{\lambda}(a+2) + \frac{1}{1-a}\right].$$

Thus, we can write that

$$\pi_2 = \left[\frac{\mu}{\lambda}(a+2) + \frac{1}{1-a}\right]^{-1}.$$

Observe that we did not make use of the equation for state 2 to determine the limiting probabilities π_k. Actually, there is always one redundant equation in the system of linear equations. We can now check that the solution obtained also satisfies the equation (2) above. We have:

$$(\lambda+\mu)\pi_2 = \lambda\pi_1 + \mu\pi_4 \quad \Longleftrightarrow \quad (\lambda+\mu)\pi_2 = \lambda\frac{\mu}{\lambda}(a+1)\pi_2 + \mu a^2\pi_2.$$

That is, we must have:

$$\mu a^2 + \mu a - \lambda = 0.$$

But this is exactly the quadratic equation satisfied by the constant a (see above).

Remarks. (i) We have obtained *a* solution of the balance equations, subject to the normalizing condition (6.5). Actually, it can be shown that there is a *unique* solution of this system of linear equations that satisfies (6.5). Therefore, we can assert that we have *the* solution to our problem.

(ii) If the server is *able* to serve two customers at the same time (also at rate μ), but begins to work as soon as there is one customer in the system, then the solution is slightly different (see [20]).

(iii) If we suppose instead that the customers always arrive two at a time, but are served only one at a time, then the balance equations become:

state j departure rate from j = arrival rate to j

$$
\begin{aligned}
0 \qquad & \lambda \pi_0 = \mu \pi_1 \\
1 \qquad & (\lambda + \mu)\pi_1 = \mu \pi_2 \\
k\ (\geq 2) \qquad & (\lambda + \mu)\pi_k = \lambda \pi_{k-2} + \mu \pi_{k+1}
\end{aligned}
$$

In such a case, we could determine at random the respective positions in the queue of the two customers who arrived together.

(iv) Finally, if the customers always arrive two at a time and are also always served two at a time, then the limiting probabilities π_n^* of the corresponding continuous-time Markov chain can be expressed in terms of the limiting probabilities π_n of the $M/M/1$ queue as follows:

$$
\pi_n^* = \pi_{n/2} = \rho^{n/2}(1 - \rho) \quad \text{for } n = 0, 2, 4, \ldots .
$$

6.2.2 The $M/M/1$ model with finite capacity

As mentioned previously, in practice the capacity of any queueing system is limited. Let c be the finite integer denoting this capacity. Suppose that we computed the limiting probabilities of a given queueing system having finite capacity and that we found that π_c is very small. Then, assuming that c is actually infinite is a valid simplifying approximation. However, if the probability that the system is saturated is far from being negligible, then we should use a finite state space.

Suppose that a certain queueing system may be adequately described by an $M/M/1$ queue having $c+1$ possible states: $0, 1, \ldots, c$. This model is often denoted by $M/M/1/c$. The balance equations of the system are then the following:

state j departure rate from j = arrival rate to j

$$
\begin{aligned}
0 \qquad & \lambda \pi_0 = \mu \pi_1 \\
k = 1, \ldots, c-1 \qquad & (\lambda + \mu)\pi_k = \lambda \pi_{k-1} + \mu \pi_{k+1} \\
c \qquad & \mu \pi_c = \lambda \pi_{c-1}
\end{aligned}
$$

Notice that the balance equations for states $j = 0, 1, \ldots, c-1$ are identical to the corresponding ones in the $M/M/1/\infty$ model. When the system has reached its maximum capacity, namely c customers, the next state visited will necessarily be $c - 1$, at an exponential rate μ. Moreover, the only way the system may enter state c is from state $c - 1$, when a new customer arrives.

Let once again $X(t)$ be the number of customers in the system at time $t \geq 0$. The stochastic process $\{X(t), t \geq 0\}$ is an irreducible birth and death process, as before. Therefore, instead of solving the previous system of linear equations, subject to $\sum_{j=0}^{c} \pi_j = 1$ [see (6.5)], we can appeal to Theorem 6.1.1. We still have:

$$\Pi_k = \left(\frac{\lambda}{\mu}\right)^k = \rho^k \quad \text{for } k = 1, 2, \ldots, c,$$

so that

$$\pi_0 = \frac{1}{1 + \sum_{k=1}^{c} \Pi_k} = \frac{1}{1 + \sum_{k=1}^{c} \rho^k}$$

and

$$\pi_j = \Pi_j \pi_0 = \frac{\rho^j}{1 + \sum_{k=1}^{c} \rho^k} = \frac{\rho^j}{\sum_{k=0}^{c} \rho^k} \quad \text{for } j = 1, 2, \ldots, c.$$

Because the state space is finite, the limiting probabilities exist for any (positive) values of the parameters λ and μ. In the particular case when $\rho = 1$, the solution is simply

$$\pi_j = \frac{1}{c + 1} \quad \text{for } j = 0, 1, \ldots, c. \tag{6.16}$$

That is, when the system is in equilibrium, the $c+1$ possible states of the Markov chain are equally likely.

When $\rho \neq 1$, we calculate

$$\sum_{k=0}^{c} \rho^k = \frac{1 - \rho^{c+1}}{1 - \rho}.$$

Hence, we can write that

$$\pi_j \stackrel{\rho \neq 1}{=} \frac{\rho^j (1 - \rho)}{1 - \rho^{c+1}} \quad \text{for } j = 0, 1, \ldots, c. \tag{6.17}$$

Remark. The probability that the system is saturated is given by

$$\pi_c = \frac{\rho^c (1 - \rho)}{1 - \rho^{c+1}}.$$

Taking the limit as ρ tends to infinity, we obtain that

$$\lim_{\rho \to \infty} \pi_c = \lim_{\rho \to \infty} \frac{\rho^c (1 - \rho)}{1 - \rho^{c+1}} = \lim_{\rho \to \infty} \frac{\rho^{-1} - 1}{\rho^{-(c+1)} - 1} = 1,$$

so that $\pi_j = 0$, for $j = 0, 1, \ldots, c-1$, as could have been expected. Conversely, we have:

$$\lim_{\rho \downarrow 0} \pi_0 = \lim_{\rho \downarrow 0} \frac{1 - \rho}{1 - \rho^{c+1}} = 1$$

and $\pi_j = 0$, for $j = 1, 2, \ldots, c$. Finally, if $\rho < 1$ and c tends to infinity, we retrieve the formula

$$\pi_j = \lim_{c \to \infty} \frac{\rho^j(1-\rho)}{1-\rho^{c+1}} = \rho^j(1-\rho) \quad \text{for } j = 0, 1, \ldots$$

obtained in the case of the $M/M/1/\infty$ model.

Making use of (6.16), we easily find that

$$\bar{N} = \frac{c}{2} \quad \text{if } \rho = 1.$$

In the general case when $\rho \neq 1$, we can show that

$$\bar{N} = \frac{\rho}{1-\rho} - \frac{(c+1)\rho^{c+1}}{1-\rho^{c+1}}.$$

Actually, when the system capacity c is small, it is a simple matter to calculate the value of $\bar{N}$ from the formula

$$\bar{N} \equiv E[N] := \sum_{k=0}^{c} k\pi_k.$$

Likewise, after having calculated the π_ks, it is not difficult to obtain the variance of the random variable N.

Next, because N_S is equal to 1 if the system in equilibrium is in any state $k \in \{1, 2, \ldots, c\}$ (and to 0 if the system is empty), the expression for the value of $\bar{N}_S$ is the same as before, namely:

$$\bar{N}_S = 1 - \pi_0,$$

which implies that

$$\bar{N}_Q = \bar{N} - 1 + \pi_0.$$

However, the limiting probability π_0 is different from the corresponding one in the $M/M/1/\infty$ model.

Finally, if we consider only the customers who actually enter the system (in equilibrium), we may write that their average entering rate is

$$\lambda_e = \lambda(1 - \pi_c).$$

We then deduce from Little's formula (6.8) that

$$\bar{T} = \frac{\bar{N}}{\lambda(1 - \pi_c)},$$

so that

$$\bar{Q} = \frac{\bar{N}}{\lambda(1 - \pi_c)} - \frac{1}{\mu}$$

because $\bar{S} \equiv E[S] = 1/\mu$, as previously.

Example 6.2.3. Consider the $M/M/1/2$ queueing system. That is, the system capacity is $c = 2$. Suppose that $\lambda = \mu$.

(a) What is the variance of the number of customers in the system in stationary regime?

(b) What is the average number of arrivals into the system (in stationary regime) during the service time of a given customer?

Solution. (a) We deduce from (6.16) that $\pi_0 = \pi_1 = \pi_2 = 1/3$. It follows that

$$E[N] = \frac{1}{3}(0 + 1 + 2) = 1 \quad \text{and} \quad E[N^2] = \frac{1}{3}(0 + 1 + 4) = \frac{5}{3},$$

so that

$$\text{VAR}[N] = \frac{5}{3} - 1^2 = \frac{2}{3}.$$

(b) Let $t_0 > 0$ be the time instant at which the customer in question begins to be served. Then, $X(t_0)$ is equal to 1 or 2. Because $\pi_k \equiv 1/3$, we can assert that

$$P[X(t_0) = 1 \mid X(t_0) \in \{1,2\}] = P[X(t_0) = 2 \mid X(t_0) \in \{1,2\}] = \frac{1}{2}.$$

Next, let K be the number of customers who enter the system while the customer of interest is being served. Because $c = 2$, the possible values of the random variable K are 0 and 1. That is, K is a Bernoulli random variable. We have, under the condition that $X(t_0) \in \{1,2\}$:

$$P[K = 0] = \frac{1}{2}\{P[K = 0 \mid X(t_0) = 1] + P[K = 0 \mid X(t_0) = 2]\}$$

$$= \frac{1}{2}\{P[K = 0 \mid X(t_0) = 1] + 1\}.$$

Moreover, we can write that

$$P[K = 0 \mid X(t_0) = 1] = P[N(t_0 + S) - N(t_0) = 0] = P[N(S) = 0],$$

where $N(t)$ is the number of arrivals in the interval $[0, t]$ and S is the service time of an arbitrary customer. Conditioning on the possible values of S, we obtain:

$$P[N(S) = 0] = \int_0^\infty P[N(S) = 0 \mid S = s] f_S(s) ds.$$

Because the arrivals of customers and the service times are, by assumption, independent random variables, we have:

$$P[N(S) = 0] = \int_0^\infty P[N(s) = 0] \mu e^{-\mu s} ds = \int_0^\infty e^{-\lambda s} \mu e^{-\mu s} ds$$

$$= \frac{\mu}{\mu + \lambda} \stackrel{\lambda = \mu}{=} \frac{1}{2}.$$

It follows that

$$P[K = 0] = \frac{1}{2}\left(\frac{1}{2} + 1\right) = \frac{3}{4},$$

which implies that $P[K = 1] = 1/4$ and

$$E[K] = 0 + 1 \times \frac{1}{4} = \frac{1}{4}.$$

Example 6.2.4. Write the balance equations for the $M/M/1/3$ queueing system if we suppose that when the server finishes serving a customer and there are two customers waiting in line, then he serves them both at the same time, at rate μ.

Solution. Here, the state $X(t)$ of the process cannot simply be the number of customers in the system at time t. Indeed, suppose that there are three customers in the system. The next state visited will not be the same if two customers are being served simultaneously or if two customers are waiting in line. In the former case, the system will go from state 3 to state 1, whereas it will move from state 3 to state 2 in the latter case. Therefore, we have to be more precise. Let (m, n) be the state of the system if there are m customers being served and n waiting to be served. The possible states are then: $(m, 0)$, for $m = 0, 1, 2$, and $(1, 1)$, $(1, 2)$, and $(2, 1)$. The balance equations of the system are the following (see Figure 6.3):

state (m, n)	departure rate from (m, n) = arrival rate to (m, n)
$(0, 0)$	$\lambda \pi_{(0,0)} = \mu(\pi_{(1,0)} + \pi_{(2,0)})$
$(1, 0)$	$(\lambda + \mu)\pi_{(1,0)} = \lambda \pi_{(0,0)} + \mu(\pi_{(1,1)} + \pi_{(2,1)})$
$(1, 1)$	$(\lambda + \mu)\pi_{(1,1)} = \lambda \pi_{(1,0)}$
$(1, 2)$	$\mu \pi_{(1,2)} = \lambda \pi_{(1,1)}$
$(2, 0)$	$(\lambda + \mu)\pi_{(2,0)} = \mu \pi_{(1,2)}$
$(2, 1)$	$\mu \pi_{(2,1)} = \lambda \pi_{(2,0)}$

To obtain the limiting probabilities, we can solve the previous system of linear equations, under the condition $\sum_{(m,n)} \pi_{(m,n)} = 1$. We express the $\pi_{(m,n)}$s in terms of $\pi_{(2,1)}$. For simplicity, we assume that $\lambda = \mu$. Then, the last equation above yields that $\pi_{(2,0)} = \pi_{(2,1)}$. Next, the equation for state $(2, 0)$ implies that $\pi_{(1,2)} = 2\pi_{(2,1)}$. It follows, using the equation for state $(1, 2)$, that we can write that $\pi_{(1,1)} = 2\pi_{(2,1)}$ as well. The equation for state $(1, 1)$ enables us to write that $\pi_{(1,0)} = 4\pi_{(2,1)}$. Finally, the first equation gives us $\pi_{(0,0)} = 5\pi_{(2,1)}$. Thus, we have:

$$(5 + 4 + 2 + 2 + 1 + 1)\pi_{(2,1)} = 1 \quad \Longrightarrow \quad \pi_{(2,1)} = \frac{1}{15},$$

so that

$$\pi_{(0,0)} = \frac{1}{3}, \quad \pi_{(1,0)} = \frac{4}{15}, \quad \pi_{(1,1)} = \pi_{(1,2)} = \frac{2}{15} \quad \text{and} \quad \pi_{(2,0)} = \frac{1}{15}.$$

Note that this solution also satisfies the equation for state $(1,0)$, which we did not use to find the limiting probabilities.

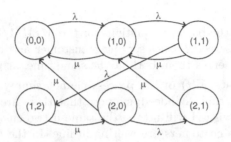

Fig. 6.3. State transition diagram for the queueing model in Example 6.2.4.

The average number of customers in the system in stationary regime is given by

$$\sum_{(m,n)} (m+n)\pi_{(m,n)} = 1 \times \pi_{(1,0)} + 2 \times (\pi_{(1,1)} + \pi_{(2,0)}) + 3 \times (\pi_{(1,2)} + \pi_{(2,1)})$$

$$= \frac{4 + 2 \times 3 + 3 \times 3}{15} = \frac{19}{15}$$

and the average entering rate of customers into the system is

$$\lambda_e = \lambda(1 - \pi_{(1,2)} - \pi_{(2,1)}) = \lambda\left(1 - \frac{3}{15}\right) = \frac{4\lambda}{5}.$$

6.3 Queueing systems with two or more servers

6.3.1 The $M/M/s$ model

Suppose that all the assumptions that were made in the formulation of the $M/M/1$ queueing model hold, but that there are actually s servers in the system, where $s \in \{2, 3, \ldots\}$. We assume that the service times of the s servers are independent $\text{Exp}(\mu)$ random variables. This model is denoted by $M/M/s$. The particular case when the number of servers tends to infinity is considered. Moreover, if the system capacity is finite, we obtain the $M/M/s/c$ model, which is treated in the next subsection.

As is generally the case in practice, we suppose that the customers wait in a single line for an idle server (or that they *take a number* when they enter the system and wait until their number is called). This means that if there are at most s customers in the system, then they are all being served, which is not necessarily the case if we assume that a queue is formed in front of each server. Furthermore, as was implicit in the previous section, the *service policy* is that of *first come, first served* (denoted by *FIFO*, for First In, First Out).

Remark. In the examples and the exercises, we often modify the basic $M/M/s$ model. For instance, we may assume that the servers do not necessarily serve at the same rate μ, or that the service policy is different from the one by default (i.e., *FIFO*), and so on.

Let $X(t)$ represent the number of customers in the system at time $t \geq 0$. The stochastic process $\{X(t), t \geq 0\}$ is a continuous-time Markov chain. The arrival process is a Poisson process with rate $\lambda > 0$. Furthermore, even though there are at least two servers, because the customers are served one at a time and the service times are exponential (thus, *continuous*) random variables, two (or more) customers cannot leave the system exactly at the same time instant. It follows that $\{X(t), t \geq 0\}$ is a birth and death process. The birth rates λ_k are all equal to λ, and the death rates μ_k are given by

$$\mu_k = \begin{cases} k\mu & \text{if } k = 1, \ldots, s - 1, \\ s\mu & \text{if } k = s, s + 1, \ldots. \end{cases}$$

Indeed, when there are k customers being served simultaneously, the time needed for a departure to take place is the minimum between k independent $\text{Exp}(\mu)$ random variables. We know that this minimum has an exponential distribution with parameter $\mu + \cdots + \mu = k\mu$ (see the remark after Proposition 5.2.1).

We deduce from what precedes that the balance equations for the $M/M/s$ queueing system are (see Figure 6.4 for the case when $s = 2$):

state j	departure rate from j = arrival rate to j
0	$\lambda \pi_0 = \mu \pi_1$
$k \in \{1, \ldots, s-1\}$	$(\lambda + k\mu)\pi_k = (k+1)\mu\pi_{k+1} + \lambda\pi_{k-1}$
$k \in \{s, s+1, \ldots\}$	$(\lambda + s\mu)\pi_k = s\mu\pi_{k+1} + \lambda\pi_{k-1}$

To solve this system of linear equations, under the condition $\sum_{k=0}^{\infty} \pi_k = 1$, we make use of Theorem 6.1.1. First, we calculate

$$\Pi_k = \frac{\lambda \times \lambda \times \cdots \times \lambda}{\mu \times 2\mu \times \cdots \times k\mu} = \frac{1}{k!}\left(\frac{\lambda}{\mu}\right)^k = \frac{1}{k!}\rho^k \quad \text{for } k = 1, 2, \ldots, s.$$

In the case when $k = s + 1, s + 2, \ldots$, we find that

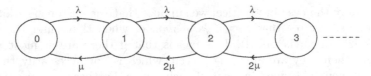

Fig. 6.4. State transition diagram for the $M/M/2$ model.

$$\Pi_k = \frac{\rho^k}{s! s^{k-s}}.$$

Next, the sum $\sum_{k=1}^{\infty} \Pi_k$ converges if and only if

$$\sum_{k=s+1}^{\infty} \Pi_k < \infty \quad \Longleftrightarrow \quad \sum_{k=s+1}^{\infty} \frac{\rho^k}{s! s^{k-s}} < \infty$$

$$\Longleftrightarrow \quad \frac{s^s}{s!} \sum_{k=s+1}^{\infty} \left(\frac{\rho}{s}\right)^k < \infty \quad \Longleftrightarrow \quad \rho < s,$$

which, again, is tantamount to saying that the arrival rate of the customers must be smaller than their (maximal) service rate.

Now, because its birth and death rates are strictly positive, the birth and death process $\{X(t), t \geq 0\}$ is irreducible and the limiting probabilities can indeed be obtained from Theorem 6.1.1. We find (after some work) that

$$\pi_0 = \left[\sum_{k=0}^{s-1} \frac{\rho^k}{k!} + \frac{\rho^s}{s!} \frac{s}{(s-\rho)} \right]^{-1} \quad \text{if } \rho < s. \tag{6.18}$$

We may then write that

$$\pi_k = \frac{\rho^k}{k!} \pi_0 \quad \text{if } k = 1, \dots, s \tag{6.19}$$

and

$$\pi_k = \frac{\rho^k}{s! s^{k-s}} \pi_0 \quad \text{if } k = s+1, s+2, \dots. \tag{6.20}$$

Obtaining the quantities $\bar{N}$ and $\bar{T}$ requires some effort. First, because the service rates are, by assumption, all equal to μ, we can write that $\bar{S} = 1/\mu$. It follows, from Little's formula (with $\lambda_e = \lambda$) that

$$\bar{N}_S = \lambda \bar{S} = \rho.$$

Next, we can show that

$$\bar{N}_Q = \frac{\rho^{s+1}}{s!} \frac{s}{(s-\rho)^2} \pi_0,$$

from which we deduce that

$$\bar{N} = \frac{\rho^{s+1}}{s!} \frac{s}{(s-\rho)^2} \pi_0 + \rho.$$

Finally, we have:

$$\bar{Q} = \frac{\bar{N}_Q}{\lambda} = \frac{\rho^{s+1}}{\lambda s!} \frac{s}{(s-\rho)^2} \pi_0 \quad \text{and} \quad \bar{T} = \bar{Q} + \frac{1}{\mu}.$$

Remark. In some applications, the number of *servers* is infinite. For example, suppose that the *customers* are people visiting a public park and that their *service time* is the random time that they spend in the park. Because people do not have to wait to be served, this situation corresponds to the case when s is infinite. We can write [see (6.18)] that

$$\lim_{s \to \infty} \pi_0 = \left(\sum_{k=0}^{\infty} \frac{\rho^k}{k!} \right)^{-1} = e^{-\rho},$$

so that

$$\lim_{s \to \infty} \pi_k = \frac{\rho^k}{k!} e^{-\rho} \quad \text{for } k = 1, 2, \dots. \tag{6.21}$$

Hence, if $K \sim \text{Poi}(\rho)$, we may state that the limiting probabilities for the $M/M/\infty$ model are given by

$$\pi_k = P[K = k] \quad \text{for } k = 0, 1, \dots. \tag{6.22}$$

It follows that $\bar{N} = \rho$. Finally, because the customers never wait to be served, we have that $\bar{N}_Q = \bar{Q} = 0$, so that $\bar{N}_S = \bar{N} = \rho$ and $\bar{T} = \bar{S} = 1/\mu$.

Now, another quantity of interest is the probability that all servers are busy, when the system is in stationary regime. We denote this probability by π_b. We can show that

$$\pi_b \equiv \sum_{k \geq s} \pi_k = \frac{s\rho^s}{s!(s-\rho)} \pi_0 \quad \text{if } \rho < s.$$

It is possible to express in terms of π_b the distribution function of the time Q that an arbitrary customer waits to be served. The random variable Q is of mixed type. We have that $P[Q = 0] = 1 - \pi_b$, because $1 - \pi_b$ is the probability that the customer in question arrives while there are fewer than s customers already present in the system. In general, we can show that, if $\rho < s$,

$$P[Q \leq t] = 1 - \pi_b e^{(\rho - s)t} \quad \text{for } t \geq 0.$$

Example 6.3.1. Consider the $M/M/2$ queueing model. Suppose that the two servers are busy. Calculate the probability that the service times of the two customers being served differ by at most one time unit.

Solution. Let S_i be the service time of the customer being served by server no. i, for $i = 1, 2$. The variables S_i are independent $\text{Exp}(\mu)$ random variables. By symmetry, we can write that

$$P[|S_1 - S_2| \le 1] = 2P[S_1 \le S_2 \le S_1 + 1]$$

$$= 2 \int_0^\infty \int_{s_1}^{s_1+1} \mu e^{-\mu s_1} \mu e^{-\mu s_2} ds_2 ds_1$$

$$= 2 \int_0^\infty \mu e^{-\mu s_1} \left\{ \int_{s_1}^{s_1+1} \mu e^{-\mu s_2} ds_2 \right\} ds_1$$

$$= 2 \int_0^\infty \mu e^{-\mu s_1} \left\{ - e^{-\mu s_2} \Big|_{s_1}^{s_1+1} \right\} ds_1$$

$$= 2 \left(1 - e^{-\mu} \right) \int_0^\infty \mu e^{-2\mu s_1} ds_1 = 2 \left(1 - e^{-\mu} \right) \frac{1}{2}.$$

Thus, the required probability is equal to $1 - e^{-\mu}$.

Remark. If we do not use symmetry, we must compute two double integrals (see Figure 6.5):

$$P[|S_1 - S_2| \le 1] = \int_0^1 \int_0^{s_1+1} \mu e^{-\mu s_1} \mu e^{-\mu s_2} ds_2 ds_1$$

$$+ \int_1^\infty \int_{s_1-1}^{s_1+1} \mu e^{-\mu s_1} \mu e^{-\mu s_2} ds_2 ds_1.$$

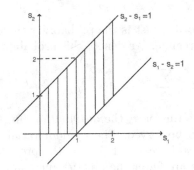

Fig. 6.5. Figure for Example 6.3.1.

Example 6.3.2. Suppose that $\lambda = 2\mu$ in an $M/M/3$ queueing system.

(a) What is the variance of the number of customers in the system at a time t_0 large enough for the system to be in stationary regime, given that nobody is waiting in line at time t_0?

(b) Knowing that the three servers are busy at time t_0, what is the probability that nobody is waiting in line?

Solution. (a) First, t_0 being large enough, we can write [see (6.19)] that

$$\pi_k^* := P[X(t_0) = k \mid X(t_0) \le 3] = \frac{\pi_k}{\pi_0 + \pi_1 + \pi_2 + \pi_3} = \frac{\rho^k/k!}{\sum_{j=0}^{3} \rho^j/j!}$$

$$\overset{\rho=2}{=} \left(\frac{3}{19}\right)\left(\frac{2^k}{k!}\right) \quad \text{for } k = 0, 1, 2, 3.$$

That is, $\pi_0^* = 3/19$, $\pi_1^* = \pi_2^* = 6/19$ and $\pi_3^* = 4/19$.

Remark. Note that because the limiting probabilities π_k are expressed in terms of π_0, we did not need to calculate the value of π_0 to obtain π_k. Actually, we have:

$$\pi_0 = \left[\sum_{k=0}^{2} \frac{2^k}{k!} + \frac{2^3}{3!}\frac{3}{(3-2)}\right]^{-1} = \frac{1}{9}.$$

Next, we calculate

$$E[X(t_0) \mid X(t_0) \le 3] = \frac{1}{19}(6 + 2 \times 6 + 3 \times 4) = \frac{30}{19}$$

and

$$E[X^2(t_0) \mid X(t_0) \le 3] = \frac{1}{19}(6 + 2^2 \times 6 + 3^2 \times 4) = \frac{66}{19},$$

so that

$$\text{VAR}[X(t_0) \mid X(t_0) \le 3] = \frac{66}{19} - \left(\frac{30}{19}\right)^2 = \frac{354}{361}.$$

(b) We look for

$$p := P[X(t_0) = 3 \mid X(t_0) \ge 3] = \frac{P[X(t_0) = 3]}{P[X(t_0) \ge 3]} = \frac{P[X(t_0) = 3]}{1 - P[X(t_0) \le 2]}$$

$$\overset{(6.19)}{=} \frac{(4/3)\pi_0}{1 - (\pi_0 + 2\pi_0 + 2\pi_0)}.$$

So, here we need the explicit value of π_0. Using the previous remark, we can write that

$$p = \frac{4/3}{\pi_0^{-1} - (1 + 2 + 2)} = \frac{4/3}{9 - (1 + 2 + 2)} = \frac{1}{3}.$$

6.3.2 The $M/M/s/c$ model

Even if the capacity, c, of an $M/M/s$ system is finite, the stochastic process $\{X(t), t \geq 0\}$ remains a birth and death process. Therefore, we can appeal to Theorem 6.1.1 to obtain the limiting probabilities of the process. One case is of special importance, namely the one for which $c = s$. The $M/M/s/s$ model is a particular case of what is known as a *loss system*, because when all the servers are busy, the arriving customers cannot (or do not want to) enter the system. Consequently, they are *lost*. An example of such a system is a parking lot. In this case, the empty parking places are the *servers*, and when the lot is full, arriving drivers must go somewhere else to park their cars.

Next, we obtain the limiting probabilities for the $M/M/s/s$ model. The balance equations of this system are the following:

state j	departure rate from j = arrival rate to j
0	$\lambda \pi_0 = \mu \pi_1$
$k \in \{1, \ldots, s-1\}$	$(\lambda + k\mu)\pi_k = (k+1)\mu \pi_{k+1} + \lambda \pi_{k-1}$
s	$s\mu \pi_s = \lambda \pi_{s-1}$

The birth and death process $\{X(t), t \geq 0\}$, where $X(t)$ represents the number of customers in the system at time t, is irreducible. Moreover, in as much as the system capacity is finite, the limiting probabilities exist for all admissible values of the parameters λ and μ. We calculate

$$\Pi_k = \frac{\lambda \times \lambda \times \cdots \times \lambda}{\mu \times 2\mu \times \cdots \times k\mu} = \frac{\rho^k}{k!} \quad \text{for } k = 1, 2, \ldots, s.$$

It follows that

$$\pi_0 = \frac{1}{1 + \sum_{k=1}^{s} \rho^k/k!} = \left(\sum_{k=0}^{s} \frac{\rho^k}{k!} \right)^{-1} \tag{6.23}$$

and

$$\pi_j = \frac{\rho^j}{j!} \pi_0 \quad \text{for } j = 1, \ldots, s. \tag{6.24}$$

Remarks. (i) If s tends to infinity, we should retrieve the results obtained for the $M/M/\infty$ queue in the previous subsection [see (6.21)]. Indeed, we have:

$$\lim_{s \to \infty} \pi_0 = \left(\sum_{k=0}^{\infty} \frac{\rho^k}{k!} \right)^{-1} = (e^\rho)^{-1} = e^{-\rho},$$

so that

$$\lim_{s \to \infty} \pi_j = \frac{\rho^j}{j!} e^{-\rho} \quad \text{for } j = 1, 2, \dots .$$

(ii) The probability π_b that all servers are busy is given by

$$\pi_b = \pi_s = \frac{\rho^s}{s!} \pi_0 = \frac{\rho^s / s!}{\sum_{j=0}^{s} \rho^j / j!}, \tag{6.25}$$

which is known as *Erlang's formula*.

(iii) Because $\rho := \lambda/\mu = \lambda E[S]$, the formulas for the limiting probabilities may be rewritten as follows:

$$\pi_0 = \left(\sum_{k=0}^{s} \frac{(\lambda E[S])^k}{k!} \right)^{-1} \quad \text{and} \quad \pi_j = \frac{(\lambda E[S])^j}{j!} \pi_0 \quad \text{for } j = 1, \dots, s. \tag{6.26}$$

A very interesting result states that the previous formulas are valid even if the random variable S is not exponentially distributed, as long as it is nonnegative. For instance, S could have a uniform distribution on the interval $(0, 1)$, or a gamma distribution, and so on. That is, (6.26) holds for the $M/G/s/s$ model (called the *Erlang loss system*), where G stands for *general*.

To complete this subsection, we calculate the various quantities of interest, which turns out to be easy, because $Q \equiv 0$. It follows at once that $\bar{Q} = \bar{N}_Q = 0$. Because $\bar{S} = 1/\mu$, as previously, we can write that

$$\bar{T} = \bar{S} = \frac{1}{\mu}$$

and

$$\bar{N} = \bar{N}_S = \lambda_e \bar{S} = \lambda (1 - \pi_s) \frac{1}{\mu} = (1 - \pi_s) \rho.$$

Example 6.3.3. The balance equations for the $M/M/2/3$ queueing system with $\lambda = 2\mu$ are:

state j	departure rate from j = arrival rate to j
0	$2\mu\pi_0 = \mu\pi_1$
1	$(2\mu + \mu)\pi_1 = 2\mu\pi_0 + (2 \times \mu)\pi_2$
2	$(2\mu + 2 \times \mu)\pi_2 = 2\mu\pi_1 + (2 \times \mu)\pi_3$
3	$(2 \times \mu)\pi_3 = 2\mu\pi_2$

That is,

$$2\pi_0 \overset{(0)}{=} \pi_1$$
$$3\pi_1 \overset{(1)}{=} 2\pi_0 + 2\pi_2$$
$$2\pi_2 \overset{(2)}{=} \pi_1 + \pi_3$$
$$\pi_3 \overset{(3)}{=} \pi_2$$

It is a simple matter to solve this system of linear equations. The equations for states 2 and 3 yield that $\pi_1 = \pi_2 = \pi_3$. Then, making use of the equation for state 0, we can write that

$$\pi_0 + 2\pi_0 + 2\pi_0 + 2\pi_0 = 1 \quad \Longrightarrow \quad \pi_0 = \frac{1}{7} \quad \text{and} \quad \pi_1 = \pi_2 = \pi_3 = \frac{2}{7}.$$

Finally, we find at once that this solution satisfies the equation for state 1 as well.

In the case of the $M/M/2/2$ queueing system (with $\lambda = 2\mu$), we deduce from (6.23) and (6.24) that

$$\pi_0 = \left(\sum_{k=0}^{2} \frac{2^k}{k!}\right)^{-1} = (1+2+2)^{-1} = \frac{1}{5} \quad \text{and} \quad \pi_1 = \pi_2 = \frac{2}{5}.$$

6.4 Exercises for Chapter 6

Solved exercises

Question no. 1

A system is made up of n components that operate independently of one another and all have an exponential lifetime, with parameter μ_k, for $k = 1, \ldots, n$. When the system breaks down, the failed components are replaced by new ones. Let $N(t)$ be the number of system breakdowns in the interval $[0, t]$. Is the stochastic process $\{N(t), t \geq 0\}$ a continuous-time Markov chain if the components are connected (a) in series? (b) in parallel? (c) in standby redundancy?

Question no. 2

The particular pure birth process known as the *Yule process* is such that $\lambda_n = n\lambda$, for $n = 0, 1, \ldots$. It can be shown that

$$p_{i,j}(t) = \binom{j-1}{i-1} e^{-i\lambda t}(1 - e^{-\lambda t})^{j-i} \quad \text{for } j \geq i \geq 1.$$

What is the expected value of $X(t)$, given that $X(0) = i > 0$?

Question no. 3

Let $\{X(t), t \geq 0\}$ be a birth and death process with state space $\{0, 1, 2\}$ and birth and death rates given by

$$\lambda_0 = \lambda, \quad \lambda_1 = 2\lambda \quad \text{and} \quad \mu_1 = \mu, \quad \mu_2 = 2\mu.$$

Find the limiting probabilities of the process from its balance equations.

Question no. 4

Find the limiting probabilities of an $M/M/1$ queue at a (large enough) time instant t_0, given that there are either two, three, or four customers in the system at t_0. Under the same condition, what is the expected time that the first customer who enters the system after t_0 will spend waiting in line if we assume that the customer who was being served at time t_0 is still present when the new one arrives?

Question no. 5

Suppose that the server in an $M/M/1$ queueing system works twice as fast when there are at least three customers in the system, so that $\mu[X(t)] = \mu$ if $X(t) = 1$ or 2, and $\mu[X(t)] = 2\mu$ if $X(t) \geq 3$. Write the balance equations for this system. What is the condition for the existence of the limiting probabilities?

Question no. 6

Consider the $M/M/1/2$ queueing system in stationary regime. Suppose that a departure took place at time t_0 and that the next two arrivals, from t_0, occurred at $t = t_0 + 1$ and $t = t_0 + 2$. What is the probability that the customer who arrived at time $t_0 + 2$ was able to enter the system?

Question no. 7

Suppose that the server in an $M/M/1/3$ queueing system decides to work twice as fast, in order to increase the profits of the system. However, after a while, the arrival rate of customers goes from λ to $\lambda/2$, because of poor service. Assume that $\lambda = \mu$. If every customer who actually enters the system pays $\$x$, what is the average amount of money that the system earns per unit of time when the service rate is μ? Is the server better off to serve at rate μ or at rate 2μ?

Question no. 8

Let $X(t)$ be the number of customers at time t in an $M/M/2$ queueing model. Suppose that $X(t_0) \geq 2$, and let τ_i be the departure time of the customer being served by server no. i, for $i = 1, 2$. Calculate the probability $P[\tau_2 \leq \tau_1 + 1]$.

Question no. 9

Write the balance equations for the $M/M/2/3$ queueing system if we suppose that the service time of server no. i has an exponential distribution with parameter μ_i, for $i = 1, 2$. That is, the two servers do not necessarily work at the same speed. Assume that, when the system is empty, an arriving customer heads for server no. 1 with probability 1. In terms of the limiting probabilities of the process, what is the average time that an entering customer spends in the system (in stationary regime)?

Question no. 10

Drivers arrive according to a Poisson process with rate λ at a service station having two gas pumps, but no waiting space for the cars. Suppose that the service time is a uniformly distributed random variable on the interval $(2,4)$ for either pump, independently from one driver to the other (and that the various service times are independent of the interarrival times of the drivers). What are the limiting probabilities of the system? What is the variance of the number of cars in the service station in equilibrium if $\lambda = 1/3$?

Exercises

Question no. 1

Let $\{N_i(t), t \geq 0\}$ be a Poisson process with rate λ_i, for $i = 1, 2$. Assume that the two Poisson processes are independent. We define

$$X(t) = N_1(t) - N_2(t) \quad \text{for } t \geq 0.$$

Is the stochastic process $\{X(t), t \geq 0\}$ a continuous-time Markov chain? Is it a birth and death process (with state space $\mathbb{Z} := \{\ldots, -2, 1, 0, 1, 2, \ldots\}$)? Justify your answers.

Question no. 2

Let $\{N(t), t \geq 0\}$ be a counting process (see Example 6.1.1) for which the time T until the first event and between two successive events has a uniform distribution on the interval $(0, 1)$. Show that the stochastic process $\{N^*(t), t \geq 0\}$, where $N^*(0) := 0$ and

$$N^*(t) := N(-\ln t) \quad \text{for } t > 0$$

is a pure birth process.

Hint. See Example 3.4.4.

Question no. 3

Consider the birth and death process $\{X(t), t \geq 0\}$ having birth and death rates $\lambda_n = \lambda$ and $\mu_n = n\mu$, for $n = 0, 1, 2, \ldots$. Calculate, if they exist, the limiting probabilities of the process.

Question no. 4

Let $\{N(t), t \geq 0\}$ be a Poisson process with rate $\lambda = \ln 2$. We define $W_i = \text{int}(\tau_i + 1)$, for $i = 0, 1, \ldots$, where τ_i is the time that the process spends in state i, and "int" designates the *integer part*. It can be shown that W_i has a geometric distribution with parameter $p = 1 - e^{-\lambda}$. Calculate the probabilities (a) $P[W_0 = W_1]$; (b) $P[W_0 > W_1]$; (c) $P[W_0 > W_1 \mid W_1 > 1]$.

Question no. 5

Suppose that $\{X(t), t \geq 0\}$ is a Yule process with $\lambda = 2$ (see solved exercise no. 2). Calculate (a) $E[\tau_1^4 + \tau_2^4 + \tau_3^4]$; (b) $E[\tau_1 + \tau_2^2 + \tau_3^3]$; (c) the correlation coefficient of τ_1^2 and τ_1^3.

Hint. We have (see Example 3.5.2) that

$$\int_0^\infty x^n \lambda e^{-\lambda x}\, dx = \frac{n!}{\lambda^n} \quad \text{for } n = 0, 1, \dots .$$

Question no. 6

Customers arrive at a certain store according to a Poisson process with rate $\lambda = 1/2$ per minute. Suppose that each customer stays exactly five minutes in the store.

(a) What is the expected number of customers in the store at time $t = 60$?

(b) What is the expected number of customers in (a), given that the store is not empty at time $t = 60$?

(c) What is the expected number of customers in (a), given that the number of customers in the store at time $t = 60$ is not equal to 1?

Question no. 7

What is the average number of customers in an $M/M/1$ queueing system in equilibrium, given that the number of customers is an odd number?

Question no. 8

Consider an $M/M/1$ queue in equilibrium, with $\lambda = \mu/2$. What is the probability that there are more than five customers in the system, given that there are at least two?

Question no. 9

Suppose that the service policy for an $M/M/1$ queue is the following: when the server finishes serving a customer, the next one to enter service is picked at random among those waiting in line. What is the expected total time that a customer who arrived while the system (in equilibrium) was in state 3 spent in the system, given that no customers who (possibly) arrived after the customer in question were served before him?

Question no. 10

Calculate the variance of the total time T that an arbitrary customer spends in an $M/M/1$ queueing system in equilibrium, given that $1 < T < 2$, if $\lambda = 1$ and $\mu = 2$.

Question no. 11

Suppose that after having been served, every customer in an $M/M/1/2$ queueing system immediately returns (exactly once) in front of the server (if there was a customer waiting to be served, then the returning customer will have to wait until the server is free). Define an appropriate state space and write the balance equations of the system.

Question no. 12

A customer who was unable to enter an $M/M/1/c$ queueing system (in equilibrium) at time t_0 decides to come back at time $t_0 + 2$. What is the probability that the customer in question is then able to enter the system, given that exactly one customer arrived in the interval $(t_0, t_0 + 2)$ and was unable to enter the system as well?

Question no. 13

Suppose that for an otherwise $M/M/1/2$ queue, a customer who enters the system, but has to wait for service, decides to leave if she is still waiting after a random time having an exponential distribution with parameter θ. Moreover, this random time is independent of the service times and the interarrival times. Let $X(t)$ be the number of customers in the system at time t. The stochastic process $\{X(t), t \geq 0\}$ is a continuous-time Markov chain (it is a birth and death process, to be precise). (a) Write the balance equations of the process. (b) Calculate the limiting probabilities in the case when $\lambda = \mu = \theta = 1$.

Question no. 14

Assume that the probability that the server, in an $M/M/1/3$ queueing system, is unable to provide the service requested by an arbitrary customer is equal to $p \in (0, 1)$, independently from one customer to the other. Assume also that the time needed by the server to decide whether he will be able or unable to service a given customer is an exponentially distributed random variable with parameter μ_0. Define an appropriate state space and write the balance equations of the system.

Question no. 15

Consider an $M/M/2$ queueing system. Suppose that any customer who accepts to pay twice as much as an ordinary customer for her service will get served at rate 2μ. If one such customer arrives while there are exactly two customers already present in the system, what is the probability that this new customer will spend less than $1/\mu$ time units (i.e., the average service time of an ordinary customer) in the system?

Question no. 16

Assume that when an $M/M/2$ queueing system is empty, an arriving customer heads for server no. 1 with probability $1/2$. What is the probability that the first two customers, from the initial time, will be served (a) by server no. 1? (b) by different servers?

Question no. 17

Let $X(t)$ be the number of customers in an $M/M/2/4$ queueing system. Suppose that t_0 is large enough for the system to be in stationary regime. Calculate the conditional expectation $E[X(t_0) \mid X(t_0) > 0]$ if $\lambda = \mu/2$.

Question no. 18

For an $M/M/4/4$ queue in equilibrium, with $\lambda = \mu$, what is the variance of the number of customers in the system, given that it is not empty?

Question no. 19

Assume that in a certain $M/M/2/3$ queueing system, the service time of an arbitrary customer is an exponentially distributed random variable with parameter μ. However, if the server has not completed his service after a random time (independent of the actual service time) having an exponential distribution with parameter μ_0, then the customer must leave the system. (a) Write the balance equations of the system. (b) What is the proportion of customers who must leave the system before having been fully serviced?

Question no. 20

For an $M/M/5/5$ queue with $\lambda = \mu$, what is the expected value of $1/N$, where N is the number of customers in the system in stationary regime, given that the system is neither empty nor full?

Multiple choice questions

Question no. 1

Let $\{X(t), t \geq 0\}$ be a birth and death process with rates $\lambda_n \equiv 1$ and $\mu_n \equiv 2$. Suppose that $X(0) = 0$. What is the probability that the process returns exactly twice to state 0 before visiting state 2?

(a) 1/27   (b) 4/27   (c) 2/9   (d) 1/3   (e) 4/9

Question no. 2

Consider the pure death process $\{X(t), t \geq 0\}$ with rates $\mu_n \equiv 1$. Calculate the probability $P[X(5) = 0 \mid X(0) = 5]$.

(a) 0.1755   (b) 0.3840   (c) 0.4405   (d) 0.5595   (e) 0.6160

Question no. 3

The continuous-time Markov chain $\{X(t), t \geq 0\}$, having state space $\{0, 1, 2\}$, is such that $\nu_i \equiv \nu$, $\rho_{0,1} = 1/2$, $\rho_{1,0} = 1/4$, and $\rho_{2,0} = 1/4$. Calculate the limiting probability that the process is not in state 0.

(a) 1/5   (b) 2/5   (c) 1/2   (d) 3/5   (e) 4/5

Question no. 4

Let T be the total time spent by an arbitrary customer in an $M/M/1$ queue with $\lambda = 1$ and $\mu = 3$. Calculate $E[T^2 \mid T > t_0]$, where $t_0 > 0$.

(a) $\frac{1}{4} + t_0 + t_0^2$ (b) $\frac{1}{2} + t_0 + t_0^2$ (c) $1 + t_0 + t_0^2$ (d) 1/4   (e) 1/2

Question no. 5

What is the probability that a customer, who left an $M/M/1$ queue with $\lambda = 1$ and $\mu = 2$ before the arrival of the next customer, spent less than one time unit in the system?

(a) $\frac{1}{2}\left(1 - e^{-1}\right)$ (b) $\frac{1}{2}\left(1 - e^{-2}\right)$ (c) $1 - e^{-1}$ (d) $1 - e^{-2}$ (e) $1 - \frac{1}{2}e^{-2}$

Question no. 6

For an $M/M/1/2$ queueing system, what is the probability that the system was full before the first departure took place, given that the second customer arrived at time $t = 2$?

(a) $\frac{1}{2\mu}\left(1 - e^{-2\mu}\right)$ (b) $\frac{1}{\mu}\left(1 - e^{-2\mu}\right)$ (c) $\frac{1}{2\mu}\left(1 - e^{-\mu}\right)$ (d) $\frac{1}{\mu}\left(1 - e^{-\mu}\right)$
(e) $\frac{1}{\mu}\left(1 - e^{-\mu} - e^{-2\mu}\right)$

Question no. 7
Suppose that the $M/M/1/2$ queueing system is modified as follows: whenever the server finishes serving a customer, he is unavailable for a random time τ (which is independent of the service times and the interarrival times) having an exponential distribution with parameter θ. Calculate the limiting probability that the server is busy serving a customer if $\lambda = \mu = 1$ and $\theta = 1/2$.

Hint. Define the following states:

> 0:   system is empty; server is available;
> 0^*:   system is empty; server is unavailable;
> 1:   one customer is being served; nobody is waiting;
> 1^*:   one customer is waiting to be served; server is unavailable;
> 2:   one customer is being served and another one is waiting;
> 2^*:   two customers are waiting to be served; server is unavailable.

(a) 3/49   (b) 13/49   (c) 16/49   (d) 19/49   (e) 20/49

Question no. 8
Suppose that there are five customers in an $M/M/2$ queueing system at a given time instant. Find the probability that the three customers waiting in line will not be served by the same server.

(a) 3/8   (b) 1/2   (c) 5/8   (d) 3/4   (e) 7/8

Question no. 9
What is the average number of customers in an $M/M/2/4$ queueing system in equilibrium, with $\lambda = \mu$, given that it is neither full nor empty?

(a) 9/7   (b) 11/7   (c) 2   (d) 15/7   (e) 17/7

Question no. 10
Suppose that the service time for an $M/G/3/3$ queueing system, with $\lambda = 1$, is a random variable S defined by $S = Z^4$, where $Z \sim N(0, 1)$. What is the average number of customers in the system in stationary regime?

Hint. The square of a $N(0, 1)$ random variable has a gamma distribution with parameters $\alpha = \lambda = 1/2$.

(a) 18/13   (b) 41/26   (c) 51/26   (d) 2   (e) 2.5

7

Time series

In many applications, in particular in economics and in hydrology, people are interested in the sequence of values of a certain variable over time. To model the variations of the variable of interest, a *time series* is often used. For example, the flow of a river on a given day may be expressed as a function of the flow on the previous days, to which a term called *noise* is added. In the first section, general properties of time series are presented. Next, various time series models are studied. Finally, the problem of modeling and using time series to forecast future values of the *state variable* is considered.

7.1 Introduction

Let $\{X_n, n = 0, 1, \ldots\}$ be a discrete-time stochastic process. If the index n represents time, it is known as a (discrete-time) *time series*. The random variable X_n is called the *state variable*. In this chapter, we are interested in *stationary* time series.

Remark. The state variable could actually be a random vector of dimension k. The time series would then be a *multivariate* time series. Likewise, if the index n is a vector, then we have a *multidimensional* or *spatial* time series. In this book, we only consider the case when the state variable is a random variable and the index n is a scalar.

Definition 7.1.1. *The stochastic process* $\{X_n, n = 0, 1, \ldots\}$ *is said to be (strictly)* **stationary** *if the distribution of the random vector* $(X_{n_1+m}, X_{n_2+m}, \ldots, X_{n_k+m})$ *is the same for all* $m \in \{0, 1, \ldots\}$ *and for all* $k = 1, 2, \ldots$, *where* $n_j \in \{0, 1, \ldots\}$ *for* $j = 1, \ldots, k$.

Remarks. (i) In words, the process $\{X_n, n = 0, 1, \ldots\}$ is stationary if the distribution of the random vector $(X_{n_1}, X_{n_2}, \ldots, X_{n_k})$ is *invariant* under any shift of length m to the right.

(ii) If we consider the process $\{X_n, n = \ldots, -1, 0, 1, \ldots\}$, then m can be any positive or negative integer in the definition, which is tantamount to saying that the time origin can be placed anywhere.

M. Lefebvre, *Basic Probability Theory with Applications*, Springer Undergraduate Texts in Mathematics and Technology, DOI: 10.1007/978-0-387-74995-2_7,
© Springer Science + Business Media, LLC 2009

Because it is generally difficult to prove that a given stochastic process is stationary, we often content ourselves with a *weaker* version of the definition. Note first that if $\{X_n, n = 0, 1, \ldots\}$ is stationary, then we may write that

$$P[X_n \leq x] = P[X_0 \leq x] \quad \text{for any } n \in \{1, 2, \ldots\},$$

which follows from the above definition with $k = 1$. Hence, we deduce that the expected value of X_n must not depend on n:

$$E[X_n] := \mu \quad \text{for all } n \in \{0, 1, 2, \ldots\}.$$

Next, if we set $k = 2$ in the definition, we obtain that

$$P[X_{n_1} \leq x_1, X_{n_2} \leq x_2] = P[X_{n_1+m} \leq x_1, X_{n_2+m} \leq x_2]$$

for any $n_1, n_2 \in \{0, 1, \ldots\}$ and for $m \in \{0, 1, \ldots\}$. This time, we deduce that the distribution of the random vector (X_{n_1}, X_{n_2}) depends only on the difference $n_2 - n_1$ (or $n_1 - n_2$), which implies that

$$\mathrm{COV}[X_{n_1}, X_{n_2}] := \gamma(n_2 - n_1) = \gamma(n_1 - n_2)$$

for all $n_1, n_2 \in \{0, 1, 2, \ldots\}$. If we choose $n_1 = i$ and $n_2 = i + j$, then we may write that

$$\mathrm{COV}[X_i, X_{i+j}] = \gamma(j) = \gamma(-j), \tag{7.1}$$

which also follows from the equation

$$\mathrm{COV}[X_i, X_{i+j}] = \mathrm{COV}[X_{i+j}, X_i].$$

Hence, $\mathrm{COV}[X_i, X_{i+j}] = \mathrm{COV}[X_i, X_{i-j}]$ if $j \leq i$.

Definition 7.1.2. *The function $\gamma(\cdot)$ is called the* (auto) covariance *function of the stationary stochastic process $\{X_n, n = 0, 1, \ldots\}$.*

Remark. If we assume that $\mu = 0$, or if we consider the centered process $\{Y_n, n = 0, 1, \ldots\}$ defined by
$$Y_n = X_n - \mu \quad \text{for } n = 0, 1, \ldots,$$

then $\gamma(\cdot)$ is also the *autocorrelation* function of the process. However, in the context of time series, this term is generally used for another function (that is defined below).

Definition 7.1.3. *If the stochastic process $\{X_n, n = 0, 1, \ldots\}$ is such that its mean $E[X_n]$ is equal to a constant $\mu \in \mathbb{R}$ for all n and Equation (7.1) is satisfied for all $i \in \{0, 1, \ldots\}$ and $j \in \{-i, -i+1, \ldots\}$, then it is said to be* **weakly** *or* **second-order stationary**.

Remarks. (i) Notice that

$$\text{VAR}[X_n] = \text{COV}[X_n, X_n] = \gamma(0),$$

which is independent of n. We denote the variance of X_n by σ^2 (or by σ_X^2 where necessary to avoid possible confusion).

(ii) Equation (7.1) implies that the function $\gamma(\cdot)$ is an *even* function. Moreover, it can be shown that

$$|\text{COV}[X_n, X_m]| \leq \text{STD}[X_n]\text{STD}[X_m]$$

(because the correlation coefficient of X_n and X_m belongs to the interval $[-1, 1]$). It follows that

$$|\text{COV}[X_n, X_{n+i}]| \stackrel{i \geq 0}{=} |\gamma(i)| \leq \text{STD}[X_n]\text{STD}[X_{n+i}] = \gamma(0) \equiv \sigma^2.$$

Therefore, the function $\gamma(\cdot)$ is bounded above by the positive constant σ^2.

(iii) In the case of a continuous-time stochastic process $\{X(t), t \geq 0\}$, the condition (7.1) becomes

$$\text{COV}[X(t), X(t+s)] = \gamma(s) = \gamma(-s) \quad \text{for all } t \geq 0 \text{ and } s \geq -t.$$

Furthermore, we may then assume that $\gamma(\cdot)$ is a *continuous* function.

Definition 7.1.4. *Let* $\{X_n, n = 0, 1, \ldots\}$ *be a weakly stationary stochastic process. The function*

$$\rho(k) = \frac{\gamma(k)}{\gamma(0)} = \frac{\gamma(k)}{\sigma^2}$$

is called the **autocorrelation function** *of the process.*

Remarks. (i) The function $\rho(\cdot)$ is sometimes denoted by ACF. Likewise, some authors write ACVF for the autocovariance function.

(ii) The autocorrelation function is such that

$$-1 \leq \rho(k) \leq 1 \quad \text{and} \quad \rho(0) = 1.$$

Moreover, because $\gamma(k) = \gamma(-k)$, we also have that $\rho(k) = \rho(-k)$. Therefore, it is not necessary to calculate the autocorrelation function for negative values of k.

Example 7.1.1. Suppose that the random variables $X_0, X_1, \ldots$ are independent and identically distributed. In addition, assume that $E[X_n] \equiv 0$ and $\text{VAR}[X_n] \equiv \sigma^2 < \infty$. Then, we have that (for $i, i+j \in \{0, 1, \ldots\}$)

$$\text{COV}[X_i, X_{i+j}] = E[X_i X_{i+j}] - 0^2 \stackrel{\text{ind.}}{=} E[X_i]E[X_{i+j}] = 0 \quad \text{if } j \neq 0.$$

In the case when $j = 0$, we can write that

$$\text{COV}[X_i, X_{i+j}] \overset{j=0}{=} \text{COV}[X_i, X_i] = \text{VAR}[X_i] = \sigma^2.$$

That is,

$$\text{COV}[X_{n_1}, X_{n_2}] = \gamma(n_2 - n_1) = \begin{cases} 0 & \text{if } n_2 - n_1 \neq 0, \\ \sigma^2 & \text{if } n_2 - n_1 = 0. \end{cases}$$

Because the mean $E[X_n]$ of the random variables is a constant and the covariance $\text{COV}[X_{n_1}, X_{n_2}]$ depends only on the difference $n_2 - n_1$, the stochastic process $\{X_n, n = 0, 1, \ldots\}$ is an example of a weakly stationary process. It is called an *i.i.d. noise* (with zero mean) and is denoted by $\text{IID}(0, \sigma^2)$.

Example 7.1.2. If the stochastic process $\{X_n, n = 0, 1, \ldots\}$ having as state space the set $\{0, 1, 2, \ldots\}$ is such that

$$P[X_{n+1} = j \mid X_n = i, X_{n-1} = i_{n-1}, \ldots, X_0 = i_0] = P[X_{n+1} = j \mid X_n = i]$$

for all states $i_0, \ldots, i_{n-1}, i, j \in \{0, 1, \ldots\}$ and for $n = 0, 1, \ldots$, then $\{X_n, n = 0, 1, \ldots\}$ is called a (discrete-time) *Markov chain*. Assume, in addition, that we can write that

$$P[X_{n+1} = j \mid X_n = i] := p_{i,j}.$$

That is, the conditional probability does not depend on n. Then, the Markov chain is said to be *time-homogeneous*. In general, the probability that the Markov chain will move from state i to state j in r *steps* is given by

$$P[X_{n+r} = j \mid X_n = i] := p_{i,j}^{(r)}$$

for any n.

Suppose that the limit

$$\pi_j := \lim_{n \to \infty} P[X_n = j \mid X_0 = i]$$

exists and does not depend on the initial state i. If we also suppose that

$$P[X_0 = i] = \pi_i \quad \text{for all } i \in \{0, 1, \ldots\},$$

then we can show that

$$P[X_n = i] = \pi_i \quad \text{for } n = 0, 1, \ldots \text{ and for all } i \in \{0, 1, \ldots\}.$$

Hence, Definition 7.1.1 is satisfied for $k = 1$. Actually, we can show that the Markov chain $\{X_n, n = 0, 1, \ldots\}$ is (strictly) stationary. In particular, if $n_2 \geq n_1$, we have that

$$P[X_{n_1+m} = i, X_{n_2+m} = j] = P[X_{n_2+m} = j \mid X_{n_1+m} = i] P[X_{n_1+m} = i]$$
$$= p_{i,j}^{(n_2-n_1)} \pi_i.$$

Note that the joint probability $P[X_{n_1+m} = i, X_{n_2+m} = j]$ does not depend on $m \in \{0, 1, \ldots\}$.

The process in Example 7.1.1 can be generalized as follows.

Definition 7.1.5. *Suppose that the random variables $X_0, X_1, \ldots$ are such that $E[X_n] \equiv 0$ and $VAR[X_n] \equiv \sigma^2 < \infty$. If the X_ns are uncorrelated and identically distributed, then the stochastic process $\{X_n, n = 0, 1, \ldots\}$ is called a **white noise** (process) and is denoted by $WN(0, \sigma^2)$.*

Remarks. (i) An i.i.d. noise is a white noise process, because if two random variables are independent, then they are also uncorrelated. However, the converse is not always true.

(ii) A very important particular case is the one when the X_ns all have a Gaussian distribution. We have seen that, in such a case, the random variables are independent *if and only if* their (covariance or) correlation coefficient is equal to 0. Hence, a *Gaussian white noise* is an i.i.d. noise. We denote this process by $\mathrm{GWN}(0, \sigma^2)$.

In Chapter 3, we defined the Gaussian or normal distribution (see p. 70). We now generalize this definition.

Definition 7.1.6. *Let $Z_1, \ldots, Z_m$ be independent $N(0, 1)$ random variables, and let μ_k be a real constant, for $k = 1, \ldots, n$. If the random variable X_k is defined by*

$$X_k = \mu_k + \sum_{j=1}^{m} c_{kj} Z_j \quad \text{for } k = 1, \ldots, n,$$

*where the c_{kj}s are real constants, then the random vector $(X_1, X_2, \ldots, X_n)$ has a **multinormal** or **multivariate normal distribution**.*

Remarks. (i) We say that X_k is a *linear combination* of the random variables $Z_1, \ldots, Z_m$.

(ii) Equivalently, $(X_1, X_2, \ldots, X_n)$ has a multinormal distribution if and only if

$$a_1 X_1 + \cdots + a_n X_n$$

has a normal distribution for all real constants $a_1, \ldots, a_n$ (with at least one $a_i \neq 0$, although we can actually take $a_i \equiv 0$ because the constant 0 can be considered as a *degenerate* normal random variable).

It can be shown that the joint density function of the random vector $\mathbf{X} := (X_1, \ldots, X_n)$ is given by

$$f_{\mathbf{X}}(\mathbf{x}) = \frac{1}{\sqrt{(2\pi)^n \det \mathbf{C}}} \exp\left\{ -\tfrac{1}{2}(\mathbf{x} - \mathbf{m})\mathbf{C}^{-1}(\mathbf{x}^T - \mathbf{m}^T) \right\} \tag{7.2}$$

for $\mathbf{x} := (x_1, \ldots, x_n) \in \mathbb{R}^n$, where

$$\mathbf{m} := (E[X_1], \ldots, E[X_n]) = (\mu_1, \ldots, \mu_n),$$

$$\mathbf{C} := (\mathrm{COV}[X_i, X_j])_{i,j=1,\ldots,n}$$

and T denotes the *transpose* of the vector.

Remarks. (i) The matrix $\mathbf{C}$ is called the *covariance matrix* of the random vector $\mathbf{X}$. Actually, it must be nonsingular for the above expression to be valid. That is, the determinant of $\mathbf{C}$ must be different from zero.

(ii) Note that the distribution of $\mathbf{X}$ depends only on the vector of means $\mathbf{m}$ and the covariance matrix $\mathbf{C}$.

(iii) We use the following notation: $\mathbf{X} \sim \mathrm{N}(\mathbf{m}, \mathbf{C})$.

Properties. (i) If the random vector $(X_1, X_2, \ldots, X_n)$ has a multinormal distribution, then any linear combination of the X_ks also has a multinormal distribution.

(ii) If, in addition, $\mathrm{COV}[X_i, X_j] = 0$, then X_i and X_j are independent random variables.

Definition 7.1.7. *A stochastic process* $\{X_n, n = 0, 1, \ldots\}$ *is said to be* **Gaussian** *if the random vector* $(X_{n_1}, \ldots, X_{n_k})$ *has a multinormal distribution for all* $n_1, \ldots, n_k \in \{0, 1, \ldots\}$ *and for any* $k \in \{1, 2, \ldots\}$.

Remarks. (i) The definition can be extended to the case when $\{X(t), t \in T\}$, where $T \subset \mathbb{R}$, is a continuous-time stochastic process.

(ii) Any *affine* transformation $\{Y_n, n = 0, 1, \ldots\}$ of a Gaussian process $\{X_n, n = 0, 1, \ldots\}$ is also a Gaussian process. That is, Y_n is defined by

$$Y_n = \alpha X_n + \beta \quad \text{for } n = 0, 1, \ldots,$$

where $\alpha \ (\neq 0)$ and β are constants. The process $\{Z_n, n = 0, 1, \ldots\}$ defined by

$$Z_n = \alpha X_{2n} + \beta \quad \text{for } n = 0, 1, \ldots,$$

for instance, is a Gaussian process as well. However, if we set

$$Z_n = X_n^2 \quad \text{for } n = 0, 1, \ldots,$$

then $\{Z_n, n = 0, 1, \ldots\}$ is clearly not a Gaussian process because $Z_n \geq 0$, which implies that Z_n does not have a Gaussian distribution.

A reason why Gaussian processes are so useful is given in the next proposition.

Proposition 7.1.1. *If a Gaussian process* $\{X_n, n = 0, 1, \ldots\}$ *is weakly stationary, then it is also (strictly) stationary.*

Proof. The proof is based on the fact that a multivariate normal distribution is completely determined by its vector of means and its covariance matrix, as mentioned above.
∎

Example 7.1.3. If (X_1, X_2) is a random vector having a multinormal distribution (i.e., a *bivariate normal* or *binormal distribution*, to be precise), then its joint probability density function can be written as follows:

$$f_{X_1, X_2}(x_1, x_2) =$$

$$c \exp\left\{ -\frac{1}{2(1 - \rho_{X_1, X_2}^2)} \left[\sum_{i=1}^{2} \left(\frac{x_i - \mu_{X_i}}{\sigma_{X_i}} \right)^2 - 2\rho_{X_1, X_2} \frac{(x_1 - \mu_{X_1})(x_2 - \mu_{X_2})}{\sigma_{X_1}\sigma_{X_2}} \right] \right\}$$

for any $(x_1, x_2) \in \mathbb{R}^2$, where $\mu_{X_i} \in \mathbb{R}$ and $\sigma_{X_i} > 0$, for $i = 1, 2$, and $-1 < \rho_{X_1, X_2} < 1$. Moreover, c is a positive constant defined by

$$c = \left[2\pi \sigma_{X_1} \sigma_{X_2} (1 - \rho_{X_1, X_2}^2)^{1/2} \right]^{-1}$$

(see Figure 7.1).

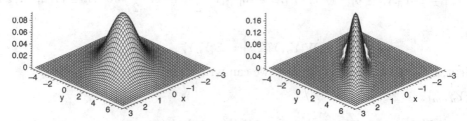

Fig. 7.1. Joint probability density functions of a random vector having a bivariate normal distribution with $\mu_{X_1} = 0$, $\mu_{X_2} = 1$, $\sigma_{X_1} = 1$, $\sigma_{X_2} = 4$, $\rho_{X_1, X_2} = 1/2$ (left), and $\rho_{X_1, X_2} = 5/6$ (right).

Because every linear combination of the random variables X_1 and X_2 has a Gaussian distribution, we can state that X_1 and X_2 themselves are Gaussian random variables. We find that $E[X_i] = \mu_{X_i}$ and $\mathrm{VAR}[X_i] = \sigma_{X_i}^2$, for $i = 1, 2$, which can be checked by integrating the joint density function $f_{X_1, X_2}(x_1, x_2)$ with respect to the appropriate variable. It follows that

$$f_{X_2}(x_2 \mid X_1 = x_1) = \frac{f_{X_1, X_2}(x_1, x_2)}{f_{X_1}(x_1)}$$

$$= c^* \exp\left\{ -\frac{1}{2(1 - \rho_{X_1, X_2}^2)} \left[\left(\frac{x_2 - \mu_{X_2}}{\sigma_{X_2}} \right) - \rho_{X_1, X_2} \left(\frac{x_1 - \mu_{X_1}}{\sigma_{X_1}} \right) \right]^2 \right\},$$

where

$$c^* = \left[\sqrt{2\pi} \sigma_{X_2} (1 - \rho_{X_1, X_2}^2)^{1/2} \right]^{-1}.$$

That is, we can write that

$$X_2 \mid \{X_1 = x_1\} \sim \mathrm{N}\left(\mu_{X_2} + \rho_{X_1,X_2}\frac{\sigma_{X_2}}{\sigma_{X_1}}(x_1 - \mu_{X_1}), \sigma_{X_2}^2(1 - \rho_{X_1,X_2}^2)\right). \qquad (7.3)$$

Remark. Note that if $\rho_{X_1,X_2} = 0$, then $f_{X_2}(x_2 \mid X_1 = x_1) \equiv f_{X_2}(x_2)$, which confirms the fact that if the correlation coefficient of two Gaussian random variables is equal to zero, then they are *independent* random variables (see p. 133).

Example 7.1.4. Let $\{X(t), t \geq 0\}$ be a Gaussian process with stationary and independent increments such that $X(0) = 0$ and $X(t) \sim \mathrm{N}(0,t)$, for $t > 0$. Define

$$Y_n = \begin{cases} 1 & \text{if } X(n) \geq 0, \\ -1 & \text{if } X(n) < 0 \end{cases}$$

for $n = 1, 2, \ldots$. Is the stochastic process $\{Y_n, n = 1, 2, \ldots\}$ weakly stationary?

Solution. We have that

$$E[Y_n] = 1 \cdot P[X_n \geq 0] + (-1) \cdot P[X_n < 0] = 1 \cdot \frac{1}{2} + (-1) \cdot \frac{1}{2} = 0 \ \forall n \in \{1, 2, \ldots\}.$$

Then,

$$\mathrm{COV}[Y_n, Y_{n+m}] = E[Y_n Y_{n+m}].$$

Next, by symmetry and continuity, we can write (for $m \in \{0, 1, \ldots\}$) that

$E[Y_n Y_{n+m}]$

$$\begin{aligned} &= 2P[X(n) \geq 0, X(n+m) \geq 0] - 2P[X(n) \geq 0, X(n+m) < 0] \\ &= 2\{P[X(n+m) \geq 0 \mid X(n) \geq 0] - P[X(n+m) < 0 \mid X(n) \geq 0]\} \\ &\quad \times \underbrace{P[X(n) \geq 0]}_{1/2} \\ &= 2P[X(n+m) \geq 0 \mid X(n) \geq 0] - 1 \\ &= 2\int_0^\infty P[X(n+m) \geq 0 \mid X(n) \geq 0, X(n) = u] f_{X(n)}(u \mid X(n) \geq 0) du - 1, \end{aligned}$$

where

$$f_{X(n)}(u \mid X(n) \geq 0) = \frac{f_{X(n)}(u)}{P[X(n) \geq 0]} = 2f_{X(n)}(u) = \frac{2}{\sqrt{2\pi n}} \exp\left\{-\frac{u^2}{2n}\right\}$$

for $u \geq 0$. It follows, using the fact that $\{X(t), t \geq 0\}$ has stationary and independent increments, that

$$E[Y_n Y_{n+m}] = 4\int_0^\infty P[\mathrm{N}(0,m) \geq -u]\frac{1}{\sqrt{2\pi n}} \exp\left\{-\frac{u^2}{2n}\right\} du - 1.$$

Now,

$$P\left[N(0,m) \geq -u\right] = 1 - \Phi(-u/\sqrt{m}) = \Phi(u/\sqrt{m}) = \int_{-\infty}^{u/\sqrt{m}} \frac{1}{\sqrt{2\pi}} e^{-z^2/2} dz$$

$$= \frac{1}{2} + \int_{0}^{u/\sqrt{m}} \frac{1}{\sqrt{2\pi}} e^{-z^2/2} dz \overset{y=z/\sqrt{2}}{=} \frac{1}{2} + \int_{0}^{u/\sqrt{2m}} \frac{1}{\sqrt{\pi}} e^{-y^2} dy$$

$$= \frac{1}{2} + \frac{1}{2} \mathrm{erf}(u/\sqrt{2m}),$$

where *erf* is called the *error function*. Hence,

$$E[Y_n Y_{n+m}] = 4 \int_0^\infty \frac{1}{2} \left[1 + \mathrm{erf}(u/\sqrt{2m})\right] \frac{1}{\sqrt{2\pi n}} \exp\left\{-\frac{u^2}{2n}\right\} du - 1$$

$$= 2 \left\{\frac{1}{2} + \int_0^\infty \mathrm{erf}(u/\sqrt{2m}) \frac{1}{\sqrt{2\pi n}} \exp\left\{-\frac{u^2}{2n}\right\} du\right\} - 1$$

$$= \frac{2}{\sqrt{2\pi n}} \int_0^\infty \mathrm{erf}(u/\sqrt{2m}) \exp\left\{-\frac{u^2}{2n}\right\} du.$$

Finally, making use of a mathematical software package, such as *Maple*, we find that the above integral is given by

$$\int_0^\infty \mathrm{erf}(u/\sqrt{2m}) \exp\left\{-\frac{u^2}{2n}\right\} du = \frac{\sqrt{2n}}{\sqrt{\pi}} \arctan(\sqrt{n}/\sqrt{m}),$$

so that

$$\mathrm{COV}[Y_n, Y_{n+m}] = E[Y_n Y_{n+m}] = \frac{2}{\pi} \arctan(\sqrt{n}/\sqrt{m}).$$

$\mathrm{COV}[Y_n, Y_{n+m}]$ depends on n, therefore the stochastic process $\{Y_n, n = 1, 2, \ldots\}$ is *not* weakly stationary.

Remark. The stochastic process $\{X(t), t \geq 0\}$ is known as a *standard Brownian motion* and is very important for the applications.

7.2 Particular time series models

7.2.1 Autoregressive processes

Let $\{Y_n, n = 0, 1, \ldots\}$ be a time series such that $E[Y_n] \equiv \mu$. We define

$$X_n = Y_n - \mu \quad \text{for } n = 0, 1, \ldots,$$

so that $E[X_n] \equiv 0$. A simple model is the one for which the state variable X_n is defined by

$$X_n = \alpha_1 X_{n-1} + \epsilon_n \quad \text{for } n = 1, 2, \ldots, \tag{7.4}$$

where $\alpha_1 \ (\neq 0)$ is a constant and $\{\epsilon_n, n = 1, 2, \ldots\}$ is a white noise $\text{WN}(0, \sigma^2)$ process. Furthermore, ϵ_n is independent of (or is at least uncorrelated with) X_{n-1}, X_{n-2}, $\ldots$, X_0. The stochastic process $\{X_n, n = 0, 1, \ldots\}$ is called an *autoregressive* process. We have that

$$X_n = \alpha_1 X_{n-1} + \epsilon_n = \alpha_1(\alpha_1 X_{n-2} + \epsilon_{n-1}) + \epsilon_n = \alpha_1^2 X_{n-2} + \alpha_1 \epsilon_{n-1} + \epsilon_n$$

$$= \cdots = \alpha_1^n X_0 + \sum_{i=1}^{n} \alpha_1^{n-i} \epsilon_i. \tag{7.5}$$

Now, if we define $\epsilon_0 = X_0$, then we can write that

$$X_n = \sum_{i=0}^{n} \alpha_1^{n-i} \epsilon_i \quad \text{for } n = 0, 1, 2, \ldots.$$

It follows (with $k \in \{0, 1, \ldots\}$) that

$$\text{COV}\left[X_n, X_{n+k}\right] = \text{COV}\left[\sum_{i=0}^{n} \alpha_1^{n-i} \epsilon_i, \sum_{i=0}^{n+k} \alpha_1^{n+k-i} \epsilon_i\right]$$

$$= \text{COV}\left[\sum_{i=0}^{n} \alpha_1^{n-i} \epsilon_i, \sum_{i=0}^{n} \alpha_1^{n+k-i} \epsilon_i + \sum_{i=n+1}^{n+k} \alpha_1^{n+k-i} \epsilon_i\right]$$

$$= \text{COV}\left[\sum_{i=0}^{n} \alpha_1^{n-i} \epsilon_i, \sum_{i=0}^{n} \alpha_1^{n+k-i} \epsilon_i\right] + \text{COV}\left[\sum_{i=0}^{n} \alpha_1^{n-i} \epsilon_i, \sum_{i=n+1}^{n+k} \alpha_1^{n+k-i} \epsilon_i\right].$$

The random variables ϵ_i being uncorrelated, the second term above is equal to 0, so that

$$\text{COV}\left[X_n, X_{n+k}\right] = \text{COV}\left[\sum_{i=0}^{n} \alpha_1^{n-i} \epsilon_i, \sum_{i=0}^{n} \alpha_1^{n+k-i} \epsilon_i\right]$$

$$= E\left[\left\{\sum_{i=0}^{n} \alpha_1^{n-i} \epsilon_i\right\}\left\{\sum_{i=0}^{n} \alpha_1^{n+k-i} \epsilon_i\right\}\right] = \sum_{i=0}^{n} \alpha_1^{n-i} \alpha_1^{n+k-i} E[\epsilon_i^2]$$

because

$$E[\epsilon_i \epsilon_j] = E[\epsilon_i] E[\epsilon_j] = 0 \quad \text{if } i \neq j.$$

Assume that $-1 < \alpha_1 < 1$ and that

$$E[\epsilon_0^2] = \text{VAR}[\epsilon_0] = \frac{\sigma^2}{1 - \alpha_1^2}.$$

Then,

$$\mathrm{COV}[X_n, X_{n+k}] = \alpha_1^{2n+k}\left(\frac{\sigma^2}{1-\alpha_1^2} + \sum_{i=1}^{n}\alpha_1^{-2i}\sigma^2\right) = \frac{\alpha_1^k}{1-\alpha_1^2}\sigma^2. \tag{7.6}$$

Indeed, we have that

$$S_n := \sum_{i=1}^{n}\alpha_1^{-2i} = \alpha_1^{-2} + \alpha_1^{-4} + \cdots + \alpha_1^{-2n}$$

$$\implies \quad \alpha_1^{-2}S_n = \alpha_1^{-4} + \cdots + \alpha_1^{-2n} + \alpha_1^{-2n-2}$$

$$\implies \quad \left(1-\alpha_1^{-2}\right)S_n = \alpha_1^{-2} - \alpha_1^{-2n-2}$$

$$\implies \quad S_n = \frac{\alpha_1^{-2} - \alpha_1^{-2n-2}}{1-\alpha_1^{-2}} = \frac{\alpha_1^{-2n} - 1}{1-\alpha_1^2}, \tag{7.7}$$

from which Equation (7.6) follows at once.

Similarly,

$$\mathrm{COV}[X_n, X_{n-k}] = \mathrm{COV}\left[\sum_{i=0}^{n}\alpha_1^{n-i}\epsilon_i, \sum_{i=0}^{n-k}\alpha_1^{n-k-i}\epsilon_i\right]$$

$$\overset{\mathrm{unc.}}{=} \mathrm{COV}\left[\sum_{i=0}^{n-k}\alpha_1^{n-i}\epsilon_i, \sum_{i=0}^{n-k}\alpha_1^{n-k-i}\epsilon_i\right] \overset{\mathrm{unc.}}{=} \sum_{i=0}^{n-k}\alpha_1^{n-i}\alpha_1^{n-k-i}E[\epsilon_i^2]$$

$$= \alpha_1^{2n-k}\sum_{i=0}^{n-k}\alpha_1^{-2i}E[\epsilon_i^2] \overset{\alpha_1^2<1}{=} \alpha_1^{2n-k}\sigma^2\left\{\frac{1}{1-\alpha_1^2} + \frac{\alpha_1^{-2(n-k)}-1}{1-\alpha_1^2}\right\}$$

$$= \frac{\sigma^2}{1-\alpha_1^2}\alpha_1^k \quad \text{for } k = 0, 1, \ldots, n.$$

Given that $\mathrm{COV}[X_n, X_{n\pm k}]$ can be written as $\gamma(k)$ (and $E[X_n] \equiv 0$), we may assert that the stochastic process $\{X_n, n = 0, 1, \ldots\}$ is a weakly stationary time series. Furthermore, we have that

$$\sigma_X^2 \equiv \mathrm{VAR}[X_n] = \mathrm{COV}[X_n, X_n] = \frac{\sigma^2}{1-\alpha_1^2} \quad \text{for } n = 0, 1, \ldots,$$

which implies that

$$\rho_{X_n, X_{n+k}} = \alpha_1^k \quad \text{for } n, k \in \{0, 1, \ldots\}. \tag{7.8}$$

Remarks. (i) Let $\mathbb{Z} = \{\ldots, -1, 0, 1, \ldots\}$. If we consider the stochastic processes $\{X_n, n \in \mathbb{Z}\}$ and $\{\epsilon_n, n \in \mathbb{Z}\}$, where $\mathrm{VAR}[\epsilon_n] \equiv \sigma^2$, then we can write that

$$X_n = \sum_{i=0}^{\infty}\alpha_1^i\epsilon_{n-i} \quad \text{for } n \in \mathbb{Z} \tag{7.9}$$

and we retrieve the formula for $\mathrm{COV}[X_n, X_{n+k}]$ *if* we assume that $|\alpha_1| < 1$, as above. Indeed, first we have that

$$E[X_n] = E\left[\sum_{i=0}^{\infty} \alpha_1^i \epsilon_{n-i}\right] = 0 \quad \text{for all } n \in \mathbb{Z}.$$

Then, using the fact that $\{\epsilon_n, n \in \mathbb{Z}\}$ is a $\mathrm{WN}(0, \sigma^2)$ process, we can write that

$$\mathrm{VAR}[X_n] = E[X_n^2] = E\left[\left(\sum_{i=0}^{\infty} \alpha_1^i \epsilon_{n-i}\right)^2\right] = \sum_{i=0}^{\infty} \alpha_1^{2i} E[\epsilon_{n-i}^2]$$

$$= \sum_{i=0}^{\infty} \alpha_1^{2i} \sigma^2 \stackrel{\alpha_1^2 < 1}{=} \left(\frac{1}{1 - \alpha_1^2}\right) \sigma^2 \quad \text{for all } n \in \mathbb{Z}. \tag{7.10}$$

Finally,

$$\mathrm{COV}[X_n, X_{n+k}] = E[X_n X_{n+k}] = E\left[\left(\sum_{i=0}^{\infty} \alpha_1^i \epsilon_{n-i}\right)\left(\sum_{j=0}^{\infty} \alpha_1^j \epsilon_{n+k-j}\right)\right]$$

$$= \sum_{i=0}^{\infty} \alpha_1^{i+(k+i)} E[\epsilon_{n-i}^2] = \sum_{i=0}^{\infty} \alpha_1^{2i+k} \sigma^2 = \left(\frac{\alpha_1^k}{1 - \alpha_1^2}\right) \sigma^2$$

if $k \in \{0, 1, \ldots\}$. When $k = -1, -2, \ldots$, we have that

$$\mathrm{COV}[X_n, X_{n+k}] = E\left[\left(\sum_{i=0}^{\infty} \alpha_1^i \epsilon_{n-i}\right)\left(\sum_{j=0}^{\infty} \alpha_1^j \epsilon_{n+k-j}\right)\right]$$

$$= E\left[\left(\sum_{i=k}^{\infty} \alpha_1^{i-k} \epsilon_{n+k-i}\right)\left(\sum_{j=0}^{\infty} \alpha_1^j \epsilon_{n+k-j}\right)\right]$$

$$= \sum_{j=0}^{\infty} \alpha_1^{2j-k} E[\epsilon_{n+k-j}^2] = \sigma^2 \left(\frac{\alpha_1^{-k}}{1 - \alpha_1^2}\right).$$

Hence, we can write that

$$\mathrm{COV}[X_n, X_{n+k}] = \left(\frac{\alpha_1^{|k|}}{1 - \alpha_1^2}\right) \sigma^2 \quad \text{for all } n, k \in \mathbb{Z}. \tag{7.11}$$

(ii) Suppose that the independent random variables ϵ_n are such that

$$P[\epsilon_n = 1] = p_0 = 1 - P[\epsilon_n = -1] \quad \text{for all } n \in \{1, 2 \ldots\},$$

where $p_0 \in (0, 1)$. Then, the stochastic process $\{X_n, n = 0, 1, \ldots\}$ defined by

$$X_n = X_{n-1} + \epsilon_n \quad \text{for } n = 1, 2, \ldots$$

(with ϵ_n independent of $X_{n-1}, \ldots, X_0$, for $n = 1, 2, \ldots$) is called a *random walk*. The initial state X_0 is often chosen to be 0. Note that if $p_0 = 1/2$, so that $E[\epsilon_n] \equiv 0$, $\{X_n, n = 0, 1, \ldots\}$ is an autoregressive process with $\alpha_1 = 1$. However, it is not stationary, which can be seen as follows:

$$X_n = X_{n-1} + \epsilon_n = X_{n-2} + \epsilon_{n-1} + \epsilon_n = \cdots = X_0 + \sum_{i=1}^{n} \epsilon_i,$$

so that

$$E[X_n] = E[X_0] + \sum_{i=1}^{n} E[\epsilon_i] = E[X_0]$$

a constant, but

$$\text{VAR}[X_n] \stackrel{\text{ind.}}{=} \text{VAR}[X_0] + \sum_{i=1}^{n} \text{VAR}[\epsilon_i] = \text{VAR}[X_0] + n\sigma^2,$$

where $\sigma^2 = 1 - 0^2 = 1$. Because the variance of X_n depends on n, the process $\{X_n, n = 0, 1, \ldots\}$ is not stationary (not even weakly).

The time series defined by Equation (7.4) is known as an autoregressive process of *order 1*, because it assumes that X_n depends only on one value of the process into the past. It can be generalized, as follows.

Definition 7.2.1. *The stochastic process* $\{X_n, n \in \mathbb{Z}\}$ *defined by*

$$X_n = \sum_{i=1}^{p} \alpha_i X_{n-i} + \epsilon_n \quad \text{for all } n \in \mathbb{Z} = \{\ldots, -1, 0, 1, \ldots\}, \tag{7.12}$$

where the $\alpha_i s$ *are constants,* $\{\epsilon_n, n \in \mathbb{Z}\}$ *is a white noise* $WN(0, \sigma^2)$ *process, and* ϵ_n *is independent of* $X_{n-1}, X_{n-2}, \ldots,$ *is called an* **autoregressive process of order** p *and is denoted by* $AR(p)$*.*

Remarks. (i) If we consider the time series from $n = 0$ only, Equation (7.12) is then valid for $n = p, p + 1, \ldots$. Alternatively, we can set $X_{n-i} = 0$ if $n - i < 0$.

(ii) We could add a constant term α_0 in (7.12).

To obtain the variance of an autoregressive process of order 1, given in Equation (7.10), we can also proceed as follows: because (by assumption) $E[X_n] \equiv 0$, we have that

$$\text{VAR}[X_n] = E[X_n^2] = E\left[(\alpha_1 X_{n-1} + \epsilon_n)^2\right]$$
$$= \alpha_1^2 E[X_{n-1}^2] + 2\alpha_1 E[X_{n-1}\epsilon_n] + E[\epsilon_n^2]$$
$$\overset{\text{ind.}}{=} \alpha_1^2 E[X_{n-1}^2] + 2\alpha_1 E[X_{n-1}]E[\epsilon_n] + \sigma^2$$
$$= \alpha_1^2 E[X_{n-1}^2] + \sigma^2.$$

Next, if we assume that the time series is weakly stationary, then we can write that $E[X_n^2] \equiv E[X_0^2]$, which implies that

$$E[X_n^2] = \alpha_1^2 E[X_n^2] + \sigma^2 \implies \text{VAR}[X_n] = \frac{\sigma^2}{1 - \alpha_1^2} \quad \text{for all } n \in \mathbb{Z}.$$

Remark. We see that the condition $|\alpha_1| < 1$ must be fulfilled, otherwise the variance of X_n would be negative (if $|\alpha_1| > 1$).

Likewise,

$$\text{COV}[X_n, X_{n-k}] = E[X_n X_{n-k}] = E\left[(\alpha_1 X_{n-1} + \epsilon_n)X_{n-k}\right]$$
$$= \alpha_1 E[X_{n-1}X_{n-k}] + E[\epsilon_n X_{n-k}]$$
$$\overset{\text{ind.}}{=} \alpha_1 \text{COV}[X_{n-1}, X_{n-k}] \quad \text{for } k = 1, 2, \ldots.$$

Hence, assuming again that the time series is weakly stationary, we obtain that

$$\gamma(k) = \alpha_1 \gamma(k-1) = \alpha_1^2 \gamma(k-2) = \cdots = \alpha_1^k \gamma(0)$$
$$\implies \gamma(k) = \alpha_1^k \frac{\sigma^2}{1 - \alpha_1^2} \quad \text{for all } k \in \mathbb{N},$$

which agrees with Equation (7.11) [because $\gamma(-k) = \gamma(k)$].

Now, for a weakly stationary AR(p) process (with $E[X_n] \equiv 0$), we have that

$$\text{COV}[X_n, X_{n-k}] = E[X_n X_{n-k}] = E\left[\left(\sum_{i=1}^{p} \alpha_i X_{n-i} + \epsilon_n\right) X_{n-k}\right]$$
$$= \sum_{i=1}^{p} \alpha_i E[X_{n-i}X_{n-k}] + E[\epsilon_n X_{n-k}]$$
$$\overset{\text{ind.}}{=} \sum_{i=1}^{p} \alpha_i \text{COV}[X_{n-i}, X_{n-k}] \quad \text{for } k = 1, 2, \ldots.$$

The previous equation can be rewritten as follows:

$$\gamma(k) = \sum_{i=1}^{p} \alpha_i \gamma(k-i) \tag{7.13}$$

$$\implies \rho(k) = \sum_{i=1}^{p} \alpha_i \rho(k-i) \quad \text{for } k = 1, 2, \ldots . \tag{7.14}$$

It can be shown that the general solution of Equation (7.14), together with $\rho(0) = 1$, is given by

$$\rho(k) = \sum_{i=1}^{p} c_i \lambda_i^k \quad \text{for } k = 0, 1, 2, \ldots, \tag{7.15}$$

where the c_is are constants and the λ_is are the roots of the equation

$$\lambda^p - \sum_{i=1}^{p} \alpha_i \lambda^{p-i} = 0. \tag{7.16}$$

Remark. We assume that the p roots of the above equation are distinct.

To determine the constants c_i, for $i = 1, \ldots, p$, we can use the fact that $\rho(0) = 1$ and the various equations obtained by setting, recursively, k equal to $1, 2, \ldots, p-1$ in (7.14) [assuming that the p roots of Equation (7.16) have been found].

Remarks. (i) It can be shown that the condition $|\lambda_i| < 1$ must be fulfilled for $i = 1, \ldots, p$ for the AR(p) process to be weakly stationary. We then find that

$$\lim_{k \to \infty} \gamma(k) = \lim_{k \to \infty} \rho(k) = 0.$$

(ii) The equations (7.14) are known as the *Yule–Walker equations.*

(iii) If we define $z = 1/\lambda$ (assuming that $\lambda \neq 0$), then Equation (7.16) becomes

$$z^{-p} - \sum_{i=1}^{p} \alpha_i z^{i-p} = 0 \quad \Longleftrightarrow \quad 1 - \sum_{i=1}^{p} \alpha_i z^i = 0. \tag{7.17}$$

The condition for the time series to be weakly stationary is now that $|z|$ must be strictly greater than 1 for all roots of the equation.

Example 7.2.1. In the case of a weakly stationary AR(1) process, Equation (7.16) becomes

$$\lambda = \alpha_1,$$

which implies that Equation (7.15) is

$$\rho(k) = c_1 \alpha_1^k \quad \text{for } k = 0, 1, 2, \ldots .$$

The condition $\rho(0) = 1$ yields at once that $c_1 = 1$. Hence, we can write that

$$\rho(k) = \alpha_1^k \quad \text{for } k = 0, 1, 2, \ldots$$

[see Equation (7.8)].

Remarks. (i) Notice that, because $|\alpha_1|$ must be strictly smaller than 1, we deduce that the autocorrelation function of a weakly stationary AR(1) process decreases geometrically (in absolute value). Furthermore, if $0 < \alpha_1 < 1$, then $\rho(k)$ is always positive, whereas in the case when $-1 < \alpha_1 < 0$, the function $\rho(k)$ alternates between geometrically decreasing positive and negative values.

(ii) Substituting $\rho(k) = \alpha_1^k$ into Equation (7.14) with $p = 1$, we obtain that

$$\alpha_1^k = \alpha_1 \alpha_1^{k-1} = \alpha_1^k,$$

so that the equation is indeed satisfied.

Example 7.2.2. To obtain the autocorrelation function of a weakly stationary AR(2) process, we must first find the two roots of the equation [see (7.16)]

$$\lambda^2 - \alpha_1 \lambda - \alpha_2 = 0.$$

We have that

$$\lambda = \frac{1}{2} \left(\alpha_1 \pm \sqrt{\alpha_1^2 + 4\alpha_2} \right).$$

Let

$$\lambda_1 = \frac{1}{2} \left(\alpha_1 - \sqrt{\alpha_1^2 + 4\alpha_2} \right) \quad \text{and} \quad \lambda_2 = \frac{1}{2} \left(\alpha_1 + \sqrt{\alpha_1^2 + 4\alpha_2} \right).$$

We can write that

$$\rho(k) = c_1 \lambda_1^k + c_2 \lambda_2^k \quad \text{for } k = 0, 1, 2, \ldots.$$

Again, $\rho(0) = 1$ yields that $c_1 + c_2 = 1$, so that

$$\rho(k) = \lambda_2^k + c_1 (\lambda_1^k - \lambda_2^k) \quad \text{for } k = 0, 1, 2, \ldots.$$

Next, making use of Equation (7.14) with $p = 2$ and $k = 1$, we get that

$$\rho(1) = \alpha_1 \rho(0) + \alpha_2 \rho(-1) = \alpha_1 + \alpha_2 \rho(1),$$

which implies that

$$\rho(1) = \frac{\alpha_1}{1 - \alpha_2}.$$

Hence (see above),

$$\rho(1) = \lambda_2 + c_1 (\lambda_1 - \lambda_2) \quad \text{and} \quad \rho(1) = \frac{\alpha_1}{1 - \alpha_2},$$

which yields

$$c_1 = \frac{[\alpha_1/(1-\alpha_2)] - \lambda_2}{\lambda_1 - \lambda_2} \quad \text{and} \quad c_2 = \frac{\lambda_1 - [\alpha_1/(1-\alpha_2)]}{\lambda_1 - \lambda_2}.$$

Now, the variance of a weakly stationary $AR(p)$ process having zero mean can be obtained as follows:

$$\text{VAR}[X_n] = E[X_n^2] = E\left[\left(\sum_{i=1}^{p} \alpha_i X_{n-i} + \epsilon_n\right) X_n\right]$$

$$= \sum_{i=1}^{p} \alpha_i E[X_{n-i} X_n] + E[\epsilon_n X_n]$$

$$\stackrel{\text{ind.}}{=} \sum_{i=1}^{p} \alpha_i \text{COV}[X_{n-i}, X_n] + E[\epsilon_n^2]$$

$$= \sum_{i=1}^{p} \alpha_i \gamma(i) + \sigma^2. \tag{7.18}$$

We can use this equation, together with (7.13), to find an explicit expression for $\text{VAR}[X_n]$.

Example 7.2.3. Consider again a weakly stationary $AR(2)$ process. From Equation (7.18),

$$\text{VAR}[X_n] = \gamma(0) = \alpha_1 \gamma(1) + \alpha_2 \gamma(2) + \sigma^2. \tag{7.19}$$

Moreover, Equation (7.13), with $p = 2$ and $k = 1$, and the fact that $\gamma(-j) = \gamma(j)$ imply that

$$\gamma(1) = \alpha_1 \gamma(0) + \alpha_2 \gamma(-1) \quad \Longrightarrow \quad \gamma(1) = \frac{\alpha_1}{1 - \alpha_2}\gamma(0),$$

whereas the same equation with $p = 2$ and $k = 2$ gives

$$\gamma(2) = \alpha_1 \gamma(1) + \alpha_2 \gamma(0) \quad \Longrightarrow \quad \gamma(2) = \left(\frac{\alpha_1^2}{1 - \alpha_2} + \alpha_2\right)\gamma(0).$$

Finally, substituting into (7.19), we find that

$$\gamma(0) = \left(\frac{\alpha_1^2}{1 - \alpha_2} + \frac{\alpha_1^2 \alpha_2}{1 - \alpha_2} + \alpha_2^2\right)\gamma(0) + \sigma^2,$$

so that

$$\gamma(0) = \left(1 - \alpha_1^2 \frac{(1 + \alpha_2)}{1 - \alpha_2} - \alpha_2^2\right)^{-1} \sigma^2.$$

Remark. If we add the constant α_0 in Equation (7.12), then the mean of a weakly stationary $AR(p)$ process is given by

$$E[X_n] = \alpha_0 + \sum_{i=1}^{p} \alpha_i E[X_{n-i}] + E[\epsilon_n] = \alpha_0 + \sum_{i=1}^{p} \alpha_i E[X_n].$$

That is,

$$E[X_n] = \frac{\alpha_0}{1 - \sum_{i=1}^{p} \alpha_i}.$$

7.2.2 Moving average processes

Definition 7.2.2. *Let* $\{\epsilon_n, n \in \mathbb{Z}\}$ *be a white noise* $WN(0, \sigma^2)$ *process. The time series* $\{X_n, n \in \mathbb{Z}\}$ *defined by*

$$X_n = \epsilon_n + \sum_{i=1}^{q} \theta_i \epsilon_{n-i} \quad \text{for all } n \in \mathbb{Z}, \tag{7.20}$$

where the $\theta_i s$ *are constants, is called a* **moving average process of order** q *and is denoted by* $MA(q)$.

Remarks. (i) We could add the constant μ in the model. However, we prefer to work with the centered process $\{X_n, n \in \mathbb{Z}\}$.
(ii) Setting $\theta_0 = 1$, Equation (7.20) can be rewritten as follows:

$$X_n = \sum_{i=0}^{q} \theta_i \epsilon_{n-i} \quad \text{for all } n \in \mathbb{Z}.$$

(iii) We deduce from Equation (7.9) that an AR(1) process can be written as an MA(q) time series with q tending to infinity and with coefficients θ_i equal to α_1^i, for $i = 0, 1, \ldots$.

The mean and the variance of an MA(q) time series are easy to calculate. First, we (indeed) have that

$$E[X_n] = \sum_{i=0}^{q} \theta_i E[\epsilon_{n-i}] = 0 \quad \text{for all } n \in \mathbb{Z}.$$

Next,

$$\text{VAR}[X_n] = E[X_n^2] = E\left[\left(\sum_{i=0}^{q} \theta_i \epsilon_{n-i}\right)^2\right].$$

Because $\{\epsilon_n, n \in \mathbb{Z}\}$ is a white noise process, we can write that

$$\text{VAR}[X_n] = \sum_{i=0}^{q} E[\theta_i^2 \epsilon_{n-i}^2] = \sum_{i=0}^{q} \theta_i^2 \sigma^2. \tag{7.21}$$

Now, the covariance of the random variables X_n and X_{n+k} is given by

$$\text{COV}[X_n, X_{n+k}] = E\left[\left(\sum_{i=0}^{q} \theta_i \epsilon_{n-i}\right)\left(\sum_{i=0}^{q} \theta_i \epsilon_{n+k-i}\right)\right].$$

Again, the fact that $\{\epsilon_n, n \in \mathbb{Z}\}$ is a white noise process implies that

$$\text{COV}[X_n, X_{n+k}] = 0 \quad \text{if } k = q+1, q+2, \ldots.$$

When $k = 1, \ldots, q$, we obtain that

$$\text{COV}[X_n, X_{n+k}] = E\left[\left(\sum_{i=0}^{q} \theta_i \epsilon_{n-i}\right)\left(\sum_{j=-k}^{q-k} \theta_{j+k} \epsilon_{n-j}\right)\right]$$

$$= \sum_{i=0}^{q-k} E\left[\theta_i \theta_{i+k} \epsilon_{n-i}^2\right] = \sum_{i=0}^{q-k} \theta_i \theta_{i+k} \sigma^2.$$

In general, we find that

$$\text{COV}[X_n, X_{n+k}] = \begin{cases} \sigma^2 \sum_{i=0}^{q-|k|} \theta_i \theta_{i+|k|} & \text{if } |k| = 0, 1, \ldots, q, \\ 0 & \text{if } |k| = q+1, q+2, \ldots. \end{cases} \tag{7.22}$$

Hence, we can state that the time series $\{X_n, n \in \mathbb{Z}\}$ is *weakly stationary*. The autocorrelation function of a moving average process of order q is therefore

$$\rho(k) = \begin{cases} \dfrac{\sum_{i=0}^{q-|k|} \theta_i \theta_{i+|k|}}{\sum_{i=0}^{q} \theta_i^2} & \text{if } |k| = 0, 1, \ldots, q, \\ 0 & \text{if } |k| = q+1, q+2, \ldots. \end{cases} \tag{7.23}$$

Particular cases. (i) The autocorrelation function of an MA(1) process is given by

$$\rho(k) = \begin{cases} 1 & \text{if } k = 0, \\ \dfrac{\theta_1}{1 + \theta_1^2} & \text{if } |k| = 1, \\ 0 & \text{if } |k| = 2, 3, \ldots. \end{cases} \tag{7.24}$$

(ii) If $q = 2$, we obtain that

$$\rho(k) = \begin{cases} 1 & \text{if } k = 0, \\ \dfrac{\theta_1(1 + \theta_2)}{1 + \theta_1^2 + \theta_2^2} & \text{if } |k| = 1, \\ \dfrac{\theta_2}{1 + \theta_1^2 + \theta_2^2} & \text{if } |k| = 2, \\ 0 & \text{if } |k| = 3, 4, \ldots. \end{cases} \tag{7.25}$$

Remarks. (i) If we replace θ_1 ($\neq 0$) by $1/\theta_1$ in $\rho(1)$ for an MA(1) time series, we obtain the same formula:

$$\rho(1) = \frac{\theta_1^{-1}}{1 + \theta_1^{-2}} = \frac{\theta_1}{\theta_1^2 + 1}. \tag{7.26}$$

Furthermore, let

$$g(x) = \frac{x}{x^2 + 1}.$$

We have that

$$g'(x) = \frac{1 - x^2}{(x^2 + 1)^2} = 0 \quad \text{if and only if } x = \pm 1.$$

Because

$$g''(x) = \frac{2x(x^2 - 3)}{(x^2 + 1)^3},$$

we deduce that $x = -1$ (resp., $x = 1$) corresponds to a minimum (resp., maximum) of the function $g(x)$. Hence, we may conclude that

$$-\frac{1}{2} \le \rho(1) \le \frac{1}{2}.$$

Conversely, solving for θ_1 in Equation (7.26), we find that

$$\theta_1 = \frac{1}{2\rho(1)} \pm \left[\frac{1}{4\rho^2(1)} - 1 \right]^{1/2}.$$

If we denote the two roots by θ_{1+} and θ_{1-}, then we have that

$$\theta_{1+} = \frac{1}{\theta_{1-}}.$$

(ii) In the case of an MA(∞) process, it can be shown that

$$\gamma(k) = \sigma^2 \sum_{i=0}^{\infty} \theta_i \theta_{i+|k|},$$

provided that the coefficients θ_i are *absolutely summable*. That is, if

$$\sum_{i=0}^{\infty} |\theta_i| < \infty.$$

The reason why time series can be used to model various processes is given in the following proposition. It is based on a result, known as *Wold's decomposition theorem*, which asserts that every weakly stationary (discrete-time) stochastic process can be written as the sum of a deterministic process and a moving average process (of infinite order).

Proposition 7.2.1. *Any weakly stationary stochastic process* $\{X_n, n \in \mathbb{Z}\}$ *can be represented as follows:*

$$X_n = \epsilon_n + \sum_{i=1}^{\infty} \theta_i \epsilon_{n-i} + \nu_n,$$

where $\{\epsilon_n, n \in \mathbb{Z}\}$ *is a white noise* $WN(0, \sigma^2)$ *process,* $\{\nu_n, n \in \mathbb{Z}\}$ *is a linearly deterministic process, and*

$$\sum_{i=1}^{\infty} \theta_i^2 < \infty.$$

Remarks. (i) If $\nu_n \equiv 0$, then the stochastic process $\{X_n, n \in \mathbb{Z}\}$ is said to be *purely nondeterministic*, which means that all linear deterministic components have been subtracted from the process. Notice that any purely nondeterministic process can be expressed as a moving average process of infinite order.

(ii) A process is *linearly deterministic* if it is perfectly predictable from its own past.

(iii) The proposition implies that any weakly stationary process has a linear structure. That is, it is a linear combination of uncorrelated random variables.

Definition 7.2.3. *The* **lag** *or* **backshift operator** *is denoted by* **L** *and is defined by*

$$LX_n = X_{n-1} \quad \text{for all } n.$$

In general, we have that

$$L^k X_n = X_{n-k} \quad \text{for all } n \text{ and for } k = 0, 1, \ldots .$$

Furthermore, the (first) **difference operator,** Δ, *is given by*

$$\Delta = 1 - L,$$

so that

$$\Delta X_n = X_n - X_{n-1} \quad \text{for all } n.$$

Remarks. (i) The operators L and Δ can be manipulated as if they were algebraic quantities. Hence,

$$\Delta^2 X_n = (1 - L)^2 X_n = (1 - 2L + L^2) X_n = X_n - 2X_{n-1} + X_{n-2}.$$

(ii) If we apply the operators L and Δ to a constant μ, we obtain that

$$L\mu = \mu \quad \text{and} \quad \Delta\mu = 0.$$

Now, using the lag operator, we can express an AR(p) process as follows:

$$\left(1 - \sum_{i=1}^{p} \alpha_i L^i\right) X_n = \epsilon_n \quad \text{for all } n \in \mathbb{Z}.$$

The time series is weakly stationary if [see Equation (7.17)] all the roots of the polynomial equation

$$1 - \sum_{i=1}^{p} \alpha_i L^i = 0 \tag{7.27}$$

lie outside the unit circle.

Next, an MA(q) time series can also be expressed by means of the lag operator:

$$X_n = \left(1 + \sum_{i=1}^{q} \theta_i L^i\right) \epsilon_n \quad \text{for all } n \in \mathbb{Z}.$$

Definition 7.2.4. *We say that an MA(q) process is* **invertible** *if all the roots of the polynomial equation*

$$1 + \sum_{i=1}^{q} \theta_i L^i = 0 \tag{7.28}$$

lie outside the unit circle.

Particular cases. (i) If $q = 1$, we simply have that

$$1 + \theta_1 L = 0 \quad \Longrightarrow \quad L = -\frac{1}{\theta_1}.$$

The root L must be such that $|L| > 1$. That is,

$$|\theta_1| < 1.$$

(ii) In the case of an MA(2) process, the two roots of

$$1 + \theta_1 L + \theta_2 L^2 = 0$$

are

$$L_+ = \frac{1}{2\theta_2}\left[-\theta_1 + (\theta_1^2 - 4\theta_2)^{1/2}\right] \quad \text{and} \quad L_- = \frac{1}{2\theta_2}\left[-\theta_1 - (\theta_1^2 - 4\theta_2)^{1/2}\right].$$

We know that a complex number $z = x + iy$ lies outside the unit circle if $x^2 + y^2 > 1$. Here, we find that L_+ and L_- both lie outside the unit circle if and only if

$$\theta_1 + \theta_2 > -1, \quad \theta_1 - \theta_2 > -1 \quad \text{and} \quad -1 < \theta_2 < 1.$$

7.2.3 Autoregressive moving average processes

To conclude this section, we define *autoregressive moving average processes* which, as their name indicates, generalize both autoregressive and moving average processes.

Definition 7.2.5. *The (zero mean) stochastic process $\{X_n, n \in \mathbb{Z}\}$ defined by*

$$X_n = \sum_{i=1}^{p} \alpha_i X_{n-i} + \epsilon_n + \sum_{j=1}^{q} \theta_j \epsilon_{n-j} \quad \text{for all } n \in \mathbb{Z},$$

*where $\{\epsilon_n, n \in \mathbb{Z}\}$ is a white noise $WN(0, \sigma^2)$ process (with ϵ_n independent of X_{n-1}, $X_{n-2}, \ldots$), is called an **autoregressive moving average process of order** (p, q) and is denoted by $ARMA(p, q)$.*

Remarks. (i) An $ARMA(p, q)$ process may be weakly stationary and invertible. It is weakly stationary if all the roots of Equation (7.27) lie outside the unit circle. That is, if the autoregressive part of the process is weakly stationary. Likewise, if the MA(q) part of the time series is invertible, then so is the $ARMA(p, q)$ process.

(ii) A weakly stationary and invertible $ARMA(p, q)$ time series can be expressed as an infinite autoregressive process or, equivalently, as a moving average process of infinite order.

(iii) For an ARMA process to have a minimal number of terms, Equations (7.27) and (7.28) should have no common roots.

Example 7.2.4. Suppose that $\{X_n, n \in \mathbb{Z}\}$ can be modeled as follows:

$$X_n = \alpha_1 X_{n-1} \quad \text{for all } n \in \mathbf{Z},$$

but cannot be observed directly. Rather, we can observe the random variable Y_n defined by

$$Y_n = X_n + \epsilon_n \quad \text{for } n \in \mathbb{Z},$$

where $\{\epsilon_n, n \in \mathbb{Z}\}$ is a $WN(0, \sigma^2)$ process (and ϵ_n is independent of $Y_{n-1}, Y_{n-2}, \ldots$).
   Let us define

$$Z_n = Y_n - \alpha_1 Y_{n-1} \quad \text{for } n \in \mathbb{Z}. \tag{7.29}$$

We have that

$$Z_n = X_n + \epsilon_n - \alpha_1(X_{n-1} + \epsilon_{n-1}).$$

That is,

$$Z_n = X_n - \alpha_1 X_{n-1} + \epsilon_n - \alpha_1 \epsilon_{n-1}$$
$$= \epsilon_n - \alpha_1 \epsilon_{n-1} \quad \text{for } n \in \mathbb{Z}.$$

We may assert that $\{Z_n, n \in \mathbb{Z}\}$ is an MA(1) process with $\theta_1 = -\alpha_1$. Therefore, we deduce from (7.29) that $\{Y_n, n \in \mathbb{Z}\}$ is actually a particular ARMA(1,1) time series.

We now calculate the autocovariance function of a weakly stationary ARMA(1,1) process, which is defined by

$$X_n = \alpha_1 X_{n-1} + \epsilon_n + \theta_1 \epsilon_{n-1} \quad \text{for all } n \in \mathbb{Z}.$$

We have that [see (7.18)]

$$
\begin{aligned}
\gamma(0) \equiv \mathrm{VAR}[X_n] &= E\left[(\alpha_1 X_{n-1} + \epsilon_n + \theta_1 \epsilon_{n-1}) X_n\right] \\
&= \alpha_1 E[X_{n-1} X_n] + E[\epsilon_n X_n] + \theta_1 E[\epsilon_{n-1} X_n] \\
&\stackrel{\text{ind.}}{=} \alpha_1 \gamma(1) + E[\epsilon_n^2] + \theta_1 \left(E[\epsilon_{n-1}\alpha_1 X_{n-1}] + E[\theta_1 \epsilon_{n-1}^2]\right) \\
&= \alpha_1 \gamma(1) + \sigma^2 + \theta_1 \left(\alpha_1 E[\epsilon_{n-1}^2] + \theta_1 \sigma^2\right) \\
&= \alpha_1 \gamma(1) + \sigma^2 \left(1 + \theta_1 \alpha_1 + \theta_1^2\right).
\end{aligned}
\tag{7.30}
$$

Likewise,

$$
\begin{aligned}
\gamma(1) = \mathrm{COV}[X_n, X_{n-1}] &= E\left[(\alpha_1 X_{n-1} + \epsilon_n + \theta_1 \epsilon_{n-1}) X_{n-1}\right] \\
&= \alpha_1 E[X_{n-1}^2] + E[\epsilon_n X_{n-1}] + \theta_1 E[\epsilon_{n-1} X_{n-1}] \\
&\stackrel{\text{ind.}}{=} \alpha_1 \gamma(0) + \theta_1 E[\epsilon_{n-1}^2] \\
&= \alpha_1 \gamma(0) + \theta_1 \sigma^2
\end{aligned}
\tag{7.31}
$$

and

$$
\begin{aligned}
\gamma(k) = \mathrm{COV}[X_n, X_{n-k}] &= E\left[(\alpha_1 X_{n-1} + \epsilon_n + \theta_1 \epsilon_{n-1}) X_{n-k}\right] \\
&= \alpha_1 E[X_{n-1} X_{n-k}] + E[\epsilon_n X_{n-k}] + \theta_1 E[\epsilon_{n-1} X_{n-k}] \\
&\stackrel{\text{ind.}}{=} \alpha_1 \gamma(k-1) \quad \text{for } k = 2, 3, \ldots.
\end{aligned}
\tag{7.32}
$$

Notice that Equations (7.30) and (7.31) enable us to write that

$$\gamma(0) = \frac{\sigma^2 \left(1 + 2\theta_1 \alpha_1 + \theta_1^2\right)}{1 - \alpha_1^2} \quad \text{and} \quad \gamma(1) = \frac{\sigma^2 \left[\theta_1 + \alpha_1(1 + \theta_1 \alpha_1 + \theta_1^2)\right]}{1 - \alpha_1^2}.$$

Finally, remember that we are interested in (weakly) stationary time series. If the time series $\{X_n, n \in \mathbb{Z}\}$ is not weakly stationary, we can consider the *differenced* process $\{Y_n, n \in \mathbb{Z}\}$ defined by

$$Y_n = \Delta X_n = X_n - X_{n-1} \quad \text{for all } n \in \mathbf{Z}.$$

If the resulting process is weakly stationary, then we can try to model it as an ARMA(p, q) process. If it is not, we can difference the $\{Y_n, n \in \mathbf{Z}\}$ process, and so on.

In general, we have the following definition.

Definition 7.2.6. *Suppose that the stochastic process* $\{Y_n, n \in \mathbf{Z}\}$ *defined by*

$$Y_n = \Delta^d X_n = X_n + \sum_{j=1}^{d} (-1)^j \binom{d}{j} X_{n-j} \quad \text{for all } n \in \mathbf{Z},$$

where $d = 1, 2, \ldots,$ *is an ARMA(p, q) process. Then,* $\{X_n, n \in \mathbf{Z}\}$ *is called an* **autoregressive integrated moving average process** *and is denoted by* ARIMA(p, d, q).

7.3 Modeling and forecasting

Let X_n be the (random) closing value of a certain stock market index, for example, the Dow Jones index, at time n. Suppose that we gathered observations $x_1, x_2, \ldots, x_N$ of this closing value over a given period of time and that we would like to model the random process $\{X_n, n = 0, 1, 2, \ldots\}$ as a particular time series. The first step is to check that the data collected constitute a weakly stationary time series. If it is not the case, we can *difference* the data until the assumption of stationarity is reasonable. In practice, one way to determine that a particular time series must be differenced is to look at the graph of the successive X_ns in time. There should be no obvious trend pattern. That is, the data should move randomly around a constant mean over time. We assume, in the sequel, that the time series considered is indeed weakly stationary.

To try to *identify* the orders p and q of an autoregressive moving average process that we would like to use as a model for a given dataset, we can look at the *sample autocorrelation function*. We have seen in the preceding section, p. 242, that the (theoretical) autocorrelation function $\rho(k)$ of a weakly stationary AR(p) process decreases geometrically (in absolute value) with $|k|$, and $\rho(k)$ is equal to 0 for $|k| > q$ in the case of an MA(q) time series [see Equation (7.23)].

To help us determine (approximately) the values of p and q, we define the *partial autocorrelation function*.

Definition 7.3.1. *Let* $\rho(k)$ *be the autocorrelation function of a weakly stationary time series* $\{X_n, n \in \mathbf{Z}\}$. *Define the determinants*

$$D_{k,1} = \begin{vmatrix} 1 & \rho(1) & \rho(2) & \ldots & \rho(k-2) & \rho(1) \\ \rho(1) & 1 & \rho(1) & \ldots & \rho(k-3) & \rho(2) \\ \rho(2) & \rho(1) & 1 & \ldots & \rho(k-4) & \rho(3) \\ \ldots & \ldots & \ldots & \ldots & \ldots & \ldots \\ \rho(k-1) & \rho(k-2) & \rho(k-3) & \ldots & \rho(1) & \rho(k) \end{vmatrix}$$

and

$$D_{k,2} = \begin{vmatrix} 1 & \rho(1) & \rho(2) & \cdots & \rho(k-2) & \rho(k-1) \\ \rho(1) & 1 & \rho(1) & \cdots & \rho(k-3) & \rho(k-2) \\ \rho(2) & \rho(1) & 1 & \cdots & \rho(k-4) & \rho(k-3) \\ \cdots & \cdots & \cdots & \cdots & \cdots & \cdots \\ \rho(k-1) & \rho(k-2) & \rho(k-3) & \cdots & \rho(1) & 1 \end{vmatrix}.$$

*The **partial autocorrelation function**, denoted by $\phi(k)$, of the process is given by*

$$\phi(k) = \frac{D_{k,1}}{D_{k,2}} \quad for \ k = 1,2,\ldots .$$

Remarks. (i) The determinants in the above definition are given by (for $k = 1$)

$$D_{1,1} = |\rho(1)| = \rho(1) \quad \text{and} \quad D_{1,2} = |1| = 1$$

and (for $k = 2$)

$$D_{2,1} = \begin{vmatrix} 1 & \rho(1) \\ \rho(1) & \rho(2) \end{vmatrix} = \rho(2) - \rho^2(1) \quad \text{and} \quad D_{2,2} = \begin{vmatrix} 1 & \rho(1) \\ \rho(1) & 1 \end{vmatrix} = 1 - \rho^2(1).$$

That is, first we write the last column of the matrix and then we add the other $k-1$ columns, starting from the first one on the left and moving right (see solved exercise no. 8 for $D_{3,1}$ and $D_{3,2}$).

(ii) The $\phi(k)$s are actually obtained by solving the system of linear equations

$$\rho(j) = \phi(1)\rho(1-j) + \phi(2)\rho(2-j) + \phi(3)\rho(3-j) + \cdots + \phi(k)\rho(k-j)$$

for $j = 1,\ldots,k$ [and using the fact that $\rho(-k) = \rho(k)$]. These equations follow from the Yule–Walker equations (7.14).

(iii) In the case of a weakly stationary AR(p) time series, $\phi(k)$ is the autocorrelation at lag k, after having removed the autocorrelation with an AR($k-1$) process. That is, it measures the correlation that is not explained by an AR($k-1$) process.

(iv) If the time series is a weakly stationary AR(p) model, we find that $\phi(k) = 0$ for $k = p+1, p+2, \ldots$, whereas $\phi(k)$ decreases (approximately) geometrically, in absolute value, in the case of an MA(q) [and an ARMA(p,q)] model (see Exercise no. 18, p. 265). Notice that the $\phi(k)$ function behaves in the exact opposite way as $\rho(k)$ does for AR(p) and MA(q) processes.

(v) We deduce from the preceding remark that if we want to use an AR(p) model for a given dataset, then the *sample* partial autocorrelations should be approximately equal to 0 from $k = p+1$. In practice, the autocorrelations $\rho(k)$ must first be estimated from the data before we can compute the sample $\phi(k)$s.

Particular cases. (i) If $\{X_n, n \in \mathbb{Z}\}$ is a weakly stationary AR(1) stochastic process, then we only have to calculate

$$\phi(1) = \frac{\rho(1)}{1} = \rho(1) \tag{7.33}$$

and we can set $\phi(k) = 0$ for $|k| = 2, 3, \ldots$.

Remark. Notice that $\phi(1)$ will always be equal to $\rho(1)$.

(ii) When the time series considered is a weakly stationary AR(2) process, we have that

$$\phi(1) = \frac{\rho(1)}{1} = \rho(1) \quad \text{and} \quad \phi(2) = \frac{\rho(2) - \rho^2(1)}{1 - \rho^2(1)} \tag{7.34}$$

[and $\phi(k) = 0$ for $|k| = 3, 4, \ldots$].

(iii) In the case of an invertible MA(1) time series, it can be shown that

$$\phi(k) = -\frac{(-\theta_1)^k}{1 + \theta_1^2 + \cdots + \theta_1^{2k}} = -\frac{(-\theta_1)^k (1 - \theta_1^2)}{1 - \theta_1^{2(k+1)}} \quad \text{for } k = 1, 2, \ldots . \tag{7.35}$$

Once a particular ARMA(p, q) process has been identified as being reasonable for the data collected, we must estimate the parameters $\alpha_1, \ldots, \alpha_p$ and $\theta_1, \ldots, \theta_q$ in the model. Finally, there exist many *statistical tests* that enable us to assess the quality of the fit of the model to the data. We do not get into these statistical questions in this book. However, we give the formula used to estimate the autocorrelation function $\rho(k)$.

The original time series $\{Y_n, n \in \mathbb{Z}\}$ is assumed to have a constant mean μ. To estimate μ, we collect some data, $y_1, y_2, \ldots, y_N$, and we calculate their *arithmetic mean*. That is, we set

$$\hat{\mu} = \frac{1}{N} \sum_{n=1}^{N} y_n.$$

Remark. The observed data $y_1, y_2, \ldots, y_N$ are particular values taken by random variables that constitute a *random sample* of the variable of interest, say Y. The quantity $\hat{\mu}$ defined above is a *point estimate* of $E[Y_n] \equiv \mu$.

Next, we set

$$x_n = y_n - \hat{\mu} \quad \text{for } n = 1, 2, \ldots, N.$$

The corresponding process $\{X_n, n \in \mathbb{Z}\}$ is supposed to have zero mean and be weakly stationary. The point estimate of the (constant) variance of the X_ns is

$$\hat{\sigma}_X^2 = \frac{1}{N} \sum_{n=1}^{N} x_n^2.$$

Remark. Remember that the variance σ_X^2 of the X_ns is not the same as the variance σ^2 of the white noise process.

More generally, to estimate the covariance of X_n and X_{n-k}, we use

$$\hat{\gamma}(k) = \frac{1}{N} \sum_{n=k+1}^{N} x_n x_{n-k}.$$

Then, the point estimate of $\rho(k)$ is given by

$$\hat{\rho}(k) = \frac{\hat{\gamma}(k)}{\hat{\gamma}(0)} \quad \text{for } k = 1, 2, \ldots .$$

Example 7.3.1. Suppose that $X_0 = 0$ and that

$$X_n = 0.5 X_{n-1} + \epsilon_n \quad \text{for } n = 1, 2, \ldots,$$

where the ϵ_ns are i.i.d. random variables having a U$(-1, 1)$ distribution (and ϵ_n is independent of $X_{n-1}, \ldots, X_1$). Therefore, $\{\epsilon_n, n = 1, 2, \ldots\}$ is an IID$(0, 1/3)$ noise and $\{X_n, n = 0, 1, \ldots\}$ is an AR(1) process.

Using a statistical software package, we generated 40 (independent) *observations* of a U$(-1, 1)$ random variable and we calculated the value x_n of X_n, for $n = 1, \ldots, 40$. The results are the following:

n	1	2	3	4	5	6	7	8	9	10
x_n	−0.41	0.70	0.76	0.38	0.60	−0.20	0.22	−0.15	−0.39	−0.76

n	11	12	13	14	15	16	17	18	19	20
x_n	0.59	−0.40	−0.05	0.39	0.42	0.38	0.47	−0.04	−0.23	−0.11

n	21	22	23	24	25	26	27	28	29	30
x_n	−0.84	−0.49	−1.23	−0.80	−0.92	−0.61	0.47	0.76	−0.44	−0.90

n	31	32	33	34	35	36	37	38	39	40
x_n	0.54	−0.12	−0.76	0.32	−0.36	0.60	0.75	1.00	0.01	−0.73

Although the number of data points is not very large, we now show whether we are able to determine that the x_ns are observations of an AR(1) process by proceeding as suggested above. First, if we look at the *scatter diagram* of the x_ns against n, we notice no obvious trend in the data (see Figure 7.2). Therefore, we may assume that the underlying stochastic process is weakly stationary.

Next, the mean of the data points is equal to -0.0395. Moreover, the point estimate of the variance σ_X^2 is approximately 0.337.

Remark. The number of observations is not large enough to obtain very accurate point estimates of the mean (which is actually equal to 0) and the variance of the X_ns, which is [see (7.10)]

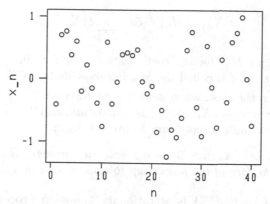

Fig. 7.2. Scatter diagram of the data in Example 7.3.1.

$$\mathrm{VAR}[X_n] \equiv \frac{\sigma^2}{1 - (0.5)^2} = 0.\bar{4}.$$

Finally, we calculate the point estimate of $\phi(k)$, for $k = 1, 2, 3$. We find that

$$\hat{\phi}(1) \simeq 0.328, \quad \hat{\phi}(2) \simeq 0.010 \quad \text{and} \quad \hat{\phi}(3) \simeq 0.057.$$

We notice that the partial autocorrelations are close to 0 for $k = 2, 3$, from which we deduce that an AR(1) model could indeed be reasonable for the data. In fact, some values of $\hat{\phi}(k)$ are larger (in absolute value) as k increases. However, the larger k is, the fewer data points are available to estimate the corresponding partial autocorrelation, so that the point estimates are less and less reliable, especially when n is small.

We now turn to the problem of using the data to *forecast* future values of the time series.

Let $\{X_n, n \in \mathbb{Z}\}$ be a time series. Suppose that we would like to forecast the value of X_{n+j}, where $j \in \{1, 2, \ldots\}$, based on the observed random variables $X_n, X_{n-1}, \ldots$. Let H_n denote the set $\{X_n, X_{n-1}, \ldots\}$. That is, H_n is the *history* of the stochastic process up to time n.

Next, denote by $g(X_{n+j} \mid H_n)$ the *predictor* of X_{n+j}, given H_n. To determine the best predictor, we need a criterion. One which is widely used is the following: we look for the function g that minimizes the *mean-square error*

$$\mathrm{MSE}(g) := E[\{X_{n+j} - g(X_{n+j} \mid H_n)\}^2].$$

It can be shown that (if the mathematical expectation exists) the *optimal predictor* is actually the conditional expectation of X_{n+j}, given H_n:

$$g^*(X_{n+j} \mid H_n) = E[X_{n+j} \mid H_n].$$

In practice, we cannot use an infinite number of random variables to forecast X_{n+j}. Therefore, the set H_n could, for example, be $\{X_n, X_{n-1}, \ldots, X_1\}$. Moreover, sometimes we look for a function $g(X_{n+j} \mid H_n)$ of the form

$$g(X_{n+j} \mid H_n) = \xi_0 + \sum_{i=1}^{n} \xi_i X_{n-i+1}. \tag{7.36}$$

The function $g(X_{n+j} \mid H_n)$ having coefficients ξ_i, for $i = 0, \dots, n$, which yield the smallest mean-square error is called the *best linear predictor* of X_{n+j}.

Now, we consider the case when we want to forecast X_{n+j} based on X_n alone. When the stochastic process $\{X_n, n \in \mathbb{Z}\}$ is *Markovian*, and X_n is known, the values of $X_{n-1}, X_{n-2}, \dots$ are actually unnecessary to predict X_{n+j}.

Proposition 7.3.1. *Let $\{X_n, n \in \mathbb{Z}\}$ be a Markovian process. The optimal predictor of X_{n+j}, based on the history of the process up to time n, is a function of X_n alone.*

Furthermore, if $\{X_n, n \in \mathbb{Z}\}$ is a stationary Gaussian process, then the optimal predictor and the best linear predictor of X_{n+j}, based on X_n, coincide, which follows from the next proposition.

Proposition 7.3.2. *Let $\{X_n, n \in \mathbb{Z}\}$ be a stationary Gaussian process. The expected value of X_{n+j}, given that $X_n = x_n$, is of the form*

$$E[X_{n+j} \mid X_n = x_n] = ax_n + b, \tag{7.37}$$

where the constants a and b are given by

$$a = \rho(j) \quad and \quad b = \mu[1 - \rho(j)]. \tag{7.38}$$

Remarks. (i) In general, if the random vector (X_1, X_2) has a bivariate normal distribution, then $E[X_2 \mid X_1 = x_1]$ is given in (7.3).

(ii) The above result can be generalized to the case when we calculate the conditional expectation $E[X_{n+j} \mid X_n, X_{n-1}, \dots, X_1]$.

(iii) Remember (see Proposition 7.1.1) that, in the case of a Gaussian process, if it is weakly stationary, then it is also strictly stationary.

(iv) We deduce from (7.3) that the conditional variance $\text{VAR}[X_{n+j} \mid X_n = x_n]$ is

$$\text{VAR}[X_{n+j} \mid X_n = x_n] = \sigma_X^2 [1 - \rho^2(j)], \tag{7.39}$$

where $\sigma_X^2 = \text{VAR}[X_n]$ for all n. Hence, the closer to one (in absolute value) the correlation coefficient $\rho(j)$ is, the more accurate is the forecast of X_{n+j} (based on X_n), which is logical.

Next, suppose that we want to forecast the value, X_{n+1}, of a weakly stationary and invertible ARMA(p, q) process at time $n + 1$, given $X_n, X_{n-1}, \dots$. We have that

$$X_{n+1} = \sum_{i=1}^{p} \alpha_i X_{n+1-i} + \epsilon_{n+1} + \sum_{j=1}^{q} \theta_j \epsilon_{n+1-j}.$$

Hence, we can write that

$$E[X_{n+1} \mid H_n] = \sum_{i=1}^{p} \alpha_i X_{n+1-i} + \sum_{j=1}^{q} \theta_j \epsilon_{n+1-j}.$$

Indeed, given the history of the process up to time n, the random variables X_{n+1-i} and ϵ_{n+1-j} are known quantities for $i, j \in \{1, 2, \ldots\}$. Moreover, by independence,

$$E[\epsilon_{n+1} \mid H_n] = E[\epsilon_{n+1}] = 0.$$

More generally, to forecast X_{n+m}, where $m \in \{1, 2, \ldots\}$, we must calculate the conditional expectation

$$E[X_{n+m} \mid H_n] = \sum_{i=1}^{p} \alpha_i E[X_{n+m-i} \mid H_n] + \sum_{j=1}^{q} \theta_j E[\epsilon_{n+m-j} \mid H_n],$$

in which

$$E[X_{n+m-i} \mid H_n] = X_{n+m-i} \quad \text{if } m \le i$$

and

$$E[\epsilon_{n+m-j} \mid H_n] = \begin{cases} 0 & \text{if } m > j, \\ \epsilon_{n+m-j} & \text{if } m \le j. \end{cases}$$

Particular case. Let $p = q = 1$, so that

$$X_{n+m} = \alpha_1 X_{n+m-1} + \epsilon_{n+m} + \theta_1 \epsilon_{n+m-1}.$$

Then,

$$E[X_{n+1} \mid H_n] = \alpha_1 X_n + \theta_1 \epsilon_n$$

and

$$E[X_{n+m} \mid H_n] = \alpha_1 E[X_{n+m-1} \mid H_n] \quad \text{for } m = 2, 3, \ldots .$$

It follows that

$$\begin{aligned} E[X_{n+2} \mid H_n] &= \alpha_1 E[X_{n+1} \mid H_n] = \alpha_1 \{\alpha_1 X_n + \theta_1 \epsilon_n\} \\ &= \alpha_1^2 X_n + \alpha_1 \theta_1 \epsilon_n \end{aligned}$$

and

$$\begin{aligned} E[X_{n+3} \mid H_n] &= \alpha_1 E[X_{n+2} \mid H_n] = \alpha_1 \{\alpha_1^2 X_n + \alpha_1 \theta_1 \epsilon_n\} \\ &= \alpha_1^3 X_n + \alpha_1^2 \theta_1 \epsilon_n, \end{aligned}$$

and so on. That is, we have that

$$E[X_{n+m} \mid H_n] = \alpha_1^m X_n + \alpha_1^{m-1}\theta_1\epsilon_n \quad \text{for } m = 2, 3, \ldots.$$

Example 7.3.2. Suppose that $\{X_n, n \in \mathbb{Z}\}$ is a weakly stationary AR(1) time series. Therefore,

$$X_{n+m} = \alpha_1 X_{n+m-1} + \epsilon_{n+m},$$

where $0 < |\alpha_1| < 1$. Proceeding as above, we find that

$$E[X_{n+m} \mid H_n] = \alpha_1^m X_n \quad \text{for } m = 1, 2, 3, \ldots.$$

This result can actually be deduced directly from the formula

$$X_{n+m} = \alpha_1 X_{n+m-1} + \epsilon_{n+m} = \alpha_1(\alpha_1 X_{n+m-2} + \epsilon_{n+m-1}) + \epsilon_{n+m}$$

$$= \cdots = \alpha_1^m X_n + \sum_{i=0}^{m-1} \alpha_1^i \epsilon_{n+m-i}$$

(because $E[\epsilon_{n+m-i} \mid H_n] = 0$ if $m > i$).

We also deduce from the previous formula that the *forecasting error* is given by

$$X_{n+m} - E[X_{n+m} \mid H_n] = \alpha_1^m X_n + \sum_{i=0}^{m-1} \alpha_1^i \epsilon_{n+m-i} - \alpha_1^m X_n = \sum_{i=0}^{m-1} \alpha_1^i \epsilon_{n+m-i}.$$

It follows that

$$\text{VAR}[X_{n+m} - E[X_{n+m} \mid H_n]] \stackrel{\text{unc.}}{=} \sum_{i=0}^{m-1} \alpha_1^{2i}\text{VAR}[\epsilon_{n+m-i}] = \sum_{i=0}^{m-1} \alpha_1^{2i}\sigma^2.$$

Example 7.3.3. In the case of an MA(q) process, we can write that

$$X_{n+m} = \epsilon_{n+m} + \sum_{i=1}^{q} \theta_i \epsilon_{n+m-i} \quad \text{for all } n \in \mathbb{Z} \text{ and for } m = 1, 2, \ldots.$$

Hence, we deduce that

$$E[X_{n+m} \mid H_n] = \begin{cases} \sum_{i=m}^{q} \theta_i \epsilon_{n+m-i} & \text{if } m = 1, \ldots, q, \\ 0 & \text{if } m = q+1, q+2, \ldots. \end{cases}$$

When $q = 1$, we simply have that

$$E[X_{n+m} \mid H_n] = \begin{cases} \theta_1 \epsilon_n & \text{if } m = 1, \\ 0 & \text{if } m = 2, 3, \ldots, \end{cases}$$

so that the variance of the forecasting error is

$$\text{VAR}\,[X_{n+m} - E[X_{n+m} \mid H_n]] = \text{VAR}[\epsilon_{n+m} + \theta_1\epsilon_{n+m-1} - \theta_1\epsilon_{n+m-1}]$$
$$= \text{VAR}[\epsilon_{n+m}] = \sigma^2 \quad \text{if } m = 1$$

and

$$\text{VAR}\,[X_{n+m} - E[X_{n+m} \mid H_n]] = \text{VAR}[\epsilon_{n+m} + \theta_1\epsilon_{n+m-1}]$$
$$\overset{\text{unc.}}{=} (1 + \theta_1^2)\sigma^2 \quad \text{if } m = 2, 3, \ldots.$$

Observe that the variance of the forecasting error is the same for any $m \in \{2, 3, \ldots\}$, whereas in the case of an AR(1) process this variance increases with m (see the previous example).

To conclude, we consider the case when we want to forecast the value of a zero mean weakly stationary time series at time $n+m$, for $m = 1, 2, \ldots$, based on a *linear* function of $X_n, X_{n-1}, \ldots, X_1$. That is, we look for the function [see (7.36)]

$$g(X_{n+m} \mid H_n) = \sum_{i=1}^{n} \xi_i X_{n-i+1}$$

that minimizes the mean-square error

$$\text{MSE}(\xi_1, \ldots, \xi_n) := E\left[\left\{X_{n+m} - \sum_{i=1}^{n} \xi_i X_{n-i+1}\right\}^2\right].$$

Remark. The fact that $E[X_n] \equiv 0$ implies that ξ_0 in (7.36) is equal to 0.

To obtain the values of the parameters ξ_i that minimize the function MSE, we differentiate this function with respect to ξ_i and we set the derivative equal to 0, for $i = 1, \ldots, n$. We find that the ξ_is must satisfy the system of linear equations

$$\sum_{i=1}^{n} \xi_i \gamma(i - r) = \gamma(m - 1 + r) \quad \text{for } r = 1, \ldots, n.$$

We can easily obtain an explicit solution when $m = 1$, that is, when the aim is to forecast the next value of the time series. We then have that

$$\sum_{i=1}^{n} \xi_i \gamma(i - r) = \gamma(r) \quad \text{for } r = 1, \ldots, n,$$

which can be rewritten as follows:

$$(\xi_1, \ldots, \xi_n) \begin{bmatrix} \gamma(1-1) & \gamma(1-2) & \ldots & \gamma(1-n) \\ \gamma(2-1) & \gamma(2-2) & \ldots & \gamma(2-n) \\ \ldots & \ldots & \ldots & \ldots \\ \gamma(n-1) & \gamma(n-2) & \ldots & \gamma(n-n) \end{bmatrix} = (\gamma(1), \ldots, \gamma(n))$$

$\Longleftrightarrow$
$$\Xi_n \Gamma_n = G_n,$$

where $\Xi_n := (\xi_1, \ldots, \xi_n)$, $G_n := (\gamma(1), \ldots, \gamma(n))$ and

$$\Gamma_n := \begin{bmatrix} \gamma(0) & \gamma(1) & \ldots & \gamma(n-1) \\ \gamma(1) & \gamma(0) & \ldots & \gamma(n-2) \\ \ldots & \ldots & \ldots & \ldots \\ \gamma(n-1) & \gamma(n-2) & \ldots & \gamma(0) \end{bmatrix}.$$

Assuming that the matrix Γ_n is invertible (or nonsingular), we can write that

$$\Xi_n = G_n \Gamma_n^{-1}$$

and

$$g(X_{n+1} \mid H_n) = \Xi_n (X_n, \ldots, X_1)^T.$$

Remark. It can be shown that the (last) coefficient ξ_n is in fact equal to the partial autocorrelation $\phi(n)$.

Finally, we find that the mean-square forecasting error is given by

$$E\left[\{X_{n+1} - g(X_{n+1} \mid H_n)\}^2\right] = \gamma(0) - G_n \Gamma_n^{-1} G_n^T.$$

Example 7.3.4. To forecast the value of X_{n+1} for a (zero mean) weakly stationary AR(1) process, based on a single observation, X_n, of the process, we calculate

$$g(X_{n+1} \mid X_n) = \xi_1 X_n = \gamma(1) \Gamma_1^{-1} X_n = \frac{\gamma(1)}{\gamma(0)} X_n.$$

Making use of (7.11), we can write that

$$g(X_{n+1} \mid X_n) = \alpha_1 X_n,$$

which is the same predictor as $E[X_{n+1} \mid H_n]$ obtained in Example 7.3.2.

Example 7.3.5. For a weakly stationary AR(2) time series (having zero mean), the best linear predictor of X_{n+1}, based on X_n, is also

$$g(X_{n+1} \mid X_n) = \frac{\gamma(1)}{\gamma(0)} X_n = \rho(1) X_n.$$

To forecast the value of X_{n+1}, based on the random variables X_n and X_{n-1}, we can show that $g(X_{n+1} \mid X_n, X_{n-1})$ is given by

$$g(X_{n+1} \mid X_n, X_{n-1}) = \alpha_1 X_n + \alpha_2 X_{n-1}.$$

Actually, we find that

$$g(X_{n+1} \mid X_n, X_{n-1}, \ldots, X_{n-l}) = \alpha_1 X_n + \alpha_2 X_{n-1}$$

for any $l \in \{1, \ldots, n-1\}$.

Remark. In general, if $\{X_n, n \in \mathbb{Z}\}$ is a weakly stationary AR(p) time series with $E[X_n] \equiv 0$, then

$$g(X_{n+1} \mid X_n, X_{n-1}, \ldots, X_1) = \alpha_1 X_n + \alpha_2 X_{n-1} + \cdots + \alpha_p X_{n-p+1}$$

for any $n \in \{p, p+1, \ldots\}$.

7.4 Exercises for Chapter 7

Solved exercises

Question no. 1

Let $\{X_n, n = 0, 1, \ldots\}$ be an IID$(0, \sigma^2)$ noise. We define

$$Y_n = \frac{X_n + X_{n-1}}{2} \quad \text{for } n = 1, 2, \ldots.$$

Is the stochastic process $\{Y_n, n = 1, 2, \ldots\}$ weakly stationary?

Question no. 2

Suppose that the random vector (X_1, X_2) has a bivariate normal distribution and that $X_1 \sim N(0,1)$ and $X_2 \sim N(0,1)$ are independent random variables. Use Proposition 4.3.1 to find the joint probability density function of the transformation

$$Y_1 = a_1 X_1 + b_1,$$

$$Y_2 = a_2 X_2 + b_2,$$

where $a_i \neq 0$ and $b_i \in \mathbb{R}$, for $i = 1, 2$.

Question no. 3

Calculate the variance of a weakly stationary AR(3) process $\{X_n, n \in \mathbb{Z}\}$ (with $E[X_n] \equiv 0$).

Question no. 4

Suppose that $X_0 = 0$ and let

$$X_n = \alpha_1 X_{n-1} + \epsilon_n \quad \text{for } n = 1, 2, \ldots,$$

where $\{\epsilon_n, n = 1, 2, \ldots\}$ is a GWN$(0, \sigma^2)$ process. We define

$$Y_n = e^{X_n} \quad \text{for } n = 1, 2, \ldots.$$

Calculate the expected value of Y_n.

Indications. (i) It can be shown that if X and Y are independent random variables, then so are $g(X)$ and $h(Y)$ for any functions g and h.

(ii) The moment-generating function of a random variable $X \sim N(\mu, \sigma^2)$ is given by

$$M_X(t) := E[e^{tX}] = \exp\left\{\mu t + \frac{\sigma^2}{2} t^2\right\}.$$

Question no. 5
Calculate the autocorrelation function $\rho(k)$, for $k = 1, 2, 3$, of an MA(3) time series.

Question no. 6
What are the possible values of the function $\rho(2)$ for an MA(2) process if (a) $\theta_1 = 1$? (b) $\theta_1 \in \mathbb{R}$?

Question no. 7
Calculate, in terms of $\gamma(1)$, the variance of a (zero mean) weakly stationary ARMA(1,2) process.

Question no. 8
Check the formula (7.35) for an invertible MA(1) time series, for $k = 1, 2, 3$.

Question no. 9
Consider the following data, denoted by $y_1, y_2, \ldots, y_{40}$:

n	1	2	3	4	5	6	7	8	9	10
y_n	−0.41	0.70	0.86	0.21	0.41	−0.30	0.07	−0.10	−0.44	−0.72

n	11	12	13	14	15	16	17	18	19	20
y_n	0.69	−0.22	−0.20	0.50	0.43	0.28	0.37	−0.13	−0.35	−0.11

n	21	22	23	24	25	26	27	28	29	30
y_n	−0.78	−0.46	−1.03	−0.68	−0.61	−0.41	0.70	0.92	−0.56	−1.09

n	31	32	33	34	35	36	37	38	39	40
y_n	0.65	0.11	−0.90	0.35	−0.17	0.52	0.84	0.85	−0.18	−0.98

Define

$$x_n = y_n - \bar{y} \quad \text{for } n = 1, \ldots, 40,$$

where $\bar{y}$ is the arithmetic mean of the data points. Calculate $\hat{\sigma}_X^2$ and the sample partial autocorrelation $\hat{\phi}(k)$ of the centered data, for $k = 1, 2, 3$. Could an MA(1) time series with $\theta_1 = 1/2$ serve as a model for $x_1, x_2, \ldots, x_{40}$ if $\epsilon_0 = 0$ and $\epsilon_n \sim U(-1, 1)$, for $n = 1, 2, \ldots, 40$? Justify.

Question no. 10

(a) Use Equation (7.3) to prove Proposition 7.3.2. (b) Calculate the mathematical expectation $E[X_{n+j}^2 \mid X_n = x_n]$.

Exercises

Question no. 1

The discrete-time stochastic process $\{X_n, n = 0, 1, \ldots\}$ is an $\text{IID}(0, \sigma^2)$ noise. Is the process $\{Y_n, n = 1, 2, \ldots\}$ defined by

$$Y_n = X_n X_{n-1} \quad \text{for } n = 1, 2, \ldots$$

weakly stationary? Justify.

Question no. 2

Suppose that $X_0 \sim N(0, 1)$ and that

$$X_n = X_0 + Y_n \quad \text{for } n = 1, 2, \ldots,$$

where $Y_n = \pm 1$ with probability $1/2$ and is independent of X_0. Is the stochastic process $\{X_n, n = 0, 1, \ldots\}$ a Gaussian process? Justify.

Question no. 3

Let $\{N(t), t \geq 0\}$ be a Poisson process with rate $\lambda > 0$ and define

$$X_n = (-1)^{N(n)} \cdot M \quad \text{for } n = 0, 1, \ldots,$$

where M is a random variable that is equal to 1 or -1 with probability $1/2$. Moreover, M is independent of $N(n)$. Calculate $E[X_n]$ and $\text{COV}[X_n, X_{n+m}]$, for $n, m \in \{0, 1, \ldots\}$. Is the stochastic process $\{X_n, n = 0, 1, \ldots\}$ weakly stationary?

Indication. We can show that $E\left[(-1)^{N(n)}\right] = e^{-2\lambda n}$, for $n = 0, 1, \ldots$.

Question no. 4

A standard Brownian motion (see p. 235) is a continuous-time Gaussian stochastic process $\{W(t), t \geq 0\}$ such that $W(0) = 0$, $E[W(t)] \equiv 0$, and $\text{COV}[W(s), W(t)] = \min\{s, t\}$ for all $s, t \geq 0$. We consider the stochastic process $\{X(t), t \geq 0\}$ for which

$$W(t) = t^{1/2} X(\ln t) \quad \text{for } t \geq 1.$$

Is $\{X(t), t \geq 0\}$ a Gaussian process? Is it (strictly) stationary? Justify.

Remark. The stochastic process $\{X(t), t \geq 0\}$ is actually a particular *Ornstein–Uhlenbeck* process.

Question no. 5

Let $\{X_n, n \in \mathbb{Z}\}$ and $\{Y_n, n \in \mathbb{Z}\}$ be two (zero mean) weakly stationary AR(1) processes defined by

$$X_n = \alpha_1 X_{n-1} + \epsilon_n \quad \text{and} \quad Y_n = \beta_1 Y_{n-1} + \eta_n \quad \forall n \in \mathbb{Z},$$

where $\{\epsilon_n, n \in \mathbb{Z}\}$ and $\{\eta_n, n \in \mathbb{Z}\}$ are independent $\mathrm{WN}(0, \sigma^2)$ processes. We define

$$Z_n = X_n + Y_n \quad \forall n \in \mathbb{Z}.$$

Calculate $E[Z_n]$ and $\mathrm{COV}[Z_n, Z_{n+m}]$, for $n, m \in \mathbb{Z}$. Is the stochastic process $\{Z_n, n \in \mathbb{Z}\}$ weakly stationary?

Question no. 6

Consider the AR(2) process $\{X_n, n = 0, 1, \ldots\}$ given by $X_0 = 0$, $X_1 = 0$, and

$$X_n = X_{n-1} + X_{n-2} + \epsilon_n \quad \text{for } n = 2, 3, \ldots .$$

Suppose that $P[\epsilon_n = 1] = P[\epsilon_n = -1] = 1/2$, for all n. Calculate the probability mass function of X_4.

Question no. 7

Let $\{X_n, n = 0, 1, \ldots\}$ and $\{Y_n, n = 0, 1, \ldots\}$ be independent random walks (see p. 239) with $p_0 = 1/2$, and define

$$Z_n = \frac{X_n + Y_n}{2} \quad \text{for } n = 0, 1, \ldots .$$

Calculate $\mathrm{COV}[Z_n, Z_{n+m}]$, for $n, m \in \{0, 1, \ldots\}$. Is the stochastic process $\{Z_n, n = 0, 1, \ldots\}$ a random walk? Justify.

Question no. 8

Suppose that $X_0 = 0$ and

$$X_n = \alpha_1 X_{n-1} + \epsilon_n \quad \text{for } n = 1, 2, \ldots,$$

where the independent random variables ϵ_n are also independent of $X_{n-1}, \ldots, X_0$ and are such that $P[\epsilon_n = 1] = P[\epsilon_n = -1] = 1/2$, for $n = 1, 2, \ldots$. Let $Y_n = X_n^2$, for all n. Calculate (a) $\mathrm{VAR}[Y_2]$ and (b) $E[Y_2^k]$, for $k \in \{1, 2, \ldots\}$.

Question no. 9

Let $\{X_n, n \in \mathbb{Z}\}$ be an MA(1) time series with $\theta_1 = 1/2$. Suppose that $P[\epsilon_n = 1] = P[\epsilon_n = -1] = 1/2$, for all $n \in \mathbb{Z}$. Calculate $E[X_n \mid X_n > 0]$, for $n \in \mathbb{Z}$.

Question no. 10

Consider the time series $\{X_n, n \in \mathbb{Z}\}$ and $\{Y_n, n \in \mathbb{Z}\}$ defined by

$$X_n = \epsilon_n + \theta_1 \epsilon_{n-1}$$

and

$$Y_n = \epsilon_n + \theta_1 \epsilon_{n-1} + \theta_2 \epsilon_{n-2}$$

for all $n \in \mathbb{Z}$. What is the correlation coefficient ρ_{X_n, Y_n}, for $n \in \mathbb{Z}$?

Question no. 11

Find, in terms of $\gamma(k)$, the variance of a (zero mean) weakly stationary ARMA(2,1) process.

Question no. 12

Is the MA(3) time series $\{X_n, n \in \mathbb{Z}\}$ defined by

$$X_n = \epsilon_n + \theta_1 \epsilon_{n-1} + \frac{1}{2}\epsilon_{n-2} + \frac{1}{4}\epsilon_{n-3} \quad \text{for all } n \in \mathbb{Z}$$

invertible if (a) $\theta_1 = \frac{1}{2}$? (b) $\theta_1 = \frac{5}{4}$? Justify.

Question no. 13

Let

$$X_n = \epsilon_n - \epsilon_{n-1} \quad \text{for all } n \in \mathbb{Z},$$

where $\{\epsilon_n, n \in \mathbb{Z}\}$ is a WN$(0, \sigma^2)$ process, so that $\{X_n, n \in \mathbb{Z}\}$ is an MA(1) time series for which $\theta_1 = -1$. Calculate VAR$[X_n \mid X_n \neq 0]$ if we assume that $P[\epsilon_n = 1] = P[\epsilon_n = -1] = 1/2$, for all $n \in \mathbb{Z}$.

Question no. 14

Let $\{X_n, n \in \mathbb{Z}\}$ be an MA(1) process and define

$$Y_n = X_n^2 \quad \text{for all } n \in \mathbb{Z}.$$

Calculate the mean and the variance of the stochastic process $\{Y_n, n \in \mathbb{Z}\}$ if $\epsilon_n \sim$ N$(0, \sigma^2)$, for all n.

Indication. The characteristic function of $Z \sim$ N$(0, 1)$ is $C_Z(\omega) = e^{-\omega^2/2}$.

Question no. 15

Calculate VAR$[X_{n+m} \mid H_n]$ for $m = 1, 2, \ldots$ if $\{X_n, n \in \mathbb{Z}\}$ is an MA(1) process and $H_n = \{X_n, X_{n-1}, \ldots\}$. Assume that the ϵ_ns are independent random variables.

Question no. 16

Let $\{X_n, n \in \mathbb{Z}\}$ be an MA(1) time series. (a) What is the best linear predictor of X_{n+1}, based on X_n. (b) Is this predictor similar to $E[X_{n+1} \mid H_n]$ (see Example 7.3.3)?

Question no. 17

Suppose that $\{X_n, n \in \mathbb{Z}\}$ is a weakly stationary ARMA(1,1) process. Calculate its partial autocorrelation function $\phi(k)$, for $k = 1, 2$, if $\alpha_1 = \theta_1 = 1/2$.

Question no. 18

Consider an invertible MA(1) time series with $\theta_1 > 0$. (a) Show that its partial autocorrelation function $\phi(k)$ decreases (approximately) geometrically, in absolute value. That is, show that

$$\left| \frac{\phi(k+1)}{\phi(k)} \right| \simeq c \quad \text{for } k \text{ large enough,}$$

where $c \in (0, 1)$ is a constant. (b) Show that

$$|\phi(k)| < \theta_1^k \quad \text{for } k = 1, 2, \dots .$$

Hint. See Equation (7.35).

Question no. 19

Let $\{\epsilon_n, n \in \mathbb{Z}\}$ be a GWN$(0,1)$ process and define

$$X_n = \frac{\epsilon_n + \epsilon_{n-1}}{2} \quad \forall n \in \mathbb{Z}.$$

Calculate the mathematical expectation $E[X_{n+j} \mid X_n = x_n]$, for $j = 1, 2, \dots$, and show that it is of the form given in Equations (7.37) and (7.38).

Question no. 20

Suppose that we want to forecast the value of X_{n+1}, based on a linear function of X_n and X_{n-1}, for a (zero mean) weakly stationary AR(2) time series. What is the corresponding mean-square forecasting error [in terms of $\gamma(0)$] if $\alpha_1 = 1/2$ and $\alpha_2 = 1/4$?

Hint. See Example 7.3.5 and Equation (7.13).

Multiple choice questions

Question no. 1

A *Bernoulli process* is a discrete-time (and discrete-state) stochastic process $\{X_n, n = 0, 1, \dots\}$, where the X_ns are i.i.d. Bernoulli random variables with parameter $p \in (0, 1)$. Calculate COV$[X_n, X_{n+m}]$, for $n, m \in \{0, 1, 2, \dots\}$.

(a) $0 \; \forall n, m$ (b) 0 if $m = 0$; $p(1-p)$ if $m \neq 0$
(c) 0 if $m \neq 0$; $p(1-p)$ if $m = 0$ (d) $p(1-p) \; \forall n, m$
(e) p^2 if $m \neq 0$; p if $m = 0$

Question no. 2

Let (X_1, X_2, X_3) have a trivariate normal distribution with $\mathbf{m} = (0, 1, -1)$ and

$$\mathbf{C} = \begin{bmatrix} 1 & 0 & -2 \\ 0 & 2 & 0 \\ -2 & 0 & 4 \end{bmatrix}.$$

Calculate $E[X_1 + X_2 \mid X_3 = 0]$.

(a) 0 (b) $1/2$ (c) 1 (d) $3/2$ (e) 2

Question no. 3

A weakly stationary AR(1) process $\{X_n, n = 0, 1, \dots\}$ is defined by $X_0 = 0$ and

$$X_n = \frac{1}{2} X_{n-1} + \epsilon_n \quad \text{for } n = 1, 2, \dots,$$

where $\{\epsilon_n, n = 1, 2, \dots\}$ is a GWN$(0, 1)$ process. Calculate $E[X_2 \mid X_2 > 0]$.

(a) $\frac{1}{2}\sqrt{5}$ (b) $\sqrt{5}$ (c) $\frac{1}{2}\sqrt{\frac{5}{\pi}}$ (d) $\sqrt{\frac{5}{2\pi}}$ (e) $\sqrt{\frac{5}{\pi}}$

Question no. 4

Suppose that $X_0 = 0$ and define the random walk (see p. 239)

$$X_n = X_{n-1} + \epsilon_n \quad \text{for } n = 1, 2, \ldots,$$

where $\{\epsilon_n, n = 1, 2, \ldots\}$ is an i.i.d. noise process such that ϵ_n takes on the value 1 or -1 with probability $1/2$, for $n = 1, 2, \ldots$. Calculate the correlation coefficient of X_n and X_{n+m}, for $n, m \in \{1, 2, \ldots\}$.

(a) $\left(\frac{n}{n+m}\right)^{1/2}$ (b) $\left(\frac{m}{n+m}\right)^{1/2}$ (c) $\frac{n}{(n+m)^{1/2}}$ (d) $\frac{n}{n+m}$ (e) $\frac{m}{n+m}$

Question no. 5

Suppose that

$$X_n = \frac{1}{2}X_{n-1} + \epsilon_n + \frac{1}{2}\epsilon_{n-1} \quad \forall n \in \mathbb{Z}.$$

That is, $\{X_n, n \in \mathbb{Z}\}$ is an ARMA(1,1) process with $\alpha_1 = \theta_1 = 1/2$. Calculate the correlation coefficient $\rho_{X_n, X_{n-2}}$.

(a) $2/7$ (b) $5/14$ (c) $1/2$ (d) $9/14$ (e) $5/7$

Question no. 6

Calculate $E[X_n \mid X_{n-1} = 2]$ for an MA(1) time series with $\theta_1 = 1$ and for which $P[\epsilon_n = 1] = P[\epsilon_n = -1] \equiv 1/2$, where the ϵ_ns are independent random variables.

(a) 0   (b) 1   (c) $3/2$ (d) 2   (e) 3

Question no. 7

What are the possible values of the autocorrelation $\rho(1)$ for an ARMA(1,1) time series with $\alpha_1 = 1/2$?

(a) $[-1, 1]$ (b) $[-3/4, 3/4]$ (c) $[-1/2, 1/2]$ (d) $[-1/4, 1/4]$
(e) $[-1/4, 3/4]$

Question no. 8

Let $\{X_n, n \in \mathbb{Z}\}$ be a stationary Gaussian process with $\mu = 3$ and $\rho(1) = 1/2$. Calculate the mathematical expectation $E[X_{n+1}X_n \mid X_n = -2]$.

(a) -3 (b) -1 (c) 0   (d) 1   (e) 3

Question no. 9

Suppose that the time series $\{X_n, n \in \mathbb{Z}\}$ is an MA(1) process with $\theta_1 = 1$ and that the independent random variables ϵ_n are such that $P[\epsilon_n = 1] = P[\epsilon_n = -1] = 1/2$, for all $n \in \mathbb{Z}$. Calculate $E[X_{n+1} \mid X_n^2 = 4]$.

(a) -1 (b) $-1/2$ (c) 0   (d) $1/2$ (e) 1

Question no. 10

We want to calculate the best linear predictor $g(X_{n+1} \mid X_n, X_{n-1})$ for an MA(1) time series with $\theta_1 = -1/2$. What is the value of the coefficient ξ_2?

(a) $-2/5$ (b) $-4/21$ (c) 0   (d) $4/21$ (e) $2/5$

A

List of symbols and abbreviations

Chapter 1

$\lim_{x \to x_0} f(x)$	limit of the function $f(x)$ as x tends to x_0
$\lim_{x \downarrow x_0} f(x)$	right-hand limit of $f(x)$ as x decreases to x_0
$\lim_{x \uparrow x_0} f(x)$	left-hand limit of $f(x)$ as x increases to x_0
$g \circ f$	composition of the functions g and f
$f'(x_0)$	derivative of $f(x)$ at x_0
$f'(x_0^+)$	right-hand derivative of $f(x)$ at x_0
$f'(x_0^-)$	left-hand derivative of $f(x)$ at x_0
$C_X(\omega)$	characteristic function of X
$F(\omega)$	Fourier transform
$M_X(t)$	moment-generating function of X
$\{a_n\}_{n=1}^\infty$	infinite sequence
$\sum_{n=1}^\infty a_n$	infinite series
$\sum_{k=1}^n a_k$	nth partial sum of the series
R	radius of convergence of the series
$f * g$	convolution of the functions f and g
$G_X(z)$	generating function of X

Chapter 2

Ω	sample space
A, B, C	events
$A \cap B = \emptyset$	incompatible or mutually exclusive events
A'	complement of event A
$P[A]$	probability of event A
$P[A \mid B]$	conditional probability of event A, given that B occurred

Chapter 3

p_X	probability (mass) function of X
F_X	distribution function of X
f_X	(probability) density function of X
$B(n, p)$	binomial distribution with parameters n and p
$Geo(p)$	geometric distribution with parameter p
$NB(r, p)$	negative binomial distribution with parameters r and p
$Hyp(N, n, d)$	hypergeometric distribution with parameters N, n, and d
$Poi(\lambda)$	Poisson distribution with parameter λ
$N(\mu, \sigma^2)$	normal or Gaussian distribution with parameters μ and σ^2
Z	random variable having a $N(0, 1)$ distribution
$\phi(z)$	density function of the $N(0, 1)$ distribution
$N(0, 1)$	standard or unit normal distribution
$\Phi(z)$	distribution function of the $N(0, 1)$ distribution
$Q(z)$	$1 - \Phi(z)$
Γ	gamma function
$G(\alpha, \lambda)$	gamma distribution with parameters α and λ
$W(\lambda, \beta)$	Weibull distribution with parameters λ and β
$Be(\alpha, \beta)$	beta distribution with parameters α and β
$LN(\mu, \sigma^2)$	lognormal distribution with parameters μ and σ^2
$E[g(X)]$	mathematical expectation of a function g of X
μ_X or $E[X]$	mean or expected value of X
x_m or $\tilde{x}$	median of X
x_p	$100(1 - p)$th quantile of X
x_p	$100(1 - p)$th percentile (if $100p$ is an integer)
σ_X^2	variance of X
μ_k'	kth-order moment about the origin or noncentral moment
μ_k	kth-order moment about the mean or central moment
β_1	skewness (coefficient)
β_2	kurtosis (coefficient)

Chapter 4

$p_{X,Y}$	joint probability function of (X, Y)
$F_{X,Y}$	joint distribution function of (X, Y)
$f_{X,Y}$	joint density function of (X, Y)
$f_X(x \mid A_Y)$	conditional density function of X, given A_Y
$E[Y \mid X = x]$	conditional expectation of Y, given that $X = x$
$X \otimes X$	convolution product of X with itself
$X \oplus X$	convolution sum of X with itself
$\text{COV}[X, Y]$	covariance of X and Y
$\text{CORR}[X, Y]$ or $\rho_{X,Y}$	correlation coefficient of X and Y
i.i.d.	independent and identically distributed
CLT	central limit theorem

Chapter 5

$R(x)$	reliability or survival function
$MTTF$	Mean Time To Failure
$MTBF$	Mean Time Between Failures
$MTTR$	Mean Time To Repair
$r(t)$	failure or hazard rate function
IFR	Increasing Failure Rate
DFR	Decreasing Failure Rate
$FR(t_1, t_2)$	interval failure rate of a system in the interval $(t_1, t_2]$
$AFR(t_1, t_2)$	average failure rate of a system over an interval $[t_1, t_2]$
$H(x_1, \ldots, x_n)$	structure function of the system
$\mathbf{x} := (x_1, \ldots, x_n)$	state vector of the system
$\mathbf{x} \geq \mathbf{y}$	$x_i \geq y_i$, $i = 1, \ldots, n$, for the vectors $\mathbf{x} = (x_1, \ldots, x_n)$ and $\mathbf{y} = (y_1, \ldots, y_n)$
$\mathbf{x} > \mathbf{y}$	$x_i \geq y_i$, $i = 1, \ldots, n$, and $x_i > y_i$ for at least one i for the vectors $\mathbf{x} = (x_1, \ldots, x_n)$ and $\mathbf{y} = (y_1, \ldots, y_n)$

Chapter 6

$\{X(t), t \in T\}$	stochastic or random process
$\rho_{i,j}$	probability that the continuous-time Markov chain $\{X(t), t \geq 0\}$, when it leaves state i, goes to state j
$\lambda_i, i = 0, 1, \ldots$	birth or arrival rates of a birth and death process
$\mu_i, i = 1, 2, \ldots$	death or departure rates of a birth and death process
π_j	limiting probability that the process will be in state j
$M/M/s$	queueing system with s servers
$X(t)$	number of customers in the system at time t
$\bar{N}$	average number of customers in the system in equilibrium
$\bar{N}_Q$	average number of customers who are waiting in line
$\bar{N}_S$	average number of customers being served
$\bar{T}$	average time that a customer spends in the system
$\bar{Q}$	average waiting time of an arbitrary customer
$\bar{S}$	average service time of an arbitrary customer
λ_a	average arrival rate
λ_e	average entering rate of customers into the system
$D(t)$	number of departures from the queueing system in $[0, t]$
N	number of customers in the system in equilibrium
N_S	number of customers being served
ρ	traffic intensity or utilization rate
$M/M/1/c$	queueing system with one server and finite capacity c
FIFO	First In, First Out
π_b	probability that all servers are busy
$M/G/s/s$	Erlang loss system

Chapter 7

$\gamma(\cdot)$	(auto) covariance function of a stationary process
$\rho(\cdot)$	autocorrelation function of a stationary process
ACF	autocorrelation function
ACVF	autocovariance function
$IID(0, \sigma^2)$	i.i.d. noise with zero mean
$WN(0, \sigma^2)$	white noise process
$GWN(0, \sigma^2)$	Gaussian white noise
$\mathbf{C}$	covariance matrix
T	transpose of a vector or a matrix
$N(\mathbf{m}, \mathbf{C})$	multinormal distribution
erf	error function
$AR(p)$	autoregressive process of order p
$MA(q)$	moving average process of order q
L	lag or backshift operator
Δ	difference operator
$ARMA(p, q)$	autoregressive moving average process of order (p, q)
$ARIMA(p, d, q)$	autoregressive integrated moving average process
$\phi(\cdot)$	partial autocorrelation function
$\hat{\theta}$	point estimate of a parameter θ
H_n	history of the stochastic process up to time n
$g(X_{n+j} \mid H_n)$	predictor of X_{n+j}, given H_n
MSE	mean-square error

B

Statistical tables

B.1 Distribution function of the binomial distribution
B.2 Distribution function of the Poisson distribution
B.3 Values of the function $\Phi(z)$
B.4 Values of the function $Q^{-1}(p)$ for some values of p

Table B.1. Distribution function of the binomial distribution

					p		
		0.05	0.10	0.20	0.25	0.40	0.50
n	x						
2	0	0.9025	0.8100	0.6400	0.5625	0.3600	0.2500
	1	0.9975	0.9900	0.9600	0.9375	0.8400	0.7500
3	0	0.8574	0.7290	0.5120	0.4219	0.2160	0.1250
	1	0.9927	0.9720	0.8960	0.8438	0.6480	0.5000
	2	0.9999	0.9990	0.9920	0.9844	0.9360	0.8750
4	0	0.8145	0.6561	0.4096	0.3164	0.1296	0.0625
	1	0.9860	0.9477	0.8192	0.7383	0.4752	0.3125
	2	0.9995	0.9963	0.9728	0.9493	0.8208	0.6875
	3	1.0000	0.9999	0.9984	0.9961	0.9744	0.9375
5	0	0.7738	0.5905	0.3277	0.2373	0.0778	0.0313
	1	0.9774	0.9185	0.7373	0.6328	0.3370	0.1875
	2	0.9988	0.9914	0.9421	0.8965	0.6826	0.5000
	3	1.0000	0.9995	0.9933	0.9844	0.9130	0.8125
	4	1.0000	1.0000	0.9997	0.9990	0.9898	0.9688
10	0	0.5987	0.3487	0.1074	0.0563	0.0060	0.0010
	1	0.9139	0.7361	0.3758	0.2440	0.0464	0.0107
	2	0.9885	0.9298	0.6778	0.5256	0.1673	0.0547
	3	0.9990	0.9872	0.8791	0.7759	0.3823	0.1719
	4	0.9999	0.9984	0.9672	0.9219	0.6331	0.3770
	5	1.0000	0.9999	0.9936	0.9803	0.8338	0.6230
	6		1.0000	0.9991	0.9965	0.9452	0.8281
	7			0.9999	0.9996	0.9877	0.9453
	8			1.0000	1.0000	0.9983	0.9893
	9					0.9999	0.9990

Table B.1. Continued

				p			
		0.05	0.10	0.20	0.25	0.40	0.50
n	x						
15	0	0.4633	0.2059	0.0352	0.0134	0.0005	0.0000
	1	0.8290	0.5490	0.1671	0.0802	0.0052	0.0005
	2	0.9638	0.8159	0.3980	0.2361	0.0271	0.0037
	3	0.9945	0.9444	0.6482	0.4613	0.0905	0.0176
	4	0.9994	0.9873	0.8358	0.6865	0.2173	0.0592
	5	0.9999	0.9977	0.9389	0.8516	0.4032	0.1509
	6	1.0000	0.9997	0.9819	0.9434	0.6098	0.3036
	7		1.0000	0.9958	0.9827	0.7869	0.5000
	8			0.9992	0.9958	0.9050	0.6964
	9			0.9999	0.9992	0.9662	0.8491
	10			1.0000	0.9999	0.9907	0.9408
	11				1.0000	0.9981	0.9824
	12					0.9997	0.9963
	13					1.0000	0.9995
	14						1.0000
20	0	0.3585	0.1216	0.0115	0.0032	0.0000	
	1	0.7358	0.3917	0.0692	0.0243	0.0005	0.0000
	2	0.9245	0.6769	0.2061	0.0913	0.0036	0.0002
	3	0.9841	0.8670	0.4114	0.2252	0.0160	0.0013
	4	0.9974	0.9568	0.6296	0.4148	0.0510	0.0059
	5	0.9997	0.9887	0.8042	0.6172	0.1256	0.0207
	6	1.0000	0.9976	0.9133	0.7858	0.2500	0.0577
	7		0.9996	0.9679	0.8982	0.4159	0.1316
	8		0.9999	0.9900	0.9591	0.5956	0.2517
	9		1.0000	0.9974	0.9861	0.7553	0.4119
	10			0.9994	0.9961	0.8725	0.5881
	11			0.9999	0.9991	0.9435	0.7483
	12			1.0000	0.9998	0.9790	0.8684
	13				1.0000	0.9935	0.9423
	14					0.9984	0.9793
	15					0.9997	0.9941
	16					1.0000	0.9987
	17						0.9998
	18						1.0000

Table B.2. Distribution function of the Poisson distribution

x	0.5	1	1.5	2	5	10	15	20
0	0.6065	0.3679	0.2231	0.1353	0.0067	0.0000		
1	0.9098	0.7358	0.5578	0.4060	0.0404	0.0005		
2	0.9856	0.9197	0.8088	0.6767	0.1247	0.0028	0.0000	
3	0.9982	0.9810	0.9344	0.8571	0.2650	0.0103	0.0002	
4	0.9998	0.9963	0.9814	0.9473	0.4405	0.0293	0.0009	0.0000
5	1.0000	0.9994	0.9955	0.9834	0.6160	0.0671	0.0028	0.0001
6		0.9999	0.9991	0.9955	0.7622	0.1301	0.0076	0.0003
7		1.0000	0.9998	0.9989	0.8666	0.2202	0.0180	0.0008
8			1.0000	0.9998	0.9319	0.3328	0.0374	0.0021
9				1.0000	0.9682	0.4579	0.0699	0.0050
10					0.9863	0.5830	0.1185	0.0108
11					0.9945	0.6968	0.1848	0.0214
12					0.9980	0.7916	0.2676	0.0390
13					0.9993	0.8645	0.3632	0.0661
14					0.9998	0.9165	0.4657	0.1049
15					0.9999	0.9513	0.5681	0.1565
16					1.0000	0.9730	0.6641	0.2211
17						0.9857	0.7489	0.2970
18						0.9928	0.8195	0.3814
19						0.9965	0.8752	0.4703
20						0.9984	0.9170	0.5591
21						0.9993	0.9469	0.6437
22						0.9997	0.9673	0.7206
23						0.9999	0.9805	0.7875
24						1.0000	0.9888	0.8432
25							0.9938	0.8878
26							0.9967	0.9221
27							0.9983	0.9475
28							0.9991	0.9657
29							0.9996	0.9782
30							0.9998	0.9865
31							0.9999	0.9919
32							1.0000	0.9953

Table B.3. Values of the function $\Phi(z)$

z	+0.00	+0.01	+0.02	+0.03	+0.04	+0.05	+0.06	+0.07	+0.08	+0.09
0.0	0.5000	0.5040	0.5080	0.5120	0.5160	0.5199	0.5239	0.5279	0.5319	0.5359
0.1	0.5398	0.5438	0.5478	0.5517	0.5557	0.5596	0.5636	0.5675	0.5714	0.5753
0.2	0.5793	0.5832	0.5871	0.5910	0.5948	0.5987	0.6026	0.6064	0.6103	0.6141
0.3	0.6179	0.6217	0.6255	0.6293	0.6331	0.6368	0.6406	0.6443	0.6480	0.6517
0.4	0.6554	0.6591	0.6628	0.6664	0.6700	0.6736	0.6772	0.6808	0.6844	0.6879
0.5	0.6915	0.6950	0.6985	0.7019	0.7054	0.7088	0.7123	0.7157	0.7190	0.7224
0.6	0.7257	0.7291	0.7324	0.7357	0.7389	0.7422	0.7454	0.7486	0.7517	0.7549
0.7	0.7580	0.7611	0.7642	0.7673	0.7704	0.7734	0.7764	0.7794	0.7823	0.7852
0.8	0.7881	0.7910	0.7939	0.7967	0.7995	0.8023	0.8051	0.8078	0.8106	0.8133
0.9	0.8159	0.8186	0.8212	0.8238	0.8264	0.8289	0.8315	0.8340	0.8365	0.8389
1.0	0.8413	0.8438	0.8461	0.8485	0.8508	0.8531	0.8554	0.8577	0.8599	0.8621
1.1	0.8643	0.8665	0.8686	0.8708	0.8729	0.8749	0.8770	0.8790	0.8810	0.8830
1.2	0.8849	0.8869	0.8888	0.8907	0.8925	0.8944	0.8962	0.8980	0.8997	0.9015
1.3	0.9032	0.9049	0.9066	0.9082	0.9099	0.9115	0.9131	0.9147	0.9162	0.9177
1.4	0.9192	0.9207	0.9222	0.9236	0.9251	0.9265	0.9279	0.9292	0.9306	0.9319
1.5	0.9332	0.9345	0.9357	0.9370	0.9382	0.9394	0.9406	0.9418	0.9429	0.9441
1.6	0.9452	0.9463	0.9474	0.9484	0.9495	0.9505	0.9515	0.9525	0.9535	0.9545
1.7	0.9554	0.9564	0.9573	0.9582	0.9591	0.9599	0.9608	0.9616	0.9625	0.9633
1.8	0.9641	0.9649	0.9656	0.9664	0.9671	0.9678	0.9686	0.9693	0.9699	0.9706
1.9	0.9713	0.9719	0.9726	0.9732	0.9738	0.9744	0.9750	0.9756	0.9761	0.9767
2.0	0.9772	0.9778	0.9783	0.9788	0.9793	0.9798	0.9803	0.9808	0.9812	0.9817
2.1	0.9821	0.9826	0.9830	0.9834	0.9838	0.9842	0.9846	0.9850	0.9854	0.9857
2.2	0.9861	0.9864	0.9868	0.9871	0.9875	0.9878	0.9881	0.9884	0.9887	0.9890
2.3	0.9893	0.9896	0.9898	0.9901	0.9904	0.9906	0.9909	0.9911	0.9913	0.9916
2.4	0.9918	0.9920	0.9922	0.9925	0.9927	0.9929	0.9931	0.9932	0.9934	0.9936
2.5	0.9938	0.9940	0.9941	0.9943	0.9945	0.9946	0.9948	0.9949	0.9951	0.9952
2.6	0.9953	0.9955	0.9956	0.9957	0.9959	0.9960	0.9961	0.9962	0.9963	0.9964
2.7	0.9965	0.9966	0.9967	0.9968	0.9969	0.9970	0.9971	0.9972	0.9973	0.9974
2.8	0.9974	0.9975	0.9976	0.9977	0.9977	0.9978	0.9979	0.9979	0.9980	0.9981
2.9	0.9981	0.9982	0.9982	0.9983	0.9984	0.9984	0.9985	0.9985	0.9986	0.9986

Table B.3. Continued

z	+0.00	+0.01	+0.02	+0.03	+0.04	+0.05	+0.06	+0.07	+0.08	+0.09
3.0	0.9987	0.9987	0.9987	0.9988	0.9988	0.9989	0.9989	0.9989	0.9990	0.9990
3.1	0.9990	0.9991	0.9991	0.9991	0.9992	0.9992	0.9992	0.9992	0.9993	0.9993
3.2	0.9993	0.9993	0.9994	0.9994	0.9994	0.9994	0.9994	0.9995	0.9995	0.9995
3.3	0.9995	0.9995	0.9995	0.9996	0.9996	0.9996	0.9996	0.9996	0.9996	0.9997
3.4	0.9997	0.9997	0.9997	0.9997	0.9997	0.9997	0.9997	0.9997	0.9997	0.9998
3.5	0.9998	0.9998	0.9998	0.9998	0.9998	0.9998	0.9998	0.9998	0.9998	0.9998
3.6	0.9998	0.9998	0.9999	0.9999	0.9999	0.9999	0.9999	0.9999	0.9999	0.9999
3.7	0.9999	0.9999	0.9999	0.9999	0.9999	0.9999	0.9999	0.9999	0.9999	0.9999
3.8	0.9999	0.9999	0.9999	0.9999	0.9999	0.9999	0.9999	0.9999	0.9999	0.9999
3.9	1.0000	1.0000	1.0000	1.0000	1.0000	1.0000	1.0000	1.0000	1.0000	1.0000

Table B.4. Values of the function $Q^{-1}(p)$ for some values of p

p	0.10	0.05	0.01	0.005	0.001	0.0001	0.00001
$Q^{-1}(p)$	1.282	1.645	2.326	2.576	3.090	3.719	4.265

C

Solutions to "Solved exercises"

Chapter 1

Question no. 1

When x decreases to 0, $1/x$ increases to ∞. The function $\sin x$ does not converge as x tends to ∞. However, because $-1 \leq \sin x \leq 1$, for any real x, we may conclude that $\lim_{x \downarrow 0} x \sin(1/x) = 0$. This result can be proved from the definition of the limit of a function. We have:

$$0 < |x| < \epsilon \quad \Longrightarrow \quad |x \sin(1/x)| \leq |x| < \epsilon.$$

Hence, we can take $\delta = \epsilon$ in Definition 1.1.1 and we can actually write that

$$\lim_{x \to 0} x \sin(1/x) = 0.$$

Note that the function $f(x) := x \sin(1/x)$ is not defined at $x = 0$.

Question no. 2

The functions $f_1(x) := \sin x$ and $f_2(x) := x$ are continuous, for any real x. Moreover, it can be shown that if $f_1(x)$ and $f_2(x)$ are continuous, then $g(x) := f_1(x)/f_2(x)$ is also a continuous function, for any x such that $f_2(x) \neq 0$. Therefore, we can assert that $f(x)$ is continuous at any $x \neq 0$.

Next, using the series expansion of the function $\sin x$:

$$\sin x = x - \frac{1}{3!}x^3 + \frac{1}{5!}x^5 - \frac{1}{7!}x^7 + \cdots,$$

we may write that

$$\frac{\sin x}{x} = 1 - \frac{1}{3!}x^2 + \frac{1}{5!}x^4 - \frac{1}{7!}x^6 + \cdots.$$

Hence, we deduce that

$$\lim_{x \to 0} \frac{\sin x}{x} = 1.$$

Remark. To obtain the previous result, we can also use l'Hospital's rule:

$$\lim_{x \to 0} \frac{\sin x}{x} = \lim_{x \to 0} \frac{\cos x}{1} = 1.$$

Because, by definition

$$f(0) = 1 = \lim_{x \to 0} \frac{\sin x}{x},$$

we conclude that the function $f(x)$ is continuous at any $x \in \mathbb{R}$.

Question no. 3

Let $g(x) = \sqrt{3x + 1}$ and $h(x) = (2x^2 + 1)^2$. We have, using the *chain rule*:

$$g'(x) = \frac{1}{2\sqrt{3x + 1}}(3) \quad \text{and} \quad h'(x) = 2(2x^2 + 1)(4x).$$

Thus,

$$f'(x) = g'(x)h(x) + g(x)h'(x) = \frac{3}{2\sqrt{3x+1}}(2x^2 + 1)^2 + \sqrt{3x+1}(8x)(2x^2 + 1)$$

$$\Longleftrightarrow \quad f'(x) = (2x^2 + 1)\left\{\frac{3(2x^2 + 1)}{2\sqrt{3x+1}} + 8x\sqrt{3x+1}\right\}.$$

Question no. 4

We have:

$$\lim_{x \downarrow 0} x \ln x = \lim_{x \downarrow 0} x \lim_{x \downarrow 0} \ln x = 0 \times (-\infty),$$

which is indeterminate. Writing that

$$x \ln x = \frac{\ln x}{1/x},$$

we obtain that

$$\lim_{x \downarrow 0} x \ln x = \lim_{x \downarrow 0} \frac{\ln x}{1/x} = \frac{-\infty}{\infty}.$$

We can then use l'Hospital's rule:

$$\lim_{x \downarrow 0} \frac{\ln x}{1/x} = \lim_{x \downarrow 0} \frac{1/x}{-1/x^2} = \lim_{x \downarrow 0} -x = 0.$$

Question no. 5

In as much as the derivative of $\ln x$ is $1/x$, we can write that

$$I_5 = \int_1^e \frac{\ln x}{x}\,dx = \frac{(\ln x)^2}{2}\Big|_1^e = \frac{(\ln e)^2 - (\ln 1)^2}{2} = \frac{1}{2}.$$

This result can be checked by using the integration by substitution method: setting $y = \ln x \Leftrightarrow x = e^y$, we deduce that

$$\int \frac{\ln x}{x}\,dx = \int \frac{y}{e^y}(e^y)'\,dy = \int y\,dy = \frac{1}{2}y^2.$$

It follows that

$$I_5 = \int_0^1 y\,dy = \frac{1}{2}y^2\Big|_0^1 = \frac{1}{2}.$$

Question no. 6

We use the integration by parts technique. We set

$$u = x^2 \quad \text{and} \quad dv = xe^{-x^2/2}\,dx.$$

Because $v = -e^{-x^2/2}$, it follows that

$$I_6 = -x^2 e^{-x^2/2}\Big|_{-\infty}^{\infty} + \int_{-\infty}^{\infty} 2xe^{-x^2/2}\,dx.$$

By l'Hospital's rule, the above constant term is equal to 0, and the integral is given by

$$\int_{-\infty}^{\infty} 2xe^{-x^2/2}\,dx = -2e^{-x^2/2}\Big|_{-\infty}^{\infty} = 0.$$

Hence, we have that $I_6 = 0$.

Remark. In probability, we deduce from this result that the *mathematical expectation* or *expected value* of the cube of a *standard normal random variable* is equal to zero.

Question no. 7

We have:

$$F(\omega) = \int_0^{\infty} e^{j\omega x} c e^{-cx}\,dx = c\int_0^{\infty} e^{(j\omega - c)x}\,dx = \frac{c}{j\omega - c} e^{(j\omega - c)x}\Big|_0^{\infty} = \frac{c}{c - j\omega}.$$

Remarks. (i) The fact that j is a *(pure) imaginary number* does not cause any problem, because it is a constant.

(ii) In probability, the function $F(\omega)$ obtained above is the *characteristic function* of a *random variable* having an *exponential distribution* with *parameter c*.

Question no. 8

We can write that

$$I_8 = \int_0^1 \int_0^{\sqrt{y}} (x+y)\,dxdy = \int_0^1 \left\{ \frac{x^2}{2}\Big|_0^{\sqrt{y}} + y\sqrt{y} \right\} dy$$

$$= \int_0^1 \frac{y}{2} + y^{3/2}\,dy = \frac{y^2}{4} + \frac{y^{5/2}}{5/2}\Big|_0^1 = \frac{1}{4} + \frac{2}{5} = \frac{13}{20}.$$

Or:

$$I_8 = \int_0^1 \int_{x^2}^1 (x+y)\,dydx = \int_0^1 \left\{ x(1-x^2) + \frac{y^2}{2}\Big|_{x^2}^1 \right\} dx$$

$$= \int_0^1 \left\{ x(1-x^2) + \frac{1}{2} - \frac{x^4}{2} \right\} dx = \frac{x^2}{2} - \frac{x^4}{4} + \frac{x}{2} - \frac{x^5}{10}\Big|_0^1$$

$$= \frac{1}{2} - \frac{1}{4} + \frac{1}{2} - \frac{1}{10} = \frac{13}{20}.$$

Remark. We easily find that

$$\int_0^1 \int_0^1 (x+y)\,dxdy = 1.$$

Hence, if $B := \{(x,y) \in \mathbb{R}^2 : 0 \le x \le 1, 0 \le y \le 1, x < y\}$, then we can write (by symmetry) that

$$\int\int_B (x+y)\,dxdy = \frac{1}{2}.$$

Question no. 9

Consider the geometric series

$$S(1/2, 1/2) := \frac{1}{2} + \frac{1}{4} + \frac{1}{8} + \cdots = \sum_{n=1}^{\infty} (1/2)^n.$$

This series converges to 1. Hence, we can write that

$$S_9 = 1 - \frac{1}{2} - \frac{1}{4} = \frac{1}{4}.$$

Or:

$$S_9 = \frac{1}{4}S(1/2, 1/2) = \frac{1}{4}.$$

Remark. The sum S_9 represents the probability that the number of tosses needed to obtain "heads" with a *fair* coin will be greater than two.

Question no. 10

We have:

$$\sum_{k=1}^{\infty}(1-p)^{k-1} = \frac{1}{1-(1-p)} = \frac{1}{p}.$$

We can differentiate this series term by term (twice). We obtain that

$$\frac{2}{p^3} = \frac{d^2}{dp^2}\sum_{k=1}^{\infty}(1-p)^{k-1} = \sum_{k=1}^{\infty}(k-1)(k-2)(1-p)^{k-3}.$$

Because

$$\sum_{k=1}^{\infty}(k-1)(k-2)(1-p)^{k-3} = \sum_{k=3}^{\infty}(k-1)(k-2)(1-p)^{k-3}$$

$$= \sum_{n=1}^{\infty}(n+1)(n)(1-p)^{n-1} = \sum_{n=1}^{\infty}n^2(1-p)^{n-1} + \sum_{n=1}^{\infty}n(1-p)^{n-1},$$

we deduce (see Example 1.4.1) that

$$S_{10} = p\left(\frac{2}{p^3} - \frac{1}{p^2}\right) = \frac{2-p}{p^2}.$$

Remark. The sum calculated above is the *average value* of the square of a *geometric random variable*.

Chapter 2

Question no. 1

We have that $\Omega = \{1, 2, 3, 4, 5, (6,1), \ldots, (6,6)\}$. Thus, there are $5+6 = 11$ elementary outcomes.

Question no. 2

Four different partitions of Ω can be formed: $\{e_1\}, \{e_2, e_3\}$, or $\{e_2\}, \{e_1, e_3\}$, or $\{e_3\}, \{e_1, e_2\}$, or $\{e_1\}, \{e_2\}, \{e_3\}$.

Question no. 3

Let S_i be the event "the sum of the two numbers obtained is equal to i" and let $D_{j,k}$ be "the number obtained on the jth roll is k." We seek

$$P[S_4 \mid D_{1,2} \cup D_{1,4} \cup D_{1,6}] = \frac{P[D_{1,2} \cap D_{2,2}]}{P[D_{1,2} \cup D_{1,4} \cup D_{1,6}]} \overset{\text{ind.}}{=} \frac{(1/6)^2}{1/2} = 1/18.$$

Question no. 4

We have that $P[A \mid B] = P[B] = P[A] = 1/4 \Rightarrow A$ and B are independent events. It follows that

$$P[A \cap B'] = P[A]P[B'] = (1/4)(3/4) = 3/16.$$

Question no. 5

Let F_i be the event "component i operates" and let F_S be "the system operates." By symmetry and incompatibility, we can write that

$$P[F_S] = 3 \times P[F_1 \cap F_2 \cap F_3'] + P[F_1 \cap F_2 \cap F_3]$$
$$\stackrel{\text{ind.}}{=} 3 \times (0.95)(0.95)(0.05) + (0.95)^3 = 0.99275.$$

Question no. 6

We have:

$$P[A] = P[A \cap B] + P[A \cap B'] = \frac{1}{4} + P[A \mid B']P[B']$$

$$= \frac{1}{4} + \left(\frac{1}{8}\right)\left(1 - \frac{1}{2}\right) = \frac{5}{16}.$$

Question no. 7

Let A_i be the event "i 'heads' were obtained." We seek

$$P[A_3 \mid A_1 \cup A_2 \cup A_3] = \frac{P[A_3]}{P[A_1 \cup A_2 \cup A_3]} \stackrel{\text{ind.}}{=} \frac{(1/2)^3}{1 - (1/2)^3} = \frac{1}{7}.$$

Question no. 8

We have:

$$P[A_1 \mid B] = \frac{P[B \mid A_1]P[A_1]}{P[B \mid A_1]P[A_1] + P[B \mid A_2]P[A_2]} \stackrel{P[A_1]=P[A_2]}{=} \frac{1/2}{1/2 + 1/4} = \frac{2}{3}.$$

Question no. 9

We have:

$$A = \{abc, acb, cab\} \quad \text{and} \quad B = \{abc, acb, bac\}.$$

(a) Because $A \cap B = \{abc, acb\} \neq \emptyset$, A and B do *not* form a partition of Ω. (In addition, $A \cup B \neq \Omega$.)

(b) $P[A] = \frac{1}{18} + \frac{1}{18} + \frac{2}{9} = P[B]$ and $P[A \cap B] = P[\{abc, acb\}] = \frac{1}{9}$. Because

$$P[A]P[B] = \frac{1}{3} \cdot \frac{1}{3} = \frac{1}{9} = P[A \cap B],$$

we can assert that A and B *are* independent events.

Question no. 10

Let A_1 be the event "A occurs on the first repetition." Likewise for B_1 and C_1. Then, we may write that

$$P[D] = P[D \mid A_1]P[A_1] + P[D \mid B_1]P[B_1] + P[D \mid C_1]P[C_1]$$
$$\Longleftrightarrow \quad P[D] = 1 \cdot P[A] + 0 + P[D]P[C]$$
$$\Longleftrightarrow \quad P[D] = \frac{P[A]}{1 - P[C]} = \frac{P[A]}{P[A] + P[B]}.$$

Question no. 11

Let A be the event "the transistor tested is defectless" and let D be "the transistor tested is defective." Then, we have that $\Omega = \{D, AD, AAD, AAA\}$.

Remark. We could also write that $\Omega = \{A', AA', AAA', AAA\}$.

Question no. 12

We have that $P[B \mid A] = 1 - P[B' \mid A] = 2/7$ and

$$P[B \mid A] = \frac{P[B \cap A]}{P[A]} = \frac{P[B]}{P[A]} \quad \text{because } B \subset A.$$

It follows that

$$P[B] = P[B \mid A]P[A] = \frac{2}{21} \simeq 0.0952.$$

Question no. 13

Let D be the event "the teacher holds a PhD," let F (resp., A, B) be "the teacher is a full (resp., associate, assistant) professor," and let L be "the teacher is a lecturer." We can write that

$$P[D] = P[D \mid F]P[F] + P[D \mid A]P[A] + P[D \mid B]P[B] + P[D \mid L]P[L]$$
$$= (0.6)(0.3) + (0.7)(0.4) + (0.9)(0.2) + (0.4)(0.1) = 0.68.$$

Question no. 14

The number of different codes is given by

$$26 \times 25 \times 24 \times 23 \times 22 = 7,893,600 \quad (= P_5^{26}).$$

Question no. 15

We have that $\Omega = \{(1,1), (1,2), \ldots, (6,6)\}$. There are 36 *equiprobable* elementary outcomes.

(a) $B = \{(1,6), (2,5), (3,4), (4,3), (5,2), (6,1)\}$ and $C = B \cup \{(5,6), (6,5)\}$. Therefore,

$$P[B \mid C] = \frac{P[B \cap C]}{P[C]} = \frac{P[B]}{P[C]} = \frac{6/36}{8/36} = 3/4.$$

(b) We have:

$$P[A \mid B] = \frac{P[A \cap B]}{P[B]} \stackrel{\text{(a)}}{=} \frac{P[\{(6,1)\}]}{6/36} = \frac{1/36}{6/36} = 1/6.$$

(c) Because the die is fair, we may write that $P[A] = 1/6 \stackrel{\text{(b)}}{=} P[A \mid B]$. Hence, A and B *are* independent events.

Question no. 16

(a) Let the events be
  $A =$ "the commuter gets home before 5:30 p.m.;"
  $B =$ "the commuter uses the compact car."
We seek

$$P[A] = P[A \mid B]P[B] + P[A \mid B']P[B']$$
$$= (0.75)(0.75) + (0.60)(0.25) = 0.7125.$$

(b) We calculate

$$P[B \mid A'] = P[A' \mid B]\frac{P[B]}{P[A']} \stackrel{\text{(a)}}{=} (1 - 0.75)\frac{0.75}{1 - 0.7125} \simeq 0.6522.$$

(c) We have that $P[A' \cap B'] = P[A' \mid B']P[B'] = (1 - 0.60)(0.25) = 0.1$.
(d) By independence, we seek

$$P[A \cap B] \cdot P[A \cap B'] + P[A \cap B'] \cdot P[A \cap B]$$

$$= 2P[A \mid B]P[B] \cdot P[A \mid B']P[B'] = 2(0.75)^2 \cdot (0.60)(0.25)$$
$$= 0.16875.$$

Question no. 17

We define the events $A =$ "it is raining," $B =$ "rain was forecast," and $C =$ "Mr. X has his umbrella." We seek

$$P[A \cap C'] = P[A \cap \underbrace{B \cap C'}_{\emptyset}] + P[A \cap B' \cap C'] = P[A \cap B' \cap C']$$

$$= P[A \mid B' \cap C']P[B' \cap C'] = \frac{1}{3}P[C' \mid B']P[B']$$
$$= \frac{1}{3} \times \frac{2}{3} \times \frac{1}{2} = \frac{1}{9}.$$

Question no. 18

We have that $\Omega = \{(1,1,1),\ldots,(1,1,6),\ldots,(6,6,6)\}$. There are $6^3 = 216$ elementary outcomes, which are all equiprobable. By *enumeration*, we find that there are 20 triples for which event F occurs:

$$(1,2,3),\ (1,2,4),\ (1,2,5),\ (1,2,6),\ (1,3,4),\ldots.$$

Consequently, the probability asked for is $\frac{20}{216} \simeq 0.0926$.

Question no. 19

We have that $\Omega = \{(G,G),(G,B),(B,G),(B,B)\}$. Moreover,

$$A_1 = \{(G,B),(B,G)\} \quad \text{and} \quad A_2 = \{(G,B),(B,G),(B,B)\}.$$

(a) Because $A_2' = \{(G,G)\}$, we have that $A_1 \cap A_2' = \emptyset$. Therefore, A_1 and A_2' are incompatible.

(b) We may write that $P[A_1] = 2/4$ and $P[A_2] = 3/4$. Given that

$$P[A_1 \cap A_2] = P[A_1] = 2/4 \neq P[A_1]P[A_2],$$

A_1 and A_2 are *not* independent.

(c) Let B_i (resp., G_i) be the event "the ith child is a boy (resp., a girl)." We first calculate

$$P[B_3] = P[B_3 \mid B_1 \cap B_2]P[B_1 \cap B_2] + P[B_3 \mid G_1 \cap G_2]P[G_1 \cap G_2]$$
$$+ P[B_3 \mid (B_1 \cap G_2) \cup (G_1 \cap B_2)]P[(B_1 \cap G_2) \cup (G_1 \cap B_2)]$$
$$= \left(\frac{11}{20}\right)\left(\frac{1}{4}\right) + \left(\frac{2}{5}\right)\left(\frac{1}{4}\right) + \left(\frac{1}{2}\right)\left(\frac{1}{2}\right) = \frac{39}{80}.$$

Hence, we may write that

$$P[B_1 \cap B_2 \mid B_3] = \frac{P[B_3 \mid B_1 \cap B_2]P[B_1 \cap B_2]}{P[B_3]} = \frac{\frac{11}{20} \cdot \frac{1}{4}}{\frac{39}{80}} \simeq 0.2821.$$

Chapter 3

Question no. 1

We must have that $1 = \frac{a}{8} + \frac{a}{4} + \frac{a}{8} = \frac{a}{2}$. Therefore, a must be equal to 2.

Question no. 2

We calculate

$$F_X(0) = \int_{-1}^{0} \frac{3}{4}(1 - x^2)\, dx = \frac{3}{4}\left(x - \frac{x^3}{3}\right)\Big|_{-1}^{0} = \frac{1}{2}.$$

Remark. This result could have been deduced from the symmetry of the function $f_X(x)$ about $x = 0$.

Question no. 3

First, we calculate $E[X] = \frac{1}{3}(1 + 2 + 3) = 2$. Next, we have:

$$E[X^2] = \frac{1}{3}(1^2 + 2^2 + 3^2) = \frac{14}{3} \implies \text{VAR}[X] = \frac{14}{3} - 2^2 = \frac{2}{3}.$$

Thus, we may write that $\text{STD}[X] = \sqrt{2/3}$.

Question no. 4

We have:

$$E[X^{1/2}] = \int_{0}^{1} x^{1/2} \cdot 2x\, dx = 2\,\frac{x^{5/2}}{5/2}\Big|_{0}^{1} = \frac{4}{5}.$$

Question no. 5

We can write that $x_{0.75}^2/4 = 0.25$ and $x_{0.75} > 0$. It follows that $x_{0.75} = 1$.

Question no. 6

We have:

$$f_Y(y) = f_X(y - 1)\left|\frac{d}{dy}(y - 1)\right| = \frac{1}{2} \quad \text{if } 1 < y < 3.$$

Question no. 7

By definition, $X \sim \text{Hyp}(N = 15, n = 2, d = 5)$.

Question no. 8

According to Table B.1, page 276, $x = 1$ is the most probable value, with $p_X(1) \simeq 0.6328 - 0.2373 = 0.3955$.

Question no. 9

We seek $P[\text{B}(10, 0.1) = 2] \overset{\text{Tab. B.1}}{\simeq} 0.9298 - 0.7361 = 0.1937$.

Question no. 10

We have:

$$P[X \geq 1 \mid X \leq 1] = \frac{P[X = 1]}{P[X \leq 1]} = \frac{e^{-5} \cdot 5}{e^{-5} + e^{-5} \cdot 5} = \frac{5}{6}.$$

Question no. 11

We want $P[\text{Poi}(2 \cdot 2) = 1] = e^{-4} \cdot 4 \simeq 0.0733$.

Question no. 12

We can write that

$$P[\text{Hyp}(N = 250, n = 5, d = 50) = 0] \simeq P\left[\text{B}\left(n = 5, p = 50/250\right) = 0\right]$$
$$= \left(\frac{4}{5}\right)^5 \simeq 0.3277.$$

Question no. 13

We have:

$$P[\text{B}(50, 0.01) \geq 4] \quad \simeq \quad P[\text{Poi}(1/2) \geq 4] = 1 - P[\text{Poi}(1/2) \leq 3]$$
$$\underset{\text{Tab. B.2}}{\simeq} 1 - 0.9982 = 0.0018.$$

Question no. 14

We seek

$$P[\text{Geo}(p = 1/2) = 5] = \left(\frac{1}{2}\right)^{5-1}\left(\frac{1}{2}\right) = \frac{1}{32} = 0.03125.$$

Question no. 15

By the memoryless property of the exponential distribution, we may write that

$$P[X > 20 \mid X > 10] = P[X > 10] = P[\text{Exp}(\lambda = 1/10) > 10] = e^{-1} \simeq 0.3679.$$

Question no. 16

We have:

$$P[X > 1] = 2\,P[X > 2] \iff e^{-\lambda} = 2\,e^{-2\lambda} \iff e^{\lambda} = 2 \iff \lambda = \ln 2.$$

Question no. 17

We can write that

$$P[\text{G}(\alpha = 2, \lambda = 1) < 4] = P[\text{Poi}(1 \cdot 4) \geq 2].$$

So, we can use a Poisson distribution with parameter 4.

Question no. 18

We can write that $X \sim \text{G}(\alpha = 10, \lambda = 2)$.

Question no. 19

We have:

$$P[|N(0,1)| < 1/2] \overset{\text{sym.}}{=} 2\,\Phi(1/2) - 1 \overset{\text{Tab. B.3}}{\simeq} 2\,(0.6915) - 1 = 0.3830.$$

Question no. 20

We have:

$$x_{0.90} = \mu_X + z_{0.90} \cdot \sigma_X \overset{\text{Tab. B.4}}{\simeq} 1 + (-1.282)\sqrt{2} \simeq -0.813.$$

Question no. 21

(a) Let X be the number of down components. Then, we have that $X \sim B(n = 5, p = 0.05)$. We seek $P[X \leq 1] \overset{\text{Tab. B.2}}{\simeq} 0.977$.

(b) Let Y be the number of devices needed to obtain a first device that does not operate. We have that $Y \sim \text{Geo}(p = P[X > 1] \overset{(a)}{\simeq} 1 - 0.977)$. Therefore, $E[Y] = 1/p \simeq 43.5$ devices.

Question no. 22

(a) Let $N(t)$ be the total number of buses passing in a time period of t hours. Then, $N(t) \sim \text{Poi}(4t)$. We seek

$$P[N(1/2) \leq 1] = P[\text{Poi}(2) \leq 1] = e^{-2} + 2e^{-2} \simeq 0.4060.$$

(b) Let T be the waiting time between the first and the third bus. Then, $T \sim G(\alpha = 2, \lambda = 4)$, so that $\text{VAR}[T] = \alpha/\lambda^2 = 1/8$ (hour)^2.

(c) Let W be the total waiting time (in minutes). Then, $W \sim \text{Exp}(1/15)$. We want

$$P[W > 20 \mid W > 5] = P[W > 15] = e^{-15/15} \simeq 0.3679.$$

Question no. 23

(a) We calculate

$$P[N(\mu, (0.1\mu)^2) \geq 1.15\mu] = 1 - \Phi\left(\frac{1.15\mu - \mu}{0.1\mu}\right)$$

$$= 1 - \Phi(1.5) \overset{\text{Tab. B.3}}{\simeq} 1 - 0.9332 = 0.0668.$$

(b) We seek $x_{0.10} = \mu + z_{0.10} \cdot \sigma \overset{\text{Tab. B.4}}{\simeq} 4 + (-1.282)(0.1)(4) \simeq 3.49.$

Question no. 24

Let X be the number of correct answers that the student gets. Then, X follows a $B(n = 10, p = 0.5)$ distribution. We want

$$P[X > 5] = 1 - P[X \le 5] \overset{\text{Tab. B.1}}{\simeq} 1 - 0.6230 = 0.3770.$$

Question no. 25

We have:

$$P[B(100, 0.1) = 15] \simeq P[\text{Poi}(10) = 15] = e^{-10} \frac{10^{15}}{15!} \simeq 0.0347.$$

Question no. 26

We can carry out the required integral to obtain the desired probability. However, we notice that the function $f_X(x)$ is symmetrical about the origin. That is, $f_X(-x) = f_X(x)$. Then, given that X is a continuous random variable that is defined in a bounded interval, we may write that $P[X < 0] = 1/2$.

Question no. 27

Let X be the IQ of the pupils. Then, $X \sim N(100, (15)^2)$. We look for

$$P[\{X < 91\} \cup \{X > 130\}] = 1 - P\left[\frac{91 - 100}{15} \le N(0, 1) \le \frac{130 - 100}{15}\right]$$

$$= 1 - [\varPhi(2) - \varPhi(-0.6)] \overset{\text{Tab. B.3}}{\simeq} 1 - [0.9772 - (1 - 0.7257)] = 0.2971.$$

Question no. 28

(a) Let X be the number of defective transistors. Then, X has a $B(n = 60, p = 0.05)$ distribution. We seek

$$P[X \le 1] = (0.95)^{60} + \binom{60}{1}(0.05)(0.95)^{59} = (0.95)^{59}(0.95 + 3) \simeq 0.1916.$$

(b) We can write that

$$P[X \le 1] \simeq P[\text{Poi}(60 \times 0.05 = 3) \le 1] = e^{-3}(1 + 3) \simeq 0.1991.$$

(c) Let W be the number of transistors that we must take to get 59 nondefective ones. Then, $W \sim NB(r = 59, p = 0.95)$. We want

$$P[W = 60] = \binom{59}{58}(0.95)^{59}(0.05) = 2.95\,(0.95)^{59} \simeq 0.1431.$$

Question no. 29

By the Bienaymé–Chebyshev inequality, we may write that

$$P[7 - k \cdot 1 \le X \le 7 + k \cdot 1] \ge 1 - \frac{1}{k^2}.$$

We want that $1 - \frac{1}{k^2} = 0.9$, from which we deduce that $k = \sqrt{10}$. Hence, the interval asked for is $[7 - \sqrt{10}, 7 + \sqrt{10}] \simeq [3.84, 10.16]$.

Question no. 30

We have:

$$f_X(x) = \frac{1}{\sqrt{2\pi}\sqrt{2}} \exp\left\{-\frac{x^2}{2(2)}\right\} \quad \Longrightarrow \quad \ln[f_X(X)] = \ln\left(\frac{1}{2\sqrt{\pi}}\right) - \frac{X^2}{4}.$$

Because $E[X^2] = \text{VAR}[X] + (E[X])^2 = 2 + 0^2 = 2$, we can then write that

$$H = E\left[-\ln\left(\frac{1}{2\sqrt{\pi}}\right) + \frac{X^2}{4}\right] = \ln(2\sqrt{\pi}) + \frac{1}{4}E[X^2]$$

$$= \ln(2\sqrt{\pi}) + \frac{2}{4} \simeq 1.766.$$

Question no. 31

(a) Let X be the total number of devices that the technician will have tried to repair at the moment of his second failure. Then, we have that $X \sim \text{NB}(r = 2, p = 0.05)$. We seek

$$P[X = 7] = \binom{6}{1}(0.05)^2(0.95)^5 \simeq 0.0116.$$

Remark. If the fact that the technician will receive at least seven out-of-order devices during this particular workday had not been mentioned in the statement of the problem, then we would have had to multiply the above probability by that of receiving at least seven out-of-order devices during an arbitrary workday, namely

$$P[N \ge 7] = P[N > 6] = (7/8)^6 \simeq 0.4488.$$

(b) We want

$$P[B(n = 10, p = 0.95) = 8] = \binom{10}{8}(0.95)^8(0.05)^2 \simeq 0.0746.$$

Remark. We can also write that

$$P[B(n = 10, p = 0.95) = 8] \quad = \quad P[B(n = 10, p = 0.05) = 2]$$

$$\stackrel{\text{Tab. B.1}}{\simeq} 0.9885 - 0.9139 = 0.0746.$$

(c) We have:

$$P[B(n = 10, p = 0.95) = 8] = P[B(n = 10, p = 0.05) = 2]$$
$$\simeq P[\text{Poi}(\lambda = 10 \times 0.05) = 2] = e^{-0.5}\frac{(0.5)^2}{2} \simeq 0.0758.$$

Remark. We could also have used Table B.2, page 278, to get $P[\text{Poi}(\lambda = 0.5) = 2]$.

(d) Let M be the number of devices that the technician could not repair, among the three taken at random. Then, $M \sim \text{Hyp}(N = 10, n = 3, d = 2)$. We seek

$$P[M = 2] = \frac{\binom{2}{2}\binom{8}{1}}{\binom{10}{3}} = \frac{1 \times 8}{\frac{10!}{3!7!}} = \frac{1}{15} \simeq 0.0667.$$

Question no. 32

Let Y be the number of cookies, among the 20, containing no raisins. If we assume that the number of raisins in a given cookie is independent of the number of raisins in the other cookies, then we may write that $Y \sim B(n = 20, p = P[X = 0])$. We have that $P[X = 0] = e^{-\lambda}$. Moreover, we find in Table B.2, page 278, that

$$P[Y \leq 2] \simeq 0.9245 \quad \text{if } p = 0.05.$$

It follows that we must take $\lambda \simeq -\ln 0.05 \simeq 3$.

Remark. We can check that with $\lambda = 3$, we have:

$$P[Y \leq 2] \simeq (0.9502)^{20} + 20(0.9502)^{19}(0.0498)$$
$$+ \underbrace{\binom{20}{2}}_{190}(0.9502)^{18}(0.0498)^2 \simeq 0.925.$$

However, λ did not have to be an integer, because it is the *average* number of raisins per cookie.

Question no. 33

Let c be the capacity of the storage tank. We want $P[X \geq c]$ to be 0.01. We have:

$$P[X \geq c] = \int_c^\infty \frac{1}{10}e^{-x/10}\,dx = e^{-c/10}.$$

Thus, the capacity must be

$$c = -10\,(\ln 0.01) \simeq 46 \quad \text{(thousands of liters)}.$$

Question no. 34

We are interested in the lifetime X (in years) of a machine. From past experience, we estimate the probability that a machine of this type lasts for more than nine years to be 0.1.

(a) We have:

$$1 = \int_0^\infty \frac{a}{(x+1)^b}\, dx \quad \Longrightarrow \quad \frac{a}{b-1} = 1.$$

Then,

$$0.1 = \int_9^\infty \frac{b-1}{(x+1)^b}\, dx \quad \Longrightarrow \quad \frac{1}{10^{b-1}} = 0.1.$$

We find that $b = 2$, which implies that $a = 1$.

(b) We want

$$P[N(7,\sigma^2) > 9] = 0.1 \quad \Longleftrightarrow \quad Q\left(\frac{2}{\sigma}\right) = 0.1 \quad \Longleftrightarrow \quad \frac{2}{\sigma} \overset{\text{Tab. B.4}}{\simeq} 1.282.$$

Therefore, σ must be approximately equal to 1.56.

(c) Let Y be the number of machines, among the ten, that will last for less than nine years. Then, $Y \sim B(n = 10, p = 0.9)$. We seek

$$
\begin{aligned}
P[Y \in \{8,9\}] \quad &= \quad P[B(n = 10, p = 0.1) \in \{1,2\}] \\
&\overset{\text{Tab. B.1}}{\simeq} 0.9298 - 0.3487 = 0.5811.
\end{aligned}
$$

Chapter 4

Question no. 1

We have:

$$p_X(x) = \sum_{y=0}^{2} p_{X,Y}(x,y) = \sum_{y=0}^{2} \frac{1}{6} = \frac{1}{2} \quad \text{if } x = 0 \text{ or } 1.$$

Question no. 2

We find that

$$f_Y(y) = \int_0^1 (x+y)\, dx = \frac{1}{2} + y \quad \text{if } 0 < y < 1.$$

Then, we may write that

$$f_X(x \mid Y = y) = \frac{x+y}{\frac{1}{2} + y} \quad \text{if } 0 < x < 1 \text{ and } 0 < y < 1.$$

Question no. 3

We first find that $\text{VAR}[X] = \text{VAR}[Y] = 1 - 0^2 = 1$. Then,

$$\text{COV}[X, Y] = \rho_{X,Y} \sigma_X \sigma_Y = 1 \cdot 1 \cdot 1 = 1.$$

Question no. 4

We can write that $P[X + Y > 1] = 1/2$, by symmetry. This result can be checked as follows:

$$P[X + Y > 1] = \int_0^1 \int_{1-x}^1 1 \, dy dx = \int_0^1 [1 - (1 - x)] \, dx = \frac{1}{2}.$$

Question no. 5

We have:

$$E[XY] = (1)(1) \left(\frac{8}{9}\right) \left(\frac{1}{2}\right)^{1+1} + (1)(2) \left(\frac{8}{9}\right) \left(\frac{1}{2}\right)^{1+2} = \left(\frac{8}{9}\right) \left(\frac{1}{4} + \frac{1}{4}\right) = \frac{4}{9}.$$

Question no. 6

We have:

$$\text{VAR}[X - 2Y] = \text{VAR}[X] + 4\,\text{VAR}[Y] - 4\,\text{COV}[X, Y] = 1 + 4 - 4(1) = 1.$$

Question no. 7

We can write that $W \sim \text{N}(0 - 1 + 2(3), 1 + 2 + 4(4)) \equiv \text{N}(5, 19)$.

Question no. 8

By the central limit theorem, we may write that

$$P[\text{Poi}(100) \leq 100] \simeq P\left[\text{N}(100, 100) \leq 100 + \frac{1}{2}\right] = \Phi(0.05) \overset{\text{Tab. B.3}}{\simeq} 0.5199$$

because if $X \sim \text{Poi}(100)$, then X has the same distribution as (for instance) $\sum_{i=1}^{100} X_i$, where the X_is are independent random variables having a Poisson distribution with parameter 1. Thus, we can use a $\text{N}(100, 100)$ distribution.

Remark. Note that not all authors make a continuity correction in this case (as in the case of the binomial distribution). If we do not make this continuity correction, then we obtain directly that

$$P[\text{Poi}(100) \leq 100] \simeq P[\text{N}(100, 100) \leq 100] = \Phi(0) = \frac{1}{2}.$$

Actually, we find, with the help of a mathematical software package, that $P[\text{Poi}(100) \leq 100] \simeq 0.5266$. Thus, here the fact of making a continuity correction does improve the approximation.

Question no. 9

We have:

$$P[X = 40] = P[39.5 \leq X \leq 40.5] \simeq P[39.5 \leq N(40, 24) \leq 40.5]$$

$$= 2\,\Phi\left(\frac{40.5 - 40}{\sqrt{24}}\right) - 1 \simeq 2\,\Phi(0.10) - 1 \stackrel{\text{Tab. B.3}}{\simeq} 0.0796.$$

Or:

$$P[X = 40] \simeq f_Y(40), \quad \text{where } Y \sim N(40, 24),$$

$$= \frac{1}{\sqrt{2\pi}\sqrt{24}} \exp\left\{-\frac{1}{2}\frac{(40-40)^2}{24}\right\} = \frac{1}{\sqrt{2\pi}\sqrt{24}} \simeq 0.0814.$$

Remark. The answer obtained by using the binomial distribution and a software package is $P[X = 40] \simeq 0.0812$, which is also the value of the probability that we find by calculating (with more accuracy than above) $\Phi((40.5-40)/\sqrt{24}) \simeq \Phi(0.102)$ rather than $\Phi(0.10)$.

Question no. 10

By the central limit theorem, we may write that

$$Y := \sum_{i=1}^{50} X_i \approx N(50(0), \sigma_Y^2),$$

so that $P[Y \geq 0] \simeq P[N(0, \sigma_Y^2) \geq 0] = 1/2$.

Question no. 11

(a) We first calculate $p_X(x) \equiv 1/3$, $p_Y(0) = 1/3$, and $p_Y(2) = 2/3$. We can check that $p_X(x)p_Y(y) = p_{X,Y}(x,y)\ \forall(x,y)$. So, X and Y *are* independent.

(b) $F_{X,Y}(0, 1/2) \equiv P[X \leq 0, Y \leq 1/2] = p_{X,Y}(-1,0) + p_{X,Y}(0,0) = 2/9$.

(c) From part (a), we have that $p_X(x) = 1/3$, for $x = -1, 0, 1$. Then,

z	0	1
$p_Z(z)$	1/3	2/3

(d) We have that $E[X^2Y^2] = (-1)^2(2)^2\,\frac{2}{9} + (1)^2(2)^2\,\frac{2}{9} + 0 = \frac{16}{9}$.

Question no. 12

We have (see Figure C.1):

$$P[X \geq Y^2] = \int_0^1 \int_x^{\sqrt{x}} 2\,dy\,dx = \int_0^1 2(\sqrt{x} - x)dx = 2\left(\frac{x^{3/2}}{3/2} - \frac{x^2}{2}\right)\Big|_0^1 = \frac{1}{3}.$$

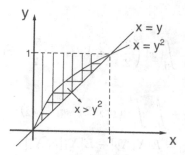

Fig. C.1. Figure for solved exercise no. 12.

Question no. 13

(a) Let $N(t)$ be the number of buses in the interval $[0, t]$, where t is in hours. We can write that $Y \equiv N(t = 12.5) \sim \text{Poi}(50)$.

(b) We have that $X_k \sim \text{Poi}(1)$, so that $E[X_k] = \text{VAR}[X_k] = 1$, for all k. Then, given that the X_ks are independent random variables, we deduce from the central limit theorem that $Y \approx N(50, 50)$.

Question no. 14

We have:

$$P[X_1 = X_2] = \sum_{i=0}^{2} P[\{X_1 = i\} \cap \{X_2 = i\}] \stackrel{\text{ind.}}{=} \sum_{i=0}^{2} P[X_1 = i]P[X_2 = i]$$
$$= (1/2)^2 + (1/4)^2 + (1/4)^2 = 3/8 = 0.375.$$

Question no. 15

We have:

$$P[\{X < 5\} \cap \{Y < 2\}] = P[\{X \le 4\} \cap \{Y \le 1\}] = P[Y = 1]$$
$$= 0.1 + 0.1 + 0.2 = 0.4.$$

Question no. 16

We first calculate

$$f_{X_1}(x_1) = \int_0^1 (2 - x_1 - x_2)\, dx_2 = \frac{3}{2} - x_1 \quad \text{if } 0 < x_1 < 1.$$

By symmetry, we can then write that

$$f_{X_2}(x_2) = \frac{3}{2} - x_2 \quad \text{if } 0 < x_2 < 1.$$

Next, we calculate

$$E[X_i] = \int_0^1 x_i \left(\frac{3}{2} - x_i\right) dx_i = \frac{5}{12} \quad \text{for } i = 1, 2$$

and

$$E[X_1 X_2] = \int_0^1 \int_0^1 x_1 x_2 \left(2 - x_1 - x_2\right) dx_1 dx_2 = \frac{1}{6}.$$

It follows that $\text{COV}[X_1, X_2] = E[X_1 X_2] - E[X_1]E[X_2] = -\frac{1}{144} \simeq -0.0069$.

Question no. 17

We first find that

y	1/2	1
$p_Y(y)$	2/3	1/3

which implies that

w	-1/2	0	1/2
$p_W(w)$	2/9	5/9	2/9

because $P[W = -1/2] = P[Y_1 = 1/2, Y_2 = 1] \overset{\text{ind.}}{=} (2/3)(1/3) = 2/9$, and so on.

Question no. 18

We can write that $\sum_{i=1}^n X_i \approx N(n(1/2), n(1/4))$, because $\mu_{X_i} = 1/2$ and $\sigma_{X_i}^2 = 1/4$, for all i. Then,

$$P\left[\sum_{i=1}^n X_i > \frac{n}{2} + 1\right] \simeq P\left[N(0,1) > \frac{\frac{n}{2} + 1 - \frac{n}{2}}{\sqrt{n}/2}\right] \simeq 0.4602$$

$$\Longleftrightarrow \quad \Phi\left(\frac{2}{\sqrt{n}}\right) \simeq 0.5398 \quad \overset{\text{Tab. B.3}}{\Longleftrightarrow} \quad \frac{2}{\sqrt{n}} \simeq 0.1 \quad \Longleftrightarrow \quad n \simeq 400.$$

Question no. 19

(a) We have that $X_1 - X_2 \sim N(0 - 0, 25 + 25) \equiv N(0, 50)$. We then calculate

$$P[X_1 - X_2 > 15] \quad = \quad P\left[N(0,1) > \frac{15 - 0}{\sqrt{50}}\right] \simeq 1 - \Phi(2.12)$$

$$\overset{\text{Tab. B.3}}{\simeq} 1 - 0.9830 = 0.0170.$$

(b) By independence, we may write that

$$f_{X_1, X_2}(x_1, x_2) = f_{X_1}(x_1) f_{X_2}(x_2) = \frac{1}{(2\pi)(25)} \exp\left\{-\frac{(x_1^2 + x_2^2)}{50}\right\}$$

for $(x_1, x_2) \in \mathbb{R}^2$.

(c) (i) We have that $P[X_1 = 2 \mid X_1 > 1] = 0$, because X_1 is a *continuous* random variable.

(ii) Because $\{X_1 = 1\} \subset \{X_1 < 2\}$, we can write that $P[X_1 < 2 \mid X_1 = 1]$ is equal to 1.

Question no. 20

Let X_i be the length of the ith section, for $i = 1, \ldots, 100$. By the central limit theorem, we may write that

$$P\left[970 \le \sum_{i=1}^{100} X_i \le 1030\right] \simeq P\left[970 \le N(\mu, \sigma^2) \le 1030\right],$$

where $\mu = 100 \times 10 = 1000$ and $\sigma^2 = 100 \times 0.9 = 90$. Thus, we seek approximately

$$1 - [\Phi(3.16) - \Phi(-3.16)] \stackrel{\text{Tab. B.3}}{\simeq} 1 - 2 \times (0.9992 - 1) = 0.0016.$$

Question no. 21

(a) We have:

$$f_X(x) = \int_0^1 3x^2 e^{-x} y(1-y)\, dy = 3x^2 e^{-x} \int_0^1 y(1-y)\, dy$$

$$= 3x^2 e^{-x} \left(\frac{y^2}{2} - \frac{y^3}{3}\right)\Big|_0^1 = \frac{x^2 e^{-x}}{2} \quad \text{if } x > 0.$$

We find that $X \sim G(\alpha = 3, \lambda = 1)$.

Next, we calculate

$$f_Y(y) = \int_0^\infty 3x^2 e^{-x} y(1-y)\, dx = 3y(1-y) \int_0^\infty x^2 e^{-x}\, dx$$

$$= 3y(1-y)\Gamma(3) = 6y(1-y) \quad \text{if } 0 < y < 1.$$

In this case, we find that $Y \sim \text{Be}(\alpha = 2, \beta = 2)$.

(b) We can check that $f_{X,Y}(x,y) \equiv f_X(x) f_Y(y)$. Therefore, by definition, X and Y are independent random variables.

(c) We have that $\text{VAR}[X] \stackrel{(a)}{=} 3/(1)^2 = 3$. Next, we calculate

$$E[X^k] = \int_0^\infty \frac{1}{2} x^{k+2} e^{-x}\, dx = \frac{1}{2}\Gamma(k+3)$$

for $k = 1, 2, \ldots$. Then, we have:

$$\beta_2 = \frac{E[X^4] - 4E[X^3]E[X] + 6E[X^2](E[X])^2 - 4E[X](E[X])^3 + (E[X])^4}{(\text{VAR}[X])^2}$$

$$= \frac{\frac{6!}{2} - 4\left(\frac{5!}{2}\right)\left(\frac{3!}{2}\right) + 6\left(\frac{4!}{2}\right)\left(\frac{3!}{2}\right)^2 - 3\left(\frac{3!}{2}\right)^4}{(3)^2}$$

$$= \frac{360 - 720 + 648 - 324 + 81}{9} = \frac{45}{9} = 5.$$

(d) First, we calculate

$$E[Y] = \int_0^1 y \cdot 6y(1-y)\, dy = \int_0^1 6y^2(1-y)\, dy = 6\left(\frac{1}{3} - \frac{1}{4}\right) = \frac{1}{2}.$$

Note that

$$f_Y\left(\tfrac{1}{2} - y\right) = f_Y\left(\tfrac{1}{2} + y\right)$$

for all $y \in (0, 1/2)$. That is, the function $f_Y(y)$ is symmetrical about the mean of Y. Because all possible values of the random variable Y are located in a bounded interval, we can then assert that $\beta_1 = 0$.

Question no. 22

(a) Summing the elements of the columns and the rows of the table, respectively, we find that

x	0	1	2
$p_X(x)$	1/3	1/3	1/3

and

y	-1	0	1
$p_Y(y)$	2/9	4/9	1/3

(b) We have, in particular, that $p_X(0)p_Y(-1) = \frac{1}{3} \times \frac{2}{9} \neq \frac{1}{9} = p_{X,Y}(0, -1)$. Therefore, X and Y are *not* independent random variables.

Remark. Because there are some 0s in the two-dimensional table, X and Y could not be independent. Indeed, a "0" in the table will never be equal to the product of the sum of the elements of the corresponding row and column.

(c) (i) By definition, we have:

$$p_Y(y \mid X = 1) = \frac{p_{X,Y}(1, y)}{p_X(1)} \overset{(a)}{=} 3\, p_{X,Y}(1, y) = \begin{cases} 0 \text{ if } y = -1, \\ 0 \text{ if } y = 0, \\ 1 \text{ if } y = 1. \end{cases}$$

That is, $Y \mid \{X = 1\}$ is the constant 1.

ii) We first find that $P[X \leq 1] \overset{(a)}{=} \frac{1}{3} + \frac{1}{3} = \frac{2}{3}$. It follows that

$$p_Y(-1 \mid X \leq 1) = \frac{3}{2} P[\{Y = -1\} \cap \{X \leq 1\}] = \frac{3}{2}\left(\frac{1}{9} + 0\right) = \frac{1}{6}.$$

Likewise, we calculate $p_Y(0 \mid X \leq 1) = \frac{3}{2}\left(\frac{2}{9} + 0\right) = \frac{1}{3}$. Thus, we have:

y	-1	0	1	Σ
$p_Y(y \mid X \leq 1)$	$1/6$	$1/3$	$1/2$	1

(d) We first calculate

$$E[X] \overset{\text{(a)}}{=} 1 \times \frac{1}{3} + 2 \times \frac{1}{3} = 1, \quad E[Y] \overset{\text{(a)}}{=} -1 \times \frac{2}{9} + 1 \times \frac{1}{3} = \frac{1}{9}$$

and

$$E[XY] = 1 \times 1 \times \frac{1}{3} + 2 \times (-1) \times \frac{1}{9} = \frac{1}{9}.$$

It follows that

$$\text{COV}[X,Y] = E[XY] - E[X]E[Y] = \frac{1}{9} - 1 \times \frac{1}{9} = 0$$

and, consequently, $\text{CORR}[X,Y] = 0$ (because $\text{STD}[X]$ and $\text{STD}[Y]$ are strictly positive).

(e) We have:

$$W = \begin{cases} 0 \text{ if } (X,Y) = (0,-1) \text{ or } (0,0), \\ 1 \text{ if } (X,Y) = (0,1) \text{ or } (1,-1) \text{ or } (1,0) \text{ or } (1,1), \\ 2 \text{ if } (X,Y) = (2,-1) \text{ or } (2,0) \text{ or } (2,1). \end{cases}$$

Making use of the function $p_{X,Y}(x,y)$, we find, by incompatibility, that the function $p_W(w)$ is given by

w	0	1	2	Σ
$p_W(w)$	$1/3$	$1/3$	$1/3$	1

Question no. 23

We have:

$$P[X + Y \leq 4 \mid X \leq 2] = \frac{P[\{X + Y \leq 4\} \cap \{X \leq 2\}]}{P[X \leq 2]}$$

$$= \frac{P[(X,Y) \in \{(1,2),(1,3),(2,2)\}]}{1 - \left(\frac{1}{12} + \frac{1}{6} + 0\right)} = \frac{\frac{1}{12} + \frac{1}{6} + \frac{1}{6}}{\frac{3}{4}} = \frac{5}{9} = 0.\bar{5}.$$

Question no. 24

Let X be the number of times that the digit "7" appears among the 10,000 random digits. Then, $X \sim \text{B}(n = 10{,}000, p = 0.1)$. We seek

$$P[X > 968] = P[X \geq 969] \simeq P\left[N(0,1) \geq \frac{969 - 0.5 - 1000}{\sqrt{900}}\right]$$

$$= Q(-1.05) = \Phi(1.05) \overset{\text{Tab. B.3}}{\simeq} 0.8531.$$

Question no. 25

(a) We have:

$$E[X^r Y^s] = \int_0^1 \int_0^x x^r y^s \frac{1}{x} \, dy \, dx = \int_0^1 x^{r-1} \left. \frac{y^{s+1}}{s+1} \right|_0^x dx$$

$$= \int_0^1 \frac{x^{r+s}}{s+1} \, dx = \frac{1}{(s+1)(r+s+1)} \qquad \text{for } r, s = 0, 1, 2, \dots .$$

(b) We calculate

$$E[X^2] = \int_0^1 x^2 \cdot 1 \, dx = \frac{1}{3}$$

and

$$E[X^2 Y^0] = \frac{1}{(0+1)(2+0+1)} = \frac{1}{3}.$$

(c) We deduce from part (a) that $E[X] = E[X^1 Y^0] = \frac{1}{2}$, $E[X^2] = E[X^2 Y^0] = \frac{1}{3}$, $E[Y] = \frac{1}{4}$, $E[Y^2] = \frac{1}{9}$, and $E[XY] = \frac{1}{6}$. Then, we have:

$$\text{VAR}[X] = \frac{1}{3} - \left(\frac{1}{2}\right)^2 = \frac{1}{12}, \quad \text{VAR}[Y] = \frac{1}{9} - \left(\frac{1}{4}\right)^2 = \frac{7}{144}$$

and

$$\rho_{X,Y} = \frac{\frac{1}{6} - \frac{1}{2} \cdot \frac{1}{4}}{\sqrt{\frac{1}{12} \cdot \frac{7}{144}}} = \sqrt{\frac{3}{7}} \simeq 0.6547.$$

Question no. 26

(a) We have:

$$\frac{1}{8} = p_Y(-1 \mid X = 2) = \frac{p_{X,Y}(2,-1)}{p_X(2)} = \frac{1/16}{P[X=2]} \quad \Longrightarrow \quad P[X=2] = \frac{1}{2}.$$

(b) We first find that

$$p_{X,Y}(2,0) = p_Y(0 \mid X = 2) p_X(2) \overset{(a)}{=} (3/8)(1/2) = 3/16.$$

Likewise, we have that $p_{X,Y}(2,1) = (1/2)(1/2) = 1/4$. We then obtain the following table:

$y \backslash x$	0	1	2	$p_Y(y)$
-1	1/16	1/8	1/16	1/4
0	3/16	1/8	3/16	1/2
1	0	0	1/4	1/4
$p_X(x)$	1/4	1/4	1/2	1

(c) We can check that W follows a binomial distribution with parameters $n = 2$ and $p = 1/2$, because [see $p_Y(y)$ in the above table]

$$p_W(w) = \binom{2}{w}(1/2)^2 \quad \text{for } w = 0, 1, 2.$$

Question no. 27

(a) We have:

$$E\left[\frac{1}{XY}\right] = \int_0^2 \int_0^x \frac{1}{xy}\frac{xy}{2}\,dy\,dx = \int_0^2 \frac{x}{2}\,dx = 1.$$

(b) We first calculate (see Figure C.2)

$$f_X(x) = \int_0^x \frac{xy}{2}\,dy = \frac{x^3}{4} \quad \text{for } 0 < x < 2.$$

Then,

$$E[X^2] = \int_0^2 x^2\frac{x^3}{4}\,dx = \frac{x^6}{24}\Big|_0^2 = \frac{8}{3} \simeq 2.67.$$

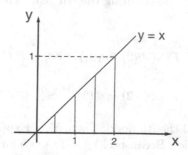

Fig. C.2. Figure for solved exercise no. 27.

(c) Making use of part (b), we may write that we seek the number x_m for which

$$\int_0^{x_m} \frac{x^3}{4}\,dx = \frac{1}{2} \quad \Longleftrightarrow \quad \frac{x_m^4}{16} = \frac{1}{2} \quad \Longrightarrow \quad x_m \simeq 1.68$$

because the median x_m must be positive, in as much as $X \in (0, 2)$.

Question no. 28

We seek

$$P[\{X < 1\} \cap \{Y < 1\}] \overset{\text{ind.}}{=} P[X < 1]P[Y < 1] \simeq 0.3402$$

because

$$P[X < 1] = \int_0^1 \frac{1}{2} e^{-x/2} \, dx = 1 - e^{-1/2}$$

and

$$P[Y < 1] = \int_0^1 4y e^{-2y^2} \, dy = 1 - e^{-2}.$$

Question no. 29

Let X_k be the kth number taken at random. We have that $E[X_k] = 1/2$ and $\mathrm{VAR}[X_k] = 1/12$, for all k. Then, by the central limit theorem, we may write that

$$P[45 \leq S < 55] \simeq P[45 \leq \mathrm{N}(100(1/2), 100(1/12)) < 55]$$
$$\simeq P[-1.73 \leq \mathrm{N}(0,1) < 1.73] = \Phi(1.73) - \Phi(-1.73)$$
$$= 2\Phi(1.73) - 1 \overset{\text{Tab. B.3}}{\simeq} 2(0.958) - 1 = 0.916.$$

Question no. 30

(a) Let X_i be the number of floods during the ith year. Then, $X_i \sim \mathrm{Poi}(\alpha = 2) \; \forall i$ and the X_is are independent. We seek

$$P\left[\sum_{i=1}^{50} X_i \geq 80\right] \overset{\text{CLT}}{\simeq} P[\mathrm{N}(50(2), 50(2)) \geq 80] = P\left[\mathrm{N}(0,1) \geq \frac{80 - 100}{10}\right]$$
$$= 1 - \Phi(-2) = \Phi(2) \overset{\text{Tab. B.3}}{\simeq} 0.9772.$$

(b) Let Y_i be the duration of the ith flood. Then, $Y_i \sim \mathrm{Exp}(\lambda = 1/5) \; \forall i$ and the Y_is are independent random variables. Because $E[Y_i] = 1/\lambda = 5$ and $\mathrm{VAR}[Y_i] = 1/\lambda^2 = 25$, we calculate

$$P\left[\sum_{i=1}^{50} Y_i < 200\right] \overset{\text{CLT}}{\simeq} P[\mathrm{N}(50(5), 50(25)) < 200] \simeq P[\mathrm{N}(0,1) < -1.41]$$
$$\overset{\text{Tab. B.3}}{\simeq} 1 - 0.9207 \simeq 0.079.$$

Chapter 5

Question no. 1

We have:

$$R(x) := P[X > x] = P[Y^2 > x] = \begin{cases} P[Y > \sqrt{x}] = 1 - \sqrt{x} \text{ if } 0 \leq x < 1, \\ \qquad\qquad 0 \qquad\qquad \text{ if } x \geq 1. \end{cases}$$

Question no. 2

We can write that $Y \mid \{X < 2\} \sim \text{Exp}(1)$ and $Y \mid \{X \geq 2\} \sim \text{Exp}(1/2)$. It follows that

$$E[Y] = E[Y \mid X < 2]P[X < 2] + E[Y \mid X \geq 2]P[X \geq 2]$$
$$= 1 \times \left[1 - e^{-(1/2)2}\right] + 2 \times e^{-(1/2)2} = 1 + e^{-1}.$$

Because $E[X] = 2$ years, we have that $MTBF = 731 + e^{-1} \simeq 731.37$ days.

Question no. 3

First, we calculate

$$F_T(t) = \int_0^t f_T(s)ds = \int_0^t \frac{1}{\lambda} \exp\left\{-\frac{e^s - 1}{\lambda} + s\right\} ds$$
$$= \int_0^t \frac{1}{\lambda} e^s \exp\left\{-\frac{e^s - 1}{\lambda}\right\} ds = -\exp\left\{-\frac{e^s - 1}{\lambda}\right\} \Big|_0^t$$
$$= 1 - \exp\left\{-\frac{e^t - 1}{\lambda}\right\} \quad \text{for } t \geq 0.$$

It follows that

$$r(t) = \frac{f_T(t)}{1 - F_T(t)} = \frac{e^t}{\lambda} \quad \text{for } t \geq 0.$$

Question no. 4

We calculate

$$r(0) = \frac{e^{-\lambda}}{\sum_{j=0}^{\infty} e^{-\lambda}\lambda^j/j!} = \frac{e^{-\lambda}}{1} = e^{-\lambda}$$

and

$$r(1) = \frac{e^{-\lambda}\lambda}{\sum_{j=1}^{\infty} e^{-\lambda}\lambda^j/j!} = \frac{e^{-\lambda}\lambda}{e^{-\lambda}(e^\lambda - 1)} = \frac{\lambda}{e^\lambda - 1}.$$

We have:

$$r(0) < r(1) \iff e^{-\lambda} < \frac{\lambda}{e^\lambda - 1} \iff 1 - e^{-\lambda} < \lambda.$$

Let

$$g(\lambda) = \lambda + e^{-\lambda} - 1.$$

Because $g(0) = 0$ and

$$g'(\lambda) = 1 - e^{-\lambda} > 0 \quad \text{for all } \lambda > 0,$$

we may assert that $g(\lambda) > 0$, for $\lambda > 0$. Hence, we conclude that $r(0) < r(1)$, so that the failure rate function is increasing at $k = 0$.

Remark. Actually, it can be shown that the function $r(k)$ is increasing at any $k \in \{0, 1, \ldots\}$.

Question no. 5

We have:

$$R(x) := P[X > x] = \int_x^1 1 \, dt = 1 - x \quad \text{if } 0 \le x \le 1.$$

Therefore,

$$AFR(0, 1/2) = \frac{\ln[R(0)] - \ln[R(1/2)]}{1 - (1/2)} = 2[\ln 1 - \ln(1/2)] = 2 \ln 2.$$

Question no. 6

The components are connected in series, therefore we indeed seek the probability $P[X_2 < X_1]$. Making use of the above formula, we can write that

$$P[X_2 < X_1] = \int_0^\infty P[X_2 < x_1 \mid X_1 = x_1] \lambda_1 e^{-\lambda_1 x_1} \, dx_1.$$

By independence, we obtain that

$$P[X_2 < X_1] = \int_0^\infty [1 - e^{-\lambda_2 x_1}] \lambda_1 e^{-\lambda_1 x_1} \, dx_1 = 1 - \lambda_1 \int_0^\infty e^{-(\lambda_1 + \lambda_2) x_1} \, dx_1$$

$$= 1 - \lambda_1 \left(\frac{1}{\lambda_1 + \lambda_2} \right) = \frac{\lambda_2}{\lambda_1 + \lambda_2}.$$

Remark. Note that $P[X_2 < X_1] = 1/2$ if $\lambda_1 = \lambda_2$, which is logical, by symmetry (because $P[X_2 = X_1] = 0$, by continuity).

Question no. 7

We have that $P[T > t_0] = P[\{T_1 > t_0\} \cup \{T_2 > t_0\}]$. Let $A_k = \{T_k > t_0\}$, for $k = 1, 2$. We seek

$$p := P[A_1 \cap A_2 \mid A_1 \cup A_2] = \frac{P[A_1 \cap A_2]}{P[A_1 \cup A_2]}$$

because $\{A_1 \cap A_2\} \subset \{A_1 \cup A_2\}$.

Next, we have:

$$P[A_1 \cap A_2] \overset{\text{ind.}}{=} P[A_1]P[A_2] = e^{-(\lambda_1+\lambda_2)t_0}$$

and

$$P[A_1 \cup A_2] = P[A_1] + P[A_2] - P[A_1 \cap A_2] = e^{-\lambda_1 t_0} + e^{-\lambda_2 t_0} - e^{-(\lambda_1+\lambda_2)t_0},$$

so that

$$p = \frac{e^{-(\lambda_1+\lambda_2)t_0}}{e^{-\lambda_1 t_0} + e^{-\lambda_2 t_0} - e^{-(\lambda_1+\lambda_2)t_0}}.$$

Question no. 8

Let $F =$ "the device operates at the initial time." In the case of a series device constituted of one brand A and one brand B component, we have that $P[F] = (0.9)^2 = 0.81$. Therefore, the probability that at least one of the two devices works is given by $1 - (1 - 0.81)^2 = 0.9639$.

If we build a single device as described above, we have:

$$P[F] = (1 - (0.1)^2)(1 - (0.1)^2) = 0.9801.$$

Thus, it is better to duplicate the components rather than to duplicate the devices.

Remark. The conclusion of this exercise can be generalized as follows: it is always better to duplicate the components in a series system than to build two distinct systems.

Question no. 9

We have:

$$H(x_1, x_2, x_3, x_4) = \max\{x_1 x_2, x_3\} x_4 = [1 - (1 - x_1 x_2)(1 - x_3)]x_4$$
$$= (x_1 x_2 + x_3 - x_1 x_2 x_3)x_4.$$

Question no. 10

The minimal path sets of the system are the following: $MP_1 = \{1, 2, 4\}$ and $MP_2 = \{3, 4\}$. Hence,

$$\pi_1(x_1, x_2, x_3, x_4) = x_1 x_2 x_4 \quad \text{and} \quad \pi_2(x_1, x_2, x_3, x_4) = x_3 x_4,$$

so that

$$H(x_1, x_2, x_3, x_4) = 1 - (1 - x_1 x_2 x_4)(1 - x_3 x_4).$$

Remark. Because $x_k = 0$ or $1 \; \forall k$, we may write that $x_k^2 = x_k$. It follows that

$$H(x_1, \ldots, x_4) = x_1 x_2 x_4 + x_3 x_4 - x_1 x_2 x_3 x_4^2 = x_1 x_2 x_4 + x_3 x_4 - x_1 x_2 x_3 x_4$$
$$= (x_1 x_2 + x_3 - x_1 x_2 x_3)x_4,$$

which agrees with the result in the previous exercise.

Chapter 6

Question no. 1

(a) Let T_k, for $k = 1, \ldots, n$, be the lifetime of component no. k. If the components are connected in series, then, by the memoryless property of the exponential distribution, we may express the time T between two system breakdowns as follows: $T = \min\{T_1, \ldots, T_n\}$. From the remark after Proposition 5.2.1, we deduce that $T \sim \text{Exp}(\mu_1 + \cdots + \mu_n)$. Hence, we may assert that $\{N(t), t \geq 0\}$ *is* a continuous-time Markov chain. Actually, it is a Poisson process with rate $\lambda = \mu_1 + \cdots + \mu_n$.

(b) When the components are connected in parallel, we have that $T = \max\{T_1, \ldots, T_n\}$. Now, the maximum $T_{1,2}$ of two independent exponential random variables does not have an exponential distribution. Indeed, we can write (see Example 5.2.2) that

$$P[T_{1,2} \leq t] = \left(1 - e^{-\mu_1 t}\right)\left(1 - e^{-\mu_2 t}\right) \quad \text{for } t \geq 0,$$

so that

$$f_{T_{1,2}}(t) = \frac{d}{dt} P[T_{1,2} \leq t] = \mu_1 e^{-\mu_1 t} + \mu_2 e^{-\mu_2 t} - (\mu_1 + \mu_2) e^{-(\mu_1 + \mu_2)t} \quad \text{for } t \geq 0.$$

Because we cannot write $f_{T_{1,2}}(t)$ in the form

$$f_{T_{1,2}}(t) = \lambda e^{-\lambda t} \quad \text{for } t \geq 0$$

for some $\lambda > 0$, we must conclude that $T_{1,2}$ is not exponentially distributed. By extension, T is not an exponential random variable either, so that $\{N(t), t \geq 0\}$ is *not* a continuous-time Markov chain.

(c) Finally, in the case when the components are placed in standby redundancy, the random variable T is not exponentially distributed either (for $n \geq 2$). Indeed, if $\mu_k \equiv \mu$, then [see (4.9)] $T \sim \text{G}(n, \mu)$. Because T does not have an exponential distribution in this particular case, it cannot be exponentially distributed for arbitrary μ_ks. Thus, $\{N(t), t \geq 0\}$ is *not* a continuous-time Markov chain.

Question no. 2

We can write that $X(t)$, given that $X(0) = i$, has a negative binomial distribution with parameters $r = i$ and $p = e^{-\lambda t}$ [see (3.2)]. From Table 3.1, p. 89, we deduce that the expected value of $X(t)$ is given by

$$E[X(t) \mid X(0) = i] = \frac{r}{p} = ie^{\lambda t}.$$

To justify this result, notice that

$$p_{1,j}(t) = e^{-\lambda t}(1 - e^{-\lambda t})^{j-1} \quad \text{for } j \geq 1.$$

That is,

$$P[X(t) = j \mid X(0) = 1] = P[\text{Geo}(p := e^{-\lambda t}) = j] \quad \text{for } j = 1, 2, \ldots .$$

Now, when $X(0) = i \geq 1$, we can represent $X(t)$ as the sum of i (independent) geometric random variables with common parameter p. Then, by linearity of the mathematical expectation, we deduce that

$$E[X(t) \mid X(0) = i] = i \times E[\text{Geo}(p := e^{-\lambda t})] = i \left(\frac{1}{e^{-\lambda t}} \right) = i e^{\lambda t}.$$

Question no. 3

The balance equations of the system are the following:

state j	departure rate from j = arrival rate to j
0	$\lambda \pi_0 = \mu \pi_1$
1	$(2\lambda + \mu)\pi_1 = \lambda \pi_0 + 2\mu \pi_2$
2	$2\mu \pi_2 = 2\lambda \pi_1$

We deduce from the first and the third equation that

$$\pi_1 = \frac{\lambda}{\mu} \pi_0 \quad \text{and} \quad \pi_2 = \frac{\lambda}{\mu} \pi_1 = \left(\frac{\lambda}{\mu} \right)^2 \pi_0.$$

Hence, we can write that

$$\pi_0 + \frac{\lambda}{\mu} \pi_0 + \left(\frac{\lambda}{\mu} \right)^2 \pi_0 = 1 \quad \Longrightarrow \quad \pi_0 = \left[1 + \frac{\lambda}{\mu} + \left(\frac{\lambda}{\mu} \right)^2 \right]^{-1},$$

from which we obtain the values of π_1 and π_2.

Question no. 4

Let $\pi_k^* = P[X(t_0) = k \mid X(t_0) \in \{2, 3, 4\}]$. We can write that

$$\pi_k^* = \frac{\pi_k}{\pi_2 + \pi_3 + \pi_4} = \frac{\rho^k}{\rho^2 + \rho^3 + \rho^4} = \frac{\rho^{k-2}}{1 + \rho + \rho^2} \quad \text{for } k = 2, 3, 4.$$

Next, let Q^* be the waiting time of the new customer. By the memoryless property of the exponential distribution, we can write that

$$E[Q^* \mid X(t_0) \in \{2, 3, 4\}] = \sum_{k=2}^{4} E[Q^* \mid \{X(t_0) = k\} \cap \{X(t_0) \in \{2, 3, 4\}\}]$$

$$\times P[X(t_0) = k \mid X(t_0) \in \{2, 3, 4\}]$$

$$= \sum_{k=2}^{4} \left(\frac{k}{\mu} \right) \pi_k^* = \frac{1}{\mu} \left(\frac{2 + 3\rho + 4\rho^2}{1 + \rho + \rho^2} \right).$$

Question no. 5

The balance equations are:

state j	departure rate from j = arrival rate to j
0	$\lambda \pi_0 = \mu \pi_1$
1	$(\lambda + \mu)\pi_1 = \lambda \pi_0 + \mu \pi_2$
2	$(\lambda + \mu)\pi_2 = \lambda \pi_1 + 2\mu \pi_3$
$n \in \{3, 4, \ldots\}$	$(\lambda + 2\mu)\pi_n = \lambda \pi_{n-1} + 2\mu \pi_{n+1}$

Next, we calculate

$$\Pi_k = \frac{\lambda \lambda \lambda \cdots \lambda}{\mu \mu (2\mu) \cdots (2\mu)} = \left(\frac{\lambda}{\mu}\right)^2 \left(\frac{\lambda}{2\mu}\right)^{k-2} \quad \text{for } k = 2, 3, \ldots \ .$$

Hence, we deduce that the sum $\sum_{k=1}^{\infty} \Pi_k$ converges if and only if

$$\sum_{k=2}^{\infty} \Pi_k < \infty \quad \Longleftrightarrow \quad \sum_{k=2}^{\infty} \left(\frac{\lambda}{\mu}\right)^2 \left(\frac{\lambda}{2\mu}\right)^{k-2} < \infty$$

$$\Longleftrightarrow \quad \left(\frac{\lambda}{\mu}\right)^2 \sum_{k=2}^{\infty} \left(\frac{\lambda}{2\mu}\right)^{k-2} < \infty \quad \Longleftrightarrow \quad \frac{\lambda}{2\mu} < 1,$$

as we could have guessed.

Question no. 6

The limiting probabilities of the system are given by [see (6.17)]

$$\pi_j = \frac{\rho^j (1 - \rho)}{1 - \rho^3} \quad \text{for } j = 0, 1, 2.$$

Let F be the event "the customer who arrived at time $t_0 + 2$ was able to enter the system." We can write that

$$P[F] = P[F \mid X(t_0^-) = 1] P[X(t_0^-) = 1 \mid X(t_0^-) \in \{1, 2\}]$$
$$+ P[F \mid X(t_0^-) = 2] P[X(t_0^-) = 2 \mid X(t_0^-) \in \{1, 2\}]$$
$$= 1 \times \frac{\pi_1}{\pi_1 + \pi_2} + P[F \mid X(t_0^-) = 2] \frac{\pi_2}{\pi_1 + \pi_2}.$$

Now, given that the system was full when a departure took place at time t_0, the customer who arrived at $t_0 + 2$ was able to enter the system if and only if the one who started being served at t_0 left the system before time $t_0 + 2$. That is, if $S \sim \text{Exp}(\mu)$,

$$P[F \mid X(t_0^-) = 2] = P[S < 2] = 1 - e^{-2\mu}.$$

Hence, the required probability is

$$P[F] = \frac{\rho}{\rho + \rho^2} + \left(1 - e^{-2\mu}\right) \frac{\rho^2}{\rho + \rho^2} = \frac{1}{1+\rho} \left[1 + \rho \left(1 - e^{-2\mu}\right)\right].$$

Question no. 7

When $\lambda_a = \lambda$ and the service rate is $\mu = \lambda$, the limiting probabilities of the system are [see (6.16)]

$$\pi_j = \frac{1}{4} \quad \text{for } j = 0, 1, 2, 3.$$

The average entering rate of customers into the system is $\lambda_e = \lambda(1 - \pi_3) = 3\lambda/4$, so that the average amount of money that the system earns per unit of time is equal to $\$x \times 3\lambda/4$.

Next, if $\lambda^* = \lambda/2$ and $\mu^* = 2\mu$, then $\rho^* = \rho/4 = 1/4$ and [see (6.17)]

$$\pi_j^* = \frac{(1/4)^j [1 - (1/4)]}{1 - (1/4)^4} \quad \text{for } j = 0, 1, 2, 3.$$

It follows that

$$\lambda_e^* = \lambda^* (1 - \pi_3^*) = \frac{\lambda}{2} \left(1 - \frac{3}{255}\right) = \frac{126\lambda}{255}$$

and, because the entering customers pay the same amount as before, we conclude that the server is better off to serve at rate μ. Indeed, we have:

$$\$x \times \frac{3\lambda}{4} > \$x \times \frac{126\lambda}{255}.$$

Question no. 8

By the memoryless property of the exponential distribution, we can write that $W_i := T_i - t_0$ is an exponentially distributed random variable with parameter μ. Indeed, we have:

$$P[W_i > t] = P[S_i - (t_0 - t_i) > t \mid S_i > t_0 - t_i] \quad \text{for } i = 1, 2,$$

where S_i is the service time of the customer being served by server no. i and $t_i > 0$ is its (known) arrival time. Because S_i is an $\text{Exp}(\mu)$ random variable, we can write that

$$P[S_i - (t_0 - t_i) > t \mid S_i > t_0 - t_i] = P[S_i > t + (t_0 - t_i) \mid S_i > t_0 - t_i]$$
$$= P[S_i > t] = e^{-\mu t}.$$

Hence, by independence, the required probability is given by

$$P[\tau_2 - t_0 \le \tau_1 - t_0 + 1] = P[W_2 \le W_1 + 1]$$
$$= P[S_2 \le S_1 + 1] = \int_0^\infty \int_0^{s_1+1} \mu e^{-\mu s_1} \mu e^{-\mu s_2} \, ds_2 ds_1$$
$$= \int_0^\infty (-1)\mu e^{-\mu(s_1+s_2)} \Big|_{s_2=0}^{s_2=s_1+1} ds_1 = \int_0^\infty \mu \left[e^{-\mu s_1} - e^{-\mu(2s_1+1)} \right] ds_1$$
$$= 1 - \frac{1}{2} e^{-\mu}.$$

Remark. Because, by symmetry and continuity, $P[S_1 < S_2] = 1/2$, we deduce that

$$P[S_1 \le S_2 \le S_1 + 1] = 1 - \frac{1}{2} e^{-\mu} - \frac{1}{2} = \frac{1}{2} \left(1 - e^{-\mu} \right),$$

from which we retrieve the result in Example 6.3.1.

Question no. 9

Because the service rates are not necessarily equal, $X(t)$ cannot simply be the number of customers in the system at time t. We define the states:

> 0: the system is empty;
> 1_1: only server no. 1 is busy;
> 1_2: only server no. 2 is busy;
> 2: the two servers are busy and nobody is waiting;
> 3: the two servers are busy and somebody is waiting.

We have:

state j	departure rate from j = arrival rate to j
0	$\lambda \pi_0 = \mu_1 \pi_{1_1} + \mu_2 \pi_{1_2}$
1_1	$(\lambda + \mu_1)\pi_{1_1} = \lambda \pi_0 + \mu_2 \pi_2$
1_2	$(\lambda + \mu_2)\pi_{1_2} = \mu_1 \pi_2$
2	$(\lambda + \mu_1 + \mu_2)\pi_2 = \lambda(\pi_{1_1} + \pi_{1_2}) + (\mu_1 + \mu_2)\pi_3$
3	$(\mu_1 + \mu_2)\pi_3 = \lambda \pi_2$

The value of $\bar{N}$ is given by

$$\bar{N} = \pi_{1_1} + \pi_{1_2} + 2\pi_2 + 3\pi_3.$$

Moreover, the average entering rate of customers is $\lambda_e = \lambda(1 - \pi_3)$. It follows, from Little's formula, that

$$\bar{T} = \frac{\bar{N}}{\lambda_e} = \frac{\pi_{1_1} + \pi_{1_2} + 2\pi_2 + 3\pi_3}{\lambda(1 - \pi_3)}.$$

Question no. 10

We have an $M/G/2/2$ queueing system for which $S \sim U(2,4)$, so that $E[S] = 3$. Making use of (6.26), we can write that

$$\pi_0 = \left(\sum_{k=0}^{2} \frac{(3\lambda)^k}{k!} \right)^{-1} = \left(1 + 3\lambda + \frac{(3\lambda)^2}{2} \right)^{-1}$$

and

$$\pi_1 = 3\lambda \pi_0 \quad \text{and} \quad \pi_2 = \frac{(3\lambda)^2}{2}\pi_0.$$

If $\lambda = 1/3$, we have:

$$\pi_0 = \left(1 + 1 + \frac{1}{2} \right)^{-1} = \frac{2}{5}, \quad \pi_1 = \pi_0 = \frac{2}{5} \quad \text{and} \quad \pi_2 = \frac{1}{2}\pi_0 = \frac{1}{5}.$$

It follows that

$$E[N] = 1 \times \pi_1 + 2 \times \pi_2 = \frac{4}{5} \quad \text{and} \quad E[N^2] = 1 \times \pi_1 + 4 \times \pi_2 = \frac{6}{5},$$

so that

$$\text{VAR}[N] = \frac{6}{5} - \left(\frac{4}{5} \right)^2 = \frac{14}{25}.$$

Chapter 7

Question no. 1

We have that

$$E[Y_n] = \frac{1}{2} \{ E[X_n] + E[X_{n-1}] \} \equiv 0.$$

Moreover, for $m \in \{ -n+1, -n+2, \ldots \}$,

$$E[Y_n Y_{n+m}] = E\left[\left(\frac{X_n + X_{n-1}}{2} \right) \left(\frac{X_{n+m} + X_{n+m-1}}{2} \right) \right]$$

$$= \frac{1}{4} \{ E[X_n X_{n+m}] + E[X_n X_{n+m-1}] $$

$$+ E[X_{n-1} X_{n+m}] + E[X_{n-1} X_{n+m-1}] \}.$$

In the case when $m = 0$, we obtain that

$$E[Y_n Y_{n+m}] = E[Y_n^2] = \frac{1}{4} \{ E[X_n^2] + 0 + 0 + E[X_{n-1}^2] \} = \frac{\sigma^2}{2}.$$

Remark. Or, because $E[Y_n] \equiv 0$:

$$E[Y_n^2] = \text{VAR}[Y_n] \stackrel{\text{ind.}}{=} \frac{1}{4}\{\text{VAR}[X_n] + \text{VAR}[X_{n-1}]\} = \frac{\sigma^2}{2}.$$

Next, if $m > 0$, we calculate

$$E[Y_n Y_{n+m}] = \frac{1}{4}\{0 + E[X_n X_{n+m-1}] + 0 + 0\} = \frac{1}{4}E[X_n X_{n+m-1}]$$
$$= \begin{cases} 0 & \text{if } m \neq 1, \\ \frac{1}{4}\sigma^2 & \text{if } m = 1. \end{cases}$$

Similarly, when $m < 0$, we find that

$$E[Y_n Y_{n+m}] = \begin{cases} 0 & \text{if } m \neq -1, \\ \frac{1}{4}\sigma^2 & \text{if } m = -1. \end{cases}$$

Hence, we can write that

$$\text{COV}[Y_n, Y_{n+m}] = \begin{cases} \frac{1}{2}\sigma^2 & \text{if } m = 0, \\[2mm] \frac{1}{4}\sigma^2 & \text{if } m = \pm 1, \\[2mm] 0 & \text{if } m \neq 0, \pm 1, \end{cases}$$

so that $\text{COV}[Y_n, Y_{n+m}] = \gamma(m) = \gamma(-m)$ (and $E[Y_n] \equiv 0$). Thus, the process $\{Y_n, n = 1, 2, \ldots\}$ *is* weakly stationary.

Question no. 2

Let

$$X_i = \frac{1}{a_i}(Y_i - b_i) := h_i(Y_1, Y_2) \quad \text{for } i = 1, 2.$$

We have that

$$\begin{vmatrix} \frac{\partial h_1(y_1, y_2)}{\partial y_1} & \frac{\partial h_1(y_1, y_2)}{\partial y_2} \\[2mm] \frac{\partial h_2(y_1, y_2)}{\partial y_1} & \frac{\partial h_2(y_1, y_2)}{\partial y_2} \end{vmatrix} = \begin{vmatrix} \frac{1}{a_1} & 0 \\[2mm] 0 & \frac{1}{a_2} \end{vmatrix} = \frac{1}{a_1 a_2} \neq 0 \quad \forall (y_1, y_2).$$

Because the partial derivatives $\partial h_i(y_1, y_2)/\partial y_j$, for $i, j = 1, 2$, are all continuous $\forall (y_i, y_2)$, we can write that

$$f_{Y_1, Y_2}(y_1, y_2) = f_{X_1, X_2}\left(\frac{1}{a_1}(y_1 - b_1), \frac{1}{a_2}(y_2 - b_2)\right)\left|\frac{1}{a_1 a_2}\right|.$$

By independence,

$$f_{X_1,X_2}(x_1,x_2) = \frac{1}{(\sqrt{2\pi})^2} e^{-(x_1^2 + x_2^2)/2} = \frac{1}{2\pi} \exp\left\{ -\frac{1}{2}(x_1^2 + x_2^2) \right\}$$

$\forall (x_1, x_2) \in \mathbb{R}^2$. It follows that

$$f_{Y_1,Y_2}(y_1, y_2) = \left| \frac{1}{a_1 a_2} \right| \frac{1}{2\pi} \exp\left\{ -\frac{1}{2} \left(\frac{1}{a_1^2}(y_1 - b_1)^2 + \frac{1}{a_2^2}(y_2 - b_2)^2 \right) \right\}$$

$\forall (y_1, y_2) \in \mathbb{R}^2$.

Remarks. (i) From Definition 7.1.6, we can assert that the random vector (Y_1, Y_2) has a bivariate normal distribution. Moreover, we have that

$$E[Y_i] = b_i, \quad \text{VAR}[Y_i] = a_i^2$$

and

$$\text{COV}[Y_1, Y_2] \stackrel{\text{ind.}}{=} a_1 a_2 \underbrace{\text{COV}[X_1, X_2]}_{0} + 0 = 0.$$

Hence, we can write that $\mathbf{m} = (b_1, b_2)$ and

$$\mathbf{C} = \begin{bmatrix} a_1^2 & 0 \\ 0 & a_2^2 \end{bmatrix}.$$

Finally, Equation (7.2) yields that

$$f_{Y_1,Y_2}(y_1, y_2) = \left\{ (2\pi)^2 a_1^2 a_2^2 \right\}^{-1/2} \exp\left\{ -\frac{1}{2}(y_1 - b_1, y_2 - b_2)\mathbf{C}^{-1} \begin{pmatrix} y_1 - b_1 \\ y_2 - b_2 \end{pmatrix} \right\},$$

where

$$\mathbf{C}^{-1} = \frac{1}{a_1^2 a_2^2} \begin{bmatrix} a_2^2 & 0 \\ 0 & a_1^2 \end{bmatrix} = \begin{bmatrix} \frac{1}{a_1^2} & 0 \\ 0 & \frac{1}{a_2^2} \end{bmatrix}.$$

It is a simple matter to check that the function $f_{Y_1,Y_2}(y_1, y_2)$ above is the same as the one obtained from Proposition 4.3.1.

(ii) We can also use the fact that $Y_1 \sim N(b_1, a_1^2)$ and $Y_2 \sim N(b_2, a_2^2)$ are independent random variables to get $f_{Y_1,Y_2}(y_1, y_2)$. Indeed, their covariance being equal to 0 (see the preceding remark), we can assert that they are independent (because they are Gaussian random variables).

(iii) If $a_1 = 0$, then $Y_1 = b_1$ is a degenerate Gaussian random variable. Its probability density function is given by

$$f_{Y_1}(y_1) = \delta(y_1 - b_1) \quad \forall y_1 \in \mathbb{R},$$

where $\delta(\cdot)$ is the Dirac delta function defined by (see p. 4)

$$\delta(x) = \begin{cases} 0 \text{ if } x \neq 0, \\ \infty \text{ if } x = 0 \end{cases} \quad \text{and} \quad \int_{-\infty}^{\infty} \delta(x)\,dx = 1.$$

Question no. 3

Equation (7.18) yields

$$\text{VAR}[X_n] \equiv \gamma(0) = \alpha_1\gamma(1) + \alpha_2\gamma(2) + \alpha_3\gamma(3) + \sigma^2 \quad (*)$$

and we deduce from (7.13) that

$$\gamma(1) \stackrel{(1)}{=} \alpha_1\gamma(0) + \alpha_2\gamma(1) + \alpha_3\gamma(2),$$

$$\gamma(2) \stackrel{(2)}{=} \alpha_1\gamma(1) + \alpha_2\gamma(0) + \alpha_3\gamma(1),$$

$$\gamma(3) \stackrel{(3)}{=} \alpha_1\gamma(2) + \alpha_2\gamma(1) + \alpha_3\gamma(0).$$

Substituting (3) into (*), we obtain that

$$\gamma(0) = \alpha_1\gamma(1) + \alpha_2\gamma(2) + \alpha_3\left[\alpha_1\gamma(2) + \alpha_2\gamma(1) + \alpha_3\gamma(0)\right] + \sigma^2$$

$$\implies \gamma(0) = \frac{1}{1 - \alpha_3^2}\left[(\alpha_1 + \alpha_2\alpha_3)\gamma(1) + (\alpha_2 + \alpha_1\alpha_3)\gamma(2) + \sigma^2\right].$$

Next, (1) and (2) imply that

$$\gamma(1) = \alpha_1\gamma(0) + \alpha_2\gamma(1) + \alpha_3\left[(\alpha_1 + \alpha_3)\gamma(1) + \alpha_2\gamma(0)\right]$$

$$\implies \gamma(1) = \gamma(0)\left\{\frac{\alpha_1 + \alpha_2\alpha_3}{1 - \alpha_2 - (\alpha_1 + \alpha_3)\alpha_3}\right\}.$$

Then, making use again of (2), we find that

$$\gamma(2) = \gamma(0)\left\{\frac{\alpha_2(1 - \alpha_2) + \alpha_1(\alpha_1 + \alpha_3)}{1 - \alpha_2 - (\alpha_1 + \alpha_3)\alpha_3}\right\}.$$

Hence, we can write that

$$(1 - \alpha_3^2)\gamma(0) = \sigma^2 + \frac{(\alpha_1 + \alpha_2\alpha_3)^2}{1 - \alpha_2 - (\alpha_1 + \alpha_3)\alpha_3}\gamma(0)$$

$$+ \frac{(\alpha_2 + \alpha_1\alpha_3)}{1 - \alpha_2 - (\alpha_1 + \alpha_3)\alpha_3}\left[\alpha_2(1 - \alpha_2) + \alpha_1(\alpha_1 + \alpha_3)\right]\gamma(0).$$

Thus, we have:

$$\gamma(0) = \sigma^2\left\{1 - \alpha_3^2\right.$$

$$\left. - \frac{(\alpha_1 + \alpha_2\alpha_3)^2 + (\alpha_2 + \alpha_1\alpha_3)\left[\alpha_2(1 - \alpha_2) + \alpha_1(\alpha_1 + \alpha_3)\right]}{1 - \alpha_2 - (\alpha_1 + \alpha_3)\alpha_3}\right\}^{-1}.$$

Remark. In the case when $\alpha_1 = \alpha_2 = \alpha_3 := \alpha$, we find that

$$\gamma(0) = \sigma^2 \left\{ \frac{1 - \alpha - 2\alpha^2}{1 - \alpha - \alpha^2 + 5\alpha^3 + 4\alpha^4} \right\}.$$

Question no. 4

We deduce from Equation (7.5) that $X_n = \sum_{i=1}^{n} \alpha_1^{n-i} \epsilon_i$, which implies that

$$
\begin{aligned}
E[Y_n] &= E\left[\exp\left\{ \sum_{i=1}^{n} \alpha_1^{n-i} \epsilon_i \right\} \right] = E\left[\prod_{i=1}^{n} \exp\left\{ \alpha_1^{n-i} \epsilon_i \right\} \right] \\
&\stackrel{\text{ind.}}{=} \prod_{i=1}^{n} E\left[e^{\alpha_1^{n-i} \epsilon_i} \right] = \prod_{i=1}^{n} \exp\left\{ \frac{\sigma^2}{2} \alpha_1^{2(n-i)} \right\} \\
&= \exp\left\{ \frac{\sigma^2}{2} \sum_{i=1}^{n} \alpha_1^{2(n-i)} \right\} = \exp\left\{ \frac{\sigma^2}{2} \alpha_1^{2n} \sum_{i=1}^{n} \alpha_1^{-2i} \right\}.
\end{aligned}
$$

Now, define

$$S_n = \sum_{i=1}^{n} \alpha_1^{-2i} = \alpha_1^{-2} + \alpha_1^{-4} + \cdots + \alpha_1^{-2n}.$$

We have [see Equation (7.7)] that

$$S_n = \frac{1 - \alpha_1^{-2n}}{\alpha_1^2 - 1}.$$

Therefore, we can write that

$$E[Y_n] = \exp\left\{ \frac{\sigma^2}{2} \left(\frac{\alpha_1^{2n} - 1}{\alpha_1^2 - 1} \right) \right\} \quad \text{for } n = 1, 2, \dots.$$

Remark. If $\{X_n, n \in \mathbb{Z}\}$ is a zero mean weakly stationary AR(1) process and $Y_n = e^{X_n}$ for all $n \in \mathbb{Z}$, then [see Equation (7.9)]

$$
\begin{aligned}
E[Y_n] &= E\left[\exp\left\{ \sum_{i=0}^{\infty} \alpha_1^{i} \epsilon_{n-i} \right\} \right] = E\left[\prod_{i=0}^{\infty} \exp\left\{ \alpha_1^{i} \epsilon_{n-i} \right\} \right] \stackrel{\text{ind.}}{=} \prod_{i=0}^{\infty} E\left[e^{\alpha_1^{i} \epsilon_{n-i}} \right] \\
&= \prod_{i=0}^{\infty} \exp\left\{ \frac{\sigma^2}{2} \alpha_1^{2i} \right\} = \exp\left\{ \frac{\sigma^2}{2} \sum_{i=0}^{\infty} \alpha_1^{2i} \right\} = \exp\left\{ \frac{\sigma^2}{2} \frac{1}{(1 - \alpha_1^2)} \right\},
\end{aligned}
$$

where we assumed that we can interchange the mathematical expectation and the infinite product.

Question no. 5

Equation (7.23) (with $\theta_0 = 1$) yields that

$$\rho(1) = \frac{\sum_{i=0}^{3-1} \theta_i \theta_{i+1}}{1 + \theta_1^2 + \theta_2^2 + \theta_3^2} = \frac{\theta_1 + \theta_1 \theta_2 + \theta_2 \theta_3}{1 + \theta_1^2 + \theta_2^2 + \theta_3^2},$$

$$\rho(2) = \frac{\sum_{i=0}^{3-2} \theta_i \theta_{i+2}}{1 + \theta_1^2 + \theta_2^2 + \theta_3^2} = \frac{\theta_2 + \theta_1 \theta_3}{1 + \theta_1^2 + \theta_2^2 + \theta_3^2}$$

and

$$\rho(3) = \frac{\sum_{i=0}^{3-3} \theta_i \theta_{i+3}}{1 + \theta_1^2 + \theta_2^2 + \theta_3^2} = \frac{\theta_3}{1 + \theta_1^2 + \theta_2^2 + \theta_3^2}.$$

Question no. 6

The function $\rho(2)$ is given by [see (7.25)]

$$\rho(2) = \frac{\theta_2}{1 + \theta_1^2 + \theta_2^2}.$$

(a) When $\theta_1 = 1$, we set

$$g(x) = \frac{x}{2 + x^2}$$

and we calculate

$$g'(x) = \frac{2 - x^2}{(2 + x^2)^2} = 0 \iff x = \pm\sqrt{2}.$$

It follows that

$$-\frac{\sqrt{2}}{4} \leq \rho(2) \leq \frac{\sqrt{2}}{4}.$$

(b) In the general case, let

$$g(x, y) = \frac{y}{1 + x^2 + y^2}.$$

We have that

$$\frac{\partial g}{\partial x} = \frac{-2xy}{(1 + x^2 + y^2)^2} = 0 \iff x = 0 \text{ or } y = 0$$

and

$$\frac{\partial g}{\partial y} = \frac{1 + x^2 - y^2}{(1 + x^2 + y^2)^2} = 0 \iff y^2 = x^2 + 1.$$

When $y = 0$, $\rho(2)$ is equal to 0, which is neither a maximum nor a minimum. Hence, the extrema of the function $g(x, y)$ are attained at $(x, y) = (0, -1)$ and $(0, 1)$. Because

$$g(0, -1) = -\frac{1}{2} \quad \text{and} \quad g(0, 1) = \frac{1}{2},$$

we conclude that

$$-\frac{1}{2} \le \rho(2) \le \frac{1}{2},$$

as the function $\rho(1)$ in the case of an MA(1) process.

Question no. 7

We can write that

$$X_n = \alpha_1 X_{n-1} + \epsilon_n + \theta_1 \epsilon_{n-1} + \theta_2 \epsilon_{n-2} \quad \forall n \in \mathbb{Z}.$$

First, we calculate

$$\gamma(0) = E[X_n^2] = E[(\alpha_1 X_{n-1} + \epsilon_n + \theta_1 \epsilon_{n-1} + \theta_2 \epsilon_{n-2}) X_n]$$
$$= \alpha_1 \gamma(1) + \underbrace{E[\epsilon_n^2]}_{\sigma^2} + \theta_1 E[\epsilon_{n-1} X_n] + \theta_2 E[\epsilon_{n-2} X_n].$$

Next, we have that

$$E[\epsilon_{n-1} X_n] = \alpha_1 E[\epsilon_{n-1} X_{n-1}] + \underbrace{E[\epsilon_{n-1}\epsilon_n]}_{0} + \theta_1 \underbrace{E[\epsilon_{n-1}^2]}_{\sigma^2} + \theta_2 \underbrace{E[\epsilon_{n-1}\epsilon_{n-2}]}_{0}$$
$$= \alpha_1 E[\epsilon_{n-1}(\alpha_1 X_{n-2} + \epsilon_{n-1} + \theta_1 \epsilon_{n-2} + \theta_2 \epsilon_{n-3})] + \theta_1 \sigma^2$$
$$= \alpha_1 E[\epsilon_{n-1}^2] + \theta_1 \sigma^2 = \sigma^2(\alpha_1 + \theta_1).$$

Similarly, we calculate

$$E[\epsilon_{n-2} X_n] = \alpha_1 E[\epsilon_{n-2} X_{n-1}] + \underbrace{E[\epsilon_{n-2}\epsilon_n]}_{0} + \theta_1 \underbrace{E[\epsilon_{n-2}\epsilon_{n-1}]}_{0} + \theta_2 \underbrace{E[\epsilon_{n-2}^2]}_{\sigma^2}$$
$$= \alpha_1 E[\epsilon_{n-2}(\alpha_1 X_{n-2} + \epsilon_{n-1} + \theta_1 \epsilon_{n-2} + \theta_2 \epsilon_{n-3})] + \theta_2 \sigma^2$$
$$= \alpha_1 \left\{ E[\epsilon_{n-2}\alpha_1 X_{n-2}] + \theta_1 \sigma^2 \right\} + \theta_2 \sigma^2$$
$$= \alpha_1 \left\{ \alpha_1 E[\epsilon_{n-2}^2] + \theta_1 \sigma^2 \right\} + \theta_2 \sigma^2 = \sigma^2(\alpha_1^2 + \alpha_1 \theta_1 + \theta_2).$$

Hence, we obtain that

$$\gamma(0) = \alpha_1 \gamma(1) + \sigma^2 + \theta_1(\alpha_1 + \theta_1)\sigma^2 + \theta_2(\alpha_1^2 + \alpha_1 \theta_1 + \theta_2)\sigma^2.$$

Remark. We see that to obtain an explicit expression for $\gamma(0)$, we must also calculate (at least) $\gamma(1)$.

Question no. 8

For $k = 1$, we have [see (7.33)] that

$$\phi(1) = \rho(1) \stackrel{(7.24)}{=} \frac{\theta_1}{1 + \theta_1^2},$$

which agrees with the formula (7.35) in the case when $k = 1$.

When $k = 2$, we can write [see (7.34)] that

$$\phi(2) = \frac{\rho(2) - \rho^2(1)}{1 - \rho^2(1)} \overset{(7.24)}{=} \frac{0 - \left(\theta_1/(1+\theta_1^2)\right)^2}{1 - \left(\theta_1/(1+\theta_1^2)\right)^2},$$

and the formula (7.35) with $k = 2$ becomes

$$\phi(2) = -\frac{(-\theta_1)^2}{1 + \theta_1^2 + \theta_1^4} = -\frac{\theta_1^2}{1 + \theta_1^2 + \theta_1^4}.$$

We indeed have that

$$\frac{-\left(\theta_1/(1+\theta_1^2)\right)^2}{1 - \left(\theta_1/(1+\theta_1^2)\right)^2} = \frac{-\theta_1^2}{(1+\theta_1^2)^2 - \theta_1^2} = -\frac{\theta_1^2}{1 + \theta_1^2 + \theta_1^4}.$$

Finally, in the case when $k = 3$, we first calculate

$$D_{3,1} = \begin{vmatrix} 1 & \rho(1) & \rho(1) \\ \rho(1) & 1 & \rho(2) \\ \rho(2) & \rho(1) & \rho(3) \end{vmatrix}$$

$$= \rho(3) - \rho(1)\rho(2) - \rho(1)[\rho(1)\rho(3) - \rho^2(2)] + \rho(1)[\rho^2(1) - \rho(2)]$$

$$= \rho^3(1) - \rho^2(1)\rho(3) - 2\rho(1)\rho(2) + \rho(1)\rho^2(2) + \rho(3)$$

and

$$D_{3,2} = \begin{vmatrix} 1 & \rho(1) & \rho(2) \\ \rho(1) & 1 & \rho(1) \\ \rho(2) & \rho(1) & 1 \end{vmatrix}$$

$$= 1 - \rho^2(1) - \rho(1)[\rho(1) - \rho(1)\rho(2)] + \rho(2)[\rho^2(1) - \rho(2)]$$

$$= 1 - 2\rho^2(1) + 2\rho^2(1)\rho(2) - \rho^2(2).$$

Because, from (7.24), $\rho(k) = 0$ for $|k| = 2, 3, \ldots$, we obtain that

$$D_{3,1} = \rho^3(1) \quad \text{and} \quad D_{3,2} = 1 - 2\rho^2(1).$$

It follows that

$$\phi(3) = \frac{\rho^3(1)}{1 - 2\rho^2(1)} \overset{(7.24)}{=} \frac{\left(\theta_1/(1+\theta_1^2)\right)^3}{1 - 2\left(\theta_1/(1+\theta_1^2)\right)^2}$$

$$= \frac{\theta_1^3}{(1+\theta_1^2)^3 - 2\theta_1^2(1+\theta_1^2)} = \frac{\theta_1^3}{1 + \theta_1^2 + \theta_1^4 + \theta_1^6},$$

which is the same formula as (7.35) with $k = 3$.

Question no. 9

We find that $\bar{y} = -0.034$. Next, we calculate

k	0	1	2	3
$\hat{\gamma}(k)$	0.3364	0.0810	−0.0347	0.0202

Hence, $\hat{\sigma}_X^2 = \hat{\gamma}(0) \simeq 0.3364$. It follows that

k	1	2	3
$\hat{\rho}(k)$	0.2408	−0.1032	0.0600

Then, we calculate (see the previous question)

$$\hat{\phi}(1) = \hat{\rho}(1) \simeq 0.2408,$$

$$\hat{\phi}(2) \simeq \frac{-0.1032 - (0.2408)^2}{1 - (0.2408)^2} \simeq -0.171$$

and

$$\hat{\phi}(3) \simeq \frac{(.2408)^3 - (.2408)^2(.06) - 2(.2408)(-.1032) + (.2408)(-.1032)^2 + .06}{1 - 2(.2408)^2 + 2(.2408)^2(-.1032) - (-.1032)^2}$$

$$\simeq 0.143.$$

Now, for an MA(1) time series with $\theta_1 = 1/2$ and $\epsilon_n \sim U(-1, 1)$, we have [see (7.21)] that

$$\sigma_X^2 = \mathrm{VAR}[X_n] = (1 + \theta_1^2)\sigma^2 = \left(1 + \frac{1}{4}\right)\frac{1}{3} = \frac{5}{12}$$

and [see (7.35) and the previous question]

$$\phi(1) = -\frac{(-1/2)}{1 + (1/2)^2} = \frac{2}{3},$$

$$\phi(2) = -\frac{(-1/2)^2}{1 + (1/2)^2 + (1/2)^4} = -\frac{4}{21} \simeq -0.190$$

and

$$\phi(3) = -\frac{(-1/2)^3}{1 + (1/2)^2 + (1/2)^4 + (1/2)^6} = \frac{8}{85} \simeq 0.094.$$

Actually, the data are indeed observations of an MA(1) time series with $\theta_1 = 1/2$ and the ϵ_ns as above. However, we see that, as in Example 7.3.1, the number of observations is too small to obtain good point estimates of σ_X^2 and the $\phi(k)$s.

Question no. 10

(a) Because, by assumption, the stochastic process $\{X_n, n \in \mathbb{Z}\}$ is stationary, we can write that

$$E[X_n] \equiv \mu \quad \text{and} \quad \text{VAR}[X_n] \equiv \sigma_X^2.$$

Therefore, Equation (7.3) implies that

$$E[X_{n+j} \mid X_n = x_n] = \mu + \rho_{X_n, X_{n+j}} \frac{\sigma_X}{\sigma_X}(x_n - \mu) = \mu + \rho(j)(x_n - \mu)$$

$$= \rho(j)x_n + \mu[1 - \rho(j)]. \quad \blacksquare$$

(b) We have that

$$E[X_{n+j}^2 \mid X_n = x_n] = \text{VAR}[X_{n+j} \mid X_n = x_n] + (E[X_{n+j} \mid X_n = x_n])^2$$

$$\overset{(7.39)}{=} \sigma_X^2[1 - \rho^2(j)] + \{\rho(j)x_n + \mu[1 - \rho(j)]\}^2.$$

Remark. We see that if $\mu = 0$ and $\sigma_X^2 = 1$, then $E[X_{n+j} \mid X_n = 0] = 0 \; \forall j$ and

$$\text{VAR}[X_{n+j} \mid X_n = 0] = E[X_{n+j}^2 \mid X_n = 0] = 1 - \rho^2(j) \quad \text{for } j = 1, 2, \ldots.$$

D

Answers to even-numbered exercises

Chapter 1

2. $F(x)$ is continuous at any $x \in \mathbb{R}$, except at $x = \{0, 1, 2\}$. At these points, it is only right-continuous.

4. Discontinuous.

6. xe^{-x}.

8. $f'(x) = \frac{2}{3}x^{-1/3}$; $x = 0$.

10. $\sqrt{2\pi}$.

12. $-\frac{1}{2}e^{-x}(\cos x + \sin x)$.

14. 3/8.

16. −

18. $\ln 2$.

20. (a) $\frac{p}{1-(1-p)z}$; (b) $k!(1-p)^k p$.

Chapter 2

2. (a) 7/12; (b) 55/72; (c) 2/9; (d) 3/8.

4. (a) $P[\{\omega_1\}] = 0.1$; $P[\{\omega_2\}] = 0.45$; $P[\{\omega_3\}] = 0.0171$; $P[\{\omega_4\}] = 0.8379$;
   (b) (i) $R = A_1 \cup (A_1^c \cap A_2) \cup (A_1^c \cap A_2^c \cap A_3)$; (ii) 0.1621; (iii) 0.2776;
   (c) (i) $B = R_1^c \cup R_2^c \cup R_3^c$; (ii) 0.9957.

6. 4/5.

8. 0.9963.

10. (a) 1/10; (b) 1/10.

12. (a) 0.325; (b) 0.05; (c) 0.25; (d) 0.66375.

14. 48/95.

16. 23/32.

18. 9.

20. 63.64%.

22. (a) 0.8829; (b) 0.9083.

24. (a) 0.9972; (b) (i) 0.8145; (ii) 0.13.
26. (a) −; (b) $\frac{1}{n-1}$; (c) 0.8585.
28. (a) 1/64; (b) 3/32.
30. (a) 86,400; (b) 3840.

Chapter 3

2. (a) Hyp($N = 100$, $n = 2$, d = number of defective devices in the batch); (b) 0.9602; (c) 0.9604; (d) 0.9608.
4. (a) Poi(3); 0.616; (b) Poi(9); 0.979; (c) Exp(3); $1 - e^{-3}$; (d) Exp(3); e^{-3}; (e) G($\alpha = 2, \lambda = 3$); (f) B($n = 100, p = 0.05$); 0.037.
6. (a) (i) B($20, \theta$); (ii) Hyp($1000, 20, 1000\theta$); (b) (i) 0.010; (ii) 0.032.
8. (a) $X \sim$ Poi(1.2); $Y \sim$ Geo(0.301); $Z \sim$ B($12, 0.301$); (b) 4th; (c) 0.166; (d) 0.218.
10. 0.7746.
12. 0.6233.
14. 0.1215.
16. 0.759.
18. 0.4.
20. $1 - e^{-5}$.
22. (a)

$$f_X(x) = \begin{cases} 1/2 & \text{if } 0 \le x \le 1, \\ 1/6 & \text{if } 1 < x < 4, \\ 0 & \text{elsewhere}; \end{cases}$$

(b) 2.5; (c) 1.5; (d) ∞; (e) (i) 1/2; (ii) 1.
24. (a) 0.657; (b) 0.663; (c) 7/17.
26. (a) 0.3754; (b) 0.3679.
28. (a) 3; (b)

$$F_X(x) = \begin{cases} 0 & \text{if } x < 0, \\ 3x^2 - 2x^3 & \text{if } 0 \le x \le 1, \\ 1 & \text{if } x > 1; \end{cases}$$

(c) $\left(\frac{27}{32}\right)2^k$, for $k = 1, 2, \ldots$; (d)

$$f_Z(z) = \begin{cases} 3(1 - \sqrt{z}) & \text{if } 0 \le z \le 1, \\ 0 & \text{elsewhere.} \end{cases}$$

30. (a) 6; (b) 7; (c) 0.8.
32. 5/8.
34. 0.00369.
36. 0.027.
38. $1 - e^{-\lambda y}$ if $y = 1, 2, \ldots$.
40. 2.963.
42. Two engines: reliability $\simeq 0.84$; four engines: reliability $\simeq 0.821$. So, here greater reliability with two engines.

44. (a) 0.75; (b) 0.2519.

46. (a) (i) 0.7833; (ii) 0.7558; (b) 0.7361.

48. (a) $E[X] = (\pi k/2)^{1/2}$; $VAR[X] = 2k\left(1 - \frac{\pi}{4}\right)$; (b) k is a scale parameter.

50. (a) $\sqrt{\pi}/8$; (b) $E[e^{X^2/2}] = 1.25$; $VAR[e^{X^2/2}] = \infty$.

52. (a)

$$Z(W) = \begin{cases} s - \alpha - \beta W & \text{if } W \geq w_0, \\ r - \alpha - \beta W & \text{if } W < w_0; \end{cases}$$

(b) $w_0 + \left[-2\sigma^2 \ln\left(\dfrac{\beta\sigma\sqrt{2\pi}}{s - r}\right)\right]^{1/2}$.

Chapter 4

2. (a) 3; (b)

$$f_X(x) = \begin{cases} 3x^2 & \text{if } 0 < x < 1, \\ 0 & \text{elsewhere;} \end{cases} \qquad f_Y(y) = \begin{cases} \frac{3}{2}(1 - y^2) & \text{if } 0 < y < 1, \\ 0 & \text{elsewhere;} \end{cases}$$

(c) $VAR[X] = 3/80$; $VAR[Y] = 19/320$; (d) 0.3974.

4. (a) $N(750\theta, 187.5\theta^2)$; (b) \$220.

6. (a) (i)

$$f_X(x) = \begin{cases} \frac{3}{4}(1 - x^2) & \text{if } -1 \leq x \leq 1, \\ 0 & \text{elsewhere;} \end{cases} \qquad f_Y(y) = \begin{cases} \frac{3}{2}\sqrt{y} & \text{if } 0 \leq y < 1, \\ 0 & \text{elsewhere;} \end{cases}$$

(ii) 0; (b) no, because $f_X(x)f_Y(y)$ is not identical to 3/4 $[\equiv f_{X,Y}(x, y)]$.

8. 1.5.

10. 1/64.

12. 0.1586.

14. −1/11.

16. 0.8665.

18. (a)

$$f_X(x) = \begin{cases} 6x(1 - x) & \text{if } 0 < x < 1, \\ 0 & \text{elsewhere;} \end{cases} \qquad f_Y(y) = \begin{cases} 3y^2 & \text{if } 0 < y < 1, \\ 0 & \text{elsewhere;} \end{cases}$$

(b) 5/16.

20. (a) 0.5987; (b) 1.608; (c) 0.0871.

22. (a) 0.6065; (b) 7.29; (c) 0.1428; (d) 0.3307; (e) 0.7291.

24. (a) 0.2048; (b) 3/5; (c) $e^{-3/5}$; (d) 0.0907.

26. (a) 0.6587; (b) 3.564.

28. (a) 0.157; (b) 0.3413; $X = \sum_{i=1}^{25} X_i$, where $X_i \sim Exp(1/2)$, for all i, and the X_is are independent random variables.

30. 10/3.

32. 18.

34. (a) $-$; (b)
$$f_X(x) = \begin{cases} \frac{2}{\pi}\sqrt{1-x^2} & \text{if } -1 \le x \le 1, \\ 0 & \text{elsewhere} \end{cases}$$

and $f_Y(y) = f_X(y)$, by symmetry; (c) no, because $f_X(x)f_Y(y)$ is not identical to $1/\pi$ [$\equiv f_{X,Y}(x,y)$]; (d) 3/4.

36. (a) 0.045; (b) 527.08.

38. (a) $T \sim \text{Exp}(1/5)$; (b)
$$f_T(t) = \begin{cases} \frac{1}{5}(1 - e^{-t/50})^9 e^{-t/50} & \text{if } t \ge 0, \\ 0 & \text{elsewhere;} \end{cases}$$

(c) $T \sim \text{G}(\alpha = 10, \lambda = 1/50)$; that is,
$$f_T(t) = \begin{cases} \dfrac{1}{50^{10}\Gamma(10)} t^9 e^{-t/50} & \text{if } t \ge 0, \\ 0 & \text{elsewhere.} \end{cases}$$

40. (a)
$$f_Y(y) = \begin{cases} 3(y-1)^2 & \text{if } 0 < y < 1, \\ 0 & \text{elsewhere;} \end{cases}$$

(b) 0.1566; (c)
$$f_Z(z) = \begin{cases} (1 - z^{-1/3})^2 & \text{if } 0 < z < 1, \\ 0 & \text{elsewhere;} \end{cases}$$

(d) 1/2.

42. (a) (i)

x_2	0	1
$p_{X_2}(x_2)$	7/12	5/12

(ii) 1/3 if $x_2 = 0$, and 2/3 if $x_2 = 1$; (b)

y	-1	0	3	8
$F_Y(y)$	1/6	1/2	5/6	1

44. (a) 0.95; (b) (i)
$$P[Y \le y \mid X_1 = x_1] = \begin{cases} 0 & \text{if } y \le x_1, \\ 1 - e^{-(y-x_1)/2} & \text{if } y > x_1; \end{cases}$$

(ii)
$$f_Y(y \mid X_1 = x_1) = \begin{cases} \frac{1}{2}e^{-(y-x_1)/2} & \text{if } y > x_1, \\ 0 & \text{elsewhere.} \end{cases}$$

46. (a) $X \sim \text{Poi}(1)$; $Y \sim \text{Poi}(1/2)$; $Y \mid \{X = 35\} \sim \text{B}(n = 35, p = 1/2)$; (b) (i) 0.5873; (ii) 0.5876; (c) 0.6970.

48. (a)

$$f_X(x) = \begin{cases} \frac{1}{8}(4-x) & \text{if } 0 \le x \le 4, \\ 0 & \text{elsewhere;} \end{cases}$$

(b) (i) 4/3; (ii) 8/9; (iii) 0.32; (iv) 2.4; (c) $-1/2$; X and Y are not independent, because $\rho_{X,Y} \ne 0$.

50. (a)

$$f_X(x) = \begin{cases} 15x^2 - 45x^4 + 30x^5 & \text{if } 0 < x < 1, \\ 0 & \text{elsewhere;} \end{cases}$$

$$f_Y(y) = \begin{cases} 30y^4(1-y) & \text{if } 0 < y < 1, \\ 0 & \text{elsewhere;} \end{cases}$$

(b) no, because, for instance, $f_{X,Y}(1, 1/2) = 22.5 \ne f_X(1)f_Y(1/2) = 0$; (c) 0.0191.

52. (a)

$$f_{X,Y}(x, y) = \begin{cases} \dfrac{1}{2(x+1)} & \text{if } -1 \le x \le 1, -1 \le y \le x, \\ 0 & \text{elsewhere;} \end{cases}$$

$$f_Y(y) = \begin{cases} \frac{1}{2}[\ln 2 - \ln(y+1)] & \text{if } -1 \le y \le 1, \\ 0 & \text{elsewhere;} \end{cases}$$

(b) (i) $-1/3$; (ii) $-1/2$; (c) 1/6.

54. (a)

$x_2 \backslash x_1$	-1	0	1
-1	1/64	6/64	1/64
0	6/64	36/64	6/64
1	1/64	6/64	1/64

(b) $-\sqrt{2}/2$; (c) no, because $\rho_{X,Y} \ne 0$.

Chapter 5

2. $r(x) = [x(1 - \ln x)]^{-1}$, for $1 \le x \le e$. We find that $r'(x) > 0$ in the interval $[1, e]$. Hence, the distribution is IFR.

4. $R(t) = (e^{-t} - e^{-2t})/(2t)$, for $t \ge 0$.

6. $\frac{71}{19}e^{-2}$.

8. $r_X(k) = \frac{1}{N-k+1}$. The distribution is IFR.

10. (a) $2\lambda/3$; (b) $\lambda/2$.

12. 1/3.

14. $R(t) = 1 - (1 - e^{-\lambda_C t})\left\{1 - [1 - (1 - e^{-\lambda_A t})^3]e^{-\lambda_B t}\right\}$.

16. $\simeq 0.3710$.

18. $0.996303 \le R(t_0) \le 0.997107$ (approximately). The exact answer is 0.996327.

20. (a) $p^3(3 - 2p)$; (b) $e^{-3\theta t}(3 - 2e^{-\theta t})$, for $t \geq 0$.

Chapter 6

2. $\{N^*(t), t \geq 0\}$ is a Poisson process with rate $\lambda = 1$. Thus, it is indeed a pure birth process.

4. (a) 1/3; (b) 1/3; (c) 2/3.

6. (a) 2.5; (b) 2.72; (c) 2.89.

8. 1/16.

10. 0.079.

12. $\left(1 - e^{-2\mu}\right)/(2\mu)$.

14. We define the states:

> 0: the system is empty;
>> n_0: there are n customers in the system and one is waiting to know whether the server can serve him or her;
>> n: there are n customers in the system and one is being served.

for $n = 1, 2, 3$. The balance equations are:

state j departure rate from j = arrival rate to j

0	$\lambda \pi_0 = \mu_0 p \pi_{1_0} + \mu \pi_1$
1_0	$(\lambda + \mu_0)\pi_{1_0} = \lambda \pi_0 + \mu_0 p \pi_{2_0} + \mu \pi_2$
2_0	$(\lambda + \mu_0)\pi_{2_0} = \lambda \pi_{1_0} + \mu_0 p \pi_{3_0} + \mu \pi_3$
3_0	$\mu_0 \pi_{3_0} = \lambda \pi_{2_0}$
1	$(\lambda + \mu)\pi_1 = \mu_0(1 - p)\pi_{1_0}$
2	$(\lambda + \mu)\pi_2 = \mu_0(1 - p)\pi_{2_0}$
3	$\mu \pi_3 = \mu_0(1 - p)\pi_{3_0}$

16. (a) $\mu/[4(\mu + \lambda)]$; (b) $1 - \{\mu/[2(\mu + \lambda)]\}$.

18. 207/256.

20. 0.77.

Chapter 7

2. Not a Gaussian process.

4. Gaussian and weakly stationary. Hence, strictly stationary.

6.

x	-4	-2	0	2	4	Σ
$p_{X_n}(x)$	1/8	1/4	1/4	1/4	1/8	1

8. (a) $4\alpha_1^2$; (b) $\sum_{j=0}^{k} \binom{2k}{2j} \alpha_1^{2j}$.

10. $\left(\frac{1 + \theta_1^2}{1 + \theta_1^2 + \theta_2^2}\right)^{1/2}$.

12. (a) Invertible; (b) not invertible.

14. $E[Y_n] \equiv \sigma^2(1 + \theta_1^2)$; $\text{VAR}[Y_n] \equiv 2\sigma^4(1 + 2\theta_1^2 + \theta_1^4)$.

16. (a) $\left(\frac{\theta_1}{1+\theta_1^2}\right) X_n$; (b) not similar.

18. $-$

20. $\simeq 0.52\gamma(0)$.

E

Answers to multiple choice questions

Chapter 1

1 b; 2 e; 3 b; 4 c; 5 a; 6 d; 7 b; 8 c; 9 e; 10 a.

Chapter 2

1 c,b,b,b,b,e; 2 c; 3 d; 4 d; 5 d; 6 c; 7 d; 8 e; 9 a; 10 c; 11 c; 12 b; 13 c; 14 a.

Chapter 3

1 d; 2 a; 3 d; 4 b; 5 c; 6 c; 7 a; 8 a; 9 e; 10 a; 11 b; 12 e; 13 a; 14 d; 15 a; 16 c; 17 c; 18 e; 19 c; 20 c.

Chapter 4

1 e; 2 b; 3 c; 4 a; 5 a; 6 e; 7 c; 8 d; 9 b; 10 c; 11 d; 12 c; 13 c; 14 e; 15 d; 16 c; 17 b; 18 e; 19 d; 20 c.

Chapter 5

1 d; 2 e; 3 c; 4 a; 5 b; 6 c; 7 c; 8 c; 9 d; 10 d.

Chapter 6

1 b; 2 d; 3 e; 4 b; 5 d; 6 a; 7 c; 8 d; 9 b; 10 c.

Chapter 7

1 c; 2 d; 3 d; 4 a; 5 b; 6 b; 7 e; 8 b; 9 c; 10 b.

References

1. Barnes, J. Wesley, *Statistical Analysis for Engineers: A Computer-Based Approach*, Prentice-Hall, Englewood Cliffs, NJ, 1988.
2. Breiman, Leo, *Probability and Stochastic Processes: With a View Toward Applications*, Houghton Mifflin, Boston, 1969.
3. Chung, Kai Lai, *Elementary Probability Theory with Stochastic Processes*, Springer-Verlag, New York, 1975.
4. Dougherty, Edward R., *Probability and Statistics for the Engineering, Computing, and Physical Sciences*, Prentice-Hall, Englewood Cliffs, NJ, 1990.
5. Feller, William, *An Introduction to Probability Theory and Its Applications*, Volume I, 3rd edition, Wiley, New York, 1968.
6. Feller, William, *An Introduction to Probability Theory and Its Applications*, Volume II, 2nd edition, Wiley, New York, 1971.
7. Hastings, Kevin J., *Probability and Statistics*, Addison-Wesley, Reading, MA, 1997.
8. Hines, William W. and Montgomery, Douglas C., *Probability and Statistics in Engineering and Management Science*, 3rd edition, Wiley, New York, 1990.
9. Hogg, Robert V. and Craig, Allen T., *Introduction to Mathematical Statistics*, 3rd edition, Macmillan, New York, 1970.
10. Hogg, Robert V. and Tanis, Elliot A., *Probability and Statistical Inference*, 6th Edition, Prentice Hall, Upper Saddle River, NJ, 2001.
11. Krée, Paul, *Introduction aux Mathématiques et à leurs Applications Fondamentales*, Dunod, Paris, 1969.
12. Lapin, Lawrence L., *Probability and Statistics for Modern Engineering*, 2nd edition, PWS-KENT, Boston, 1990.
13. Leon-Garcia, Alberto, *Probability and Random Processes for Electrical Engineering*, 2nd edition, Addison-Wesley, Reading, MA, 1994.
14. Lindgren, Bernard W., *Statistical Theory*, 3rd edition, Macmillan, New York, 1976.
15. Maksoudian, Y. Leon, *Probability and Statistics with Applications*, International, Scranton, PA, 1969.
16. Miller, Irwin, Freund, John E., and Johnson, Richard A., *Probability and Statistics for Engineers*, 4th edition, Prentice-Hall, Englewood Cliffs, NJ, 1990.
17. Papoulis, Athanasios, *Probability, Random Variables, and Stochastic Processes*, 3rd edition, McGraw-Hill, New York, 1991.

18. Roberts, Richard A., *An Introduction to Applied Probability*, Addison-Wesley, Reading, MA, 1992.
19. Ross, Sheldon M., *Introduction to Probability and Statistics for Engineers and Scientists*, Wiley, New York, 1987.
20. Ross, Sheldon M., *Introduction to Probability Models*, 7th edition, Academic Press, San Diego, 2000.
21. Spiegel, Murray R., *Theory and Problems of Advanced Calculus*, Schaum's Outline Series, McGraw-Hill, New York, 1973.
22. Walpole, Ronald E., Myers, Raymond H., Myers, Sharon L., and Ye, Keying, *Probability and Statistics for Engineers and Scientists*, 7th edition, Prentice Hall, Upper Saddle River, NJ, 2002.

Index